95

Structure and Bonding

Springer-Verlag Berlin Heidelberg GmbH

Liquid Crystals II

Volume Editor: D.M.P. Mingos

With contributions by
D.W. Bruce, P. Davidson, B. Donnio, J.W. Goodby
D. Guillon, C.T. Imrie

 Springer

In references Structure and Bonding is abbreviated
Struct.Bond. and is cited as a journal.

Springer WWW home page: HTTP://www.springer.de

ISSN 0081-5993

CIP Data applied for

ISBN 978-3-662-14714-6   ISBN 978-3-540-68118-2 (eBook)
DOI 10.1007/978-3-540-68118-2
© Springer-Verlag Berlin Heidelberg 1999
Originally published by Springer-Verlag Berlin Heidelberg New York in 1999.
Softcover reprint of the hardcover 1st edition 1999

Typesetting: Scientific Publishing Services (P) Ltd, Madras
Production editor: Christiane Messerschmidt, Rheinau
Cover: Medio V. Leins, Berlin
SPIN: 10649393   66/3020 – 5 4 3 2 1 0 – Printed on acid-free paper

Volume Editor

Editorial Statement

It is difficult to accept that the first volume of *Structure and Bonding* was published more than 30 years ago. The growing series of volumes with their characteristic green and white covers has provided a personal landmark throughout the scientific career of many of us. When you visit a new library, the green and white bands not only guide you to the inorganic section, but also provide some stimulating reading. For some of us, our views on ligand field theory, hard and soft acids and bases, and bonding in cluster compounds were structured by articles in these volumes. The series has almost reached one hundred, and it is perhaps appropriate to reconsider its aims and scope.

The distinguished original Editorial Board consisted of C. K. Jørgenson, J. B. Neilands, R. S. Nyholm, D. Reinen and R. J. P. Williams – names that we clearly associate with the renaissance of modern inorganic chemistry. Not surprisingly the Preface in the first volume showed the foresight and imagination that one would expect of such pioneers. They argued that a new review series was justified because "A valuable service is performed by bringing together up-to-date authoritative reviews from the different fields of modern inorganic chemistry, chemical physics and biochemistry, where the general subject of chemical bonding involves (usually) a metal and a small number of associated atoms... We are especially interested in the role of the 'complex metal-ligand' moiety... and wish to direct attention towards borderline subjects... We hope that this series may help to bridge the gaps between some of these different fields and perhaps provide in the process some stimulation and scientific profit to the reader."

Since that time progress in all the scientific disciplines that they highlighted in their Preface has been enormous, and the volume of primary publications has increased exponentially. At the same time powerful new tools have become available to chemists to enable them to tackle problems that were unthinkable only a few years ago. They were right to stress the importance of inter-disciplinary studies, and indeed the growth of bioinorganic chemistry has been so dramatic that it now justifies its own review series, *Topics in Biological Inorganic Chemistry*, which has been initiated by some of the former *Structure and Bonding* editors. In the flush of success associated with the development of ligand field theory, the founding editors perhaps stressed the importance of the metal-ligand bond too much for modern tastes. Inorganic chemists are now freer to exploit all parts of the periodic table and tackle problems in materials chemistry, biology and medicine. Also the traditional barriers between organic and inorganic chemistry now seem an irrelevance. This catholic approach to the subject is held together by the basic belief that an understanding of the structure and dynamics of the chemical interactions at the molecular level will lead to great insights into the observations made in the laboratory or observed in nature.

We expect the scope of the *Structure and Bonding* series to span the entire periodic table and address structure and bonding issues wherever they may be relevant. Therefore, it is anticipated that there will be reviews dealing not only with the traditional areas of chemical bonding based on valence problems and dynamics, but also nanostructures, molecular electronics, supramolecular structure, surfaces and clusters. These represent new and important developing areas of chemistry, but others will no doubt also emerge. Physical and spectroscopic techniques used to determine, examine and model structures will fall within the purview of *Structure and Bonding* to the extent that the focus is on the scientific results obtained and not on specialist information concerning the techniques themselves. Issues associated with the development of bonding models and generalisations that illuminate not only the structures of molecules, but also their reactivities and the rates of their chemical reactions will also be considered relevant.

Previously, *Structure and Bonding* brought together not only volumes dealing with specific topics, but also published unsolicited articles. In the future, we shall only be publishing volumes on specific topics. It is our goal to provide the scientific community with thematic volumes that contain critical and timely reviews across a wide range of subjects. However, all members of the Editorial Board welcome suggestions for future volumes from readers.

We join Springer-Verlag in thanking the editors now leaving the Editorial Board, M. J. Clarke, J. B. Goodenough, C. K. Jørgensen, G. A. Palmer, P. J. Sadler, R. Weiss, and R. J. P. Williams, whose efforts over the last decades served to maintain the high standards of the series.

Therefore, with the twenty-first century near at hand, we hope that you will agree that the changes we are making not only represent a cosmetic change of cover, but a change of emphasis which accurately reflects the way in which the science is developing.

Allen J. Bard, Austin
Ian G. Dance, Sydney
Peter Day, London
James A. Ibers, Evanston
Toyohi Kunitake, Fukuoka
Thomas J. Meyer, Chapel Hill
D.M.P. Mingos, London
Herbert W. Roesky, Göttingen
Jean-Pierre Sauvage, Strasbourg
Arndt Simon, Stuttgart
Fred Wudl, Los Angeles   October 1998

Preface

The liquid crystalline state may be identified as a distinct and unique state of matter which is characterised by properties which resembles those of both solids and liquids. It was first recognised in the middle of the last century through the study of nerve myelin and derivatives of cholesterol. The research in the area really gathered momentum, however, when as a result of the pioneering work of Gray in the early 1970's organic compounds showing liquid crystalline properties were shown to be suitable to form the basis of display devices in the electronic products.

The study of liquid crystals is truly multidisciplinary and has attached the attention of physicists, biologists, chemists, mathematicians and electronics engineers. It is therefore impossible to cover all these aspects fully in two small volumes and therefore it was decided in view of the overall title of the series to concentrate on the structural and bonding aspects of the subject. The Chapters presented in these two volumes have been organised to cover the following fundamental aspects of the subject. The calculation of the structures of liquid crystals, an account of their dynamical properties and a discussion of computer simulations of liquid crystalline phases formed by Gay-Berne mesogens. The relationships between molecular conformation and packing are analysed in some detail. The crystal structures of liquid crystal mesogens and the importance of their X-ray scattering properties for characterisational purposes are discussed. These are followed by chapters on columnar phases, dimeric liquid crystals, twist grain boundary phases and metallomesogens. I should like to thank the authors for providing such a detailed analysis of these important aspects of the subject and doing it in a way that I think will be intelligible to the general reader. I should also like to thank Dr. John Seddon for assisting me in making the choice of authors and topics for this volume.

For the general reader the following brief summary and basic definitions may provide a helpful introduction:

A compound which displays liquid crystal properties is referred to as a *mesogen* and said to exhibit *mesomorphism*. Liquid crystals may be considered either as disordered solids or ordered liquids, and their properties are very dependent on temperature and the presence or absence of solvent. In *thermotropic liquid crystals* the phases of the liquid crystals may be observed to change as the temperature is increased. In *lyotropic liquid crystals* the ordered crystalline state is disrupted by the addition of a solvent, which is very commonly water. For these systems temperature changes may also be responsible for phase changes, but only when used in combination with solvent.

The conventional liquid state is described as the *isotropic phase*. The temperature at which the compound passes from the solid phase into a mesophase is described as the *melting point* and the transition temperature between a mesophase and an isotropic liquid is described as the *clearing point*.

The mesophases of thermotropic liquid crystals are described as *calamitic* if the constituent molecules are rod-like and *columnar*, if the constituent molecules, which often have a disc like shape(discotic), stack into columns.

The simplest mesophase is the nematic phase. It is very fluid and involves highly disordered molecules having only short-range positional order, but with the molecules preferentially aligned on average in a particular direction (the director). If the constituent compound is racemic then it is possible to form a phase from the enantiomerically pure compound which is a *chiral nematic phase*.

Smectic phases are more highly ordered than nematic phases, and with an ordering of the molecules into layers. There are a number of different smectic phases which reflect differing degree of ordering. *Crystal smectic phases* are characterised by the appearance of inter-layer structural correlations and may in some cases be accompanied by a loss of molecular rotational freedom.

Columnar mesophases are observed generally for disc shaped molecules and in these phases the molecules show some orientational order with respect to the short molecular axis. Nematic phases which are directly related to calamitic nematic phases are observed and are similarly very fluid. Many of the columnar phases have symmetry resulting from ordered side-to-side packing – with or without ordering in the columns. In the disordered hexagonal phase the molecules define columns which are packed together in a hexagonal manner. Within the columns there is a variable degree of ordering. The ordered phases resemble the crystal smectic phases. Rectangular, oblique and tilted phases have also been identified.

Micelles are formed when surfactant molecules are dissolved in water. The precise concentration for micelle formation is described as the *critical micelle concentration*. Micelle formation results because of the disparate nature of the two ends of the surfactant molecules, i.e. there are hydrophobic and hydrophilic ends. If the concentration of surfactant is increased above the critical concentration then the density of micelles becomes sufficiently large that they begin to form ordered arrays. These ordered arrays correspond to *lyotropic liquid crystals*. Different mesophases result from different micelle types. Spherical micelles in general give rise to cubic close packed phases. Rod-like micelles give rise to hexagonal mesophases and disc mesophases give rise to lamellar phases. A variety of further phases are based on bilayer structures, ranging from the flat lamellar phases, to the highly convoluted bicontinuous phases, many of cubic symmetry, in which the interface is curved.

Many polymers are capable of forming mesophases in either aqueous or non-aqueous solvents. Furthermore, liquid crystal phases may form for pure block-copolymers through a tendency for the different polymer blocks to separate on a microscopic scale.

Hopefully with this brief introduction the reader will be able to appreciate fully the chapters which follow and which have been written by experts in the field. The complexity and beauty of the liquid crystalline phase has attracted many able scientists and the applications of liquid crystals in the electronics industry have provided a secure funding base for the subject. This is therefore still a field which is expanding rapidly and many research avenues remain to be explored by newcomers. Perhaps after reading these volumes of *Structure and Bonding* you will be tempted to join this exciting endeavour.

D.M.P. Mingos

Contents

Contents of Volume 94

Liquid Crystals I

Volume Editor: D.M.P. Mingos

Selected Topics in X-Ray Scattering by Liquid-Crystalline Polymers

Patrick Davidson

Laboratoire de Physique des Solides, Bât. 510, Université Paris Sud,
91405 Orsay cedex, France
E-mail: davidson@lps.u-psud.fr

This article presents several topics in the field of X-ray scattering by liquid-crystalline polymers. Through the description of these topics, a comprehensive overview of the experimental and theoretical methods used in this field is sketched. A first section briefly describes the main features of the experimental set-ups. Then three different sections describe the scattering by the three main types of liquid-crystalline polymers. A section on thermotropic main-chain polymers deals with the measurement of the nematic order parameter and with the local order in the nematic phase such as the chain-like correlations and the smectic fluctuations. This section also shows how the existence of hairpin defects was proved experimentally. The next section describes the scattering by thermotropic side-chain polymers. It deals with the positional long-range order in the SmA phase and with the various fluctuations which affect it. Moreover, the scattering by edge-dislocations as an example of localized defect is also discussed. The last section describes the scattering by lyotropic liquid-crystalline polymers. The Tobacco Mosaic Virus is a typical example of this class of materials and its scattering pattern is first detailed. Then some statistical physics models of the nematic and hexagonal phases are summarized. The industrially important phenomenon of flow alignment is also discussed. Finally, two examples of mineral liquid-crystalline polymers are described.

Keywords: Liquid crystals, Polymers, X-ray scattering

1
Introduction

The field of liquid crystalline polymers (LCPs) covers extremely widespread and diverse materials ranging from Kevlar, a synthetic LCP used as a structural polymer in the industry, to DNA which is a polymer of utmost biological importance as is well known. Furthermore, spurred by industrial applications, this field has grown at a tremendous pace in the last decade. Synthetic chemists have come up with a wealth of new and original architectures for compounds that combine both the mechanical properties of polymers and the properties of liquid crystals as anisotropic fluids. Therefore, it has become almost impossible to write a completely exhaustive review of the structural studies of these materials. Recent reviews cover some more restricted areas such as comb-like polymers [1] and main-chain polymers [2] which represent the two main classes of thermotropic LCPs. The aim of this paper is not to try to review this broad field but rather to give an overview of the state of the art through the study of several particular topics.

Two main classes of LCPs can be defined depending on whether their mesophases are formed by solutions or by the pure compounds: the first one is that of the lyotropic LCPs [3] the polymorphism of which depends on the concentration of the polymer in the solvent. Examples of this class include Kevlar [4] and DNA [5]. The second class is that of the thermotropic LCPs the polymorphism of which depends on temperature. It is essentially based upon rod-like mesogenic moieties polymerized together with small flexible chains called spacers. This class is further divided into two main types according to their chemical architecture. The rod-like mesogenic cores may be polymerized in a chain giving rise to main-chain polymers [6] (Fig. 1) or they may be grafted as side-chains on a polymer backbone giving rise to side-chain polymers [7] (Fig. 1).

It should be noted here that there exists nowadays a large number of thermotropic LCPs that do not belong to these two kinds. They may be based on disk-like mesogenic moieties such as main-chain and side-chain discotic LCPs. They may still be based on rod-like mesogenic moieties but they may have different chemical architectures. Copolymers of all kinds have also been synthesized. It is not our goal here to review all the structures of these compounds but rather to convey the main ideas that should guide one in the study of these new materials. It is indeed very

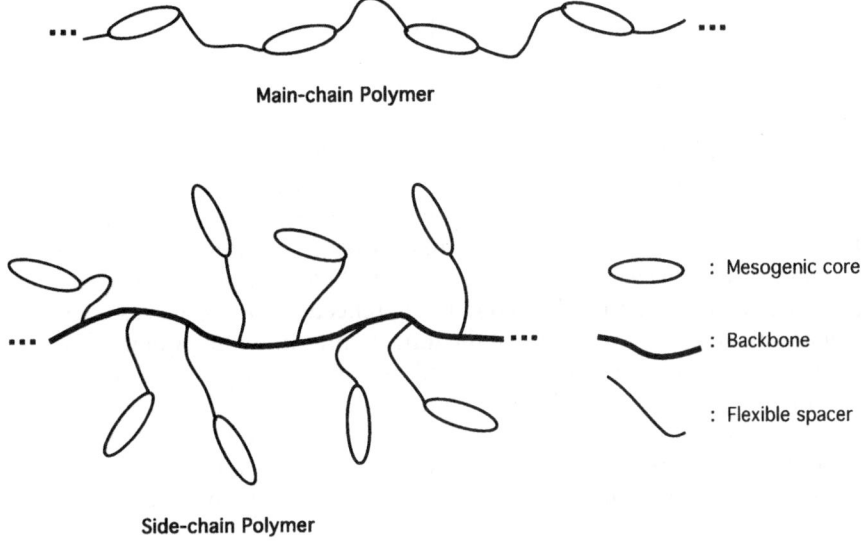

Fig. 1. Chemical architectures of thermotropic LCPs

important to determine precisely the structures of new materials in order to understand their macroscopic properties. This applies from the simple identification of the various types of liquid crystalline ordering to more elaborate investigations such as quantifying the degree of orientational order, i.e. the nematic order parameter, or the anisotropy of the polymer chain. More sophisticated studies can even give access to physical constants such as elastic moduli which in turn give information about the thermodynamics of the phase.

Altogether, the X-ray scattering by LCPs is similar to that of low molar mass liquid crystals (LMMLCs). Actually, the nature of a given mesophase only depends on its symmetry, i.e. the set of long-range orders it shows, so that the polymeric nature of the material does not change the way we consider this phase. In particular, all the methods devised to analyze the structures of LMMLCs also apply to the mesophases of LCPs. Nonetheless, the X-ray scattering patterns of LCPs are often richer than those of the corresponding monomers. This is due to the short range correlations brought about by physically connecting the monomers. These correlations often give rise to a large additional diffuse scattering. From another point of view, another specificity of LCPs is that small angle neutron scattering (SANS) has proved an extremely helpful tool to study their structures. There are two main reasons for this: first, SANS techniques probe matter at a length scale (10–1000 Å) which usually corresponds to the dimensions of the polymer chain. Second, neutrons are sensitive to the isotopic nature of the nuclei which allows one to observe directly the conformation of selectively deuterated polymer chains. Since this technique and its applications in the field of LCPs have been reviewed recently [8], we shall only mention it briefly when necessary in the course of this article.

2
Description of X-Ray Scattering Set-Ups

X-rays are electromagnetic radiations of the same nature as light but of much larger energy and smaller wavelength. The X-rays used in crystallography have a wavelength of about 1 Å which is comparable to atomic dimensions. X-rays interact essentially with the electrons so that information is obtained about the organization of the electron density which of course follows closely the atomic density. The most convenient and commonly used radiation is that emitted by Cu anodes at a wavelength $\lambda_{CuK\alpha} = 1.54$ Å. An "ideal" crystalline organization gives rise to highly localized discrete Bragg reflections, a special case of highly cooperative scattering in which all the units scatter in phase. A more distorted "imperfect" organization such as a liquid-like order only gives rise to a weaker continuous scattering spread over a large angular range. This is called a diffuse scattering. In general, the Bragg reflections indicate the existence of some positional long-range order whereas the diffuse scattering points to local orders, fluctuations and defects. Several classical textbooks describe X-ray scattering in detail [9].

In order to perform X-ray scattering studies, samples of LCPs often need to be held in special vials well suited to X-ray scattering. The most commonly used vials are cylindrical capillary tubes made of boron glass called Lindemann glass capillaries which are commercially available. The state of the sample, powder or single domain, prompts the kind of studies that can be achieved. Consequently, spending time to produce a single domain of a given mesophase is quite worthwhile because a lot of valuable information is lost through powder averaging. For instance, the powder pattern of a nematic phase is very similar to that of an isotropic liquid. Therefore, one cannot detect the existence of a nematic phase from its scattering pattern in the powder form. Also, the intensity scattered by a single domain lies in a smaller solid angle and has a larger signal-over-noise ratio. Single domains of a mesophase can be obtained in several ways. The most convenient and usual one is by application of a magnetic field. The use of LCPs requires some specific caution due to their high viscosity. Although LMMLCs usually align in magnetic fields as low as 0.1 Tesla, larger field intensities (1–2 Teslas) are recommended in the case of LCPs. Thermotropic LCPs are most conveniently aligned by slowly cooling samples through the nematic/isotropic (N/I) phase transition at a rate of 0.1 K/min. Most of the orientation takes place in the biphasic region and the cooling rate can be increased inside the nematic range. Surface anchoring may also hamper the alignment by a magnetic field so that no sample dimension should be smaller than 100 μm. Nevertheless, the state of alignment will still depend on the shape of the vial holding the sample. Aligned samples may also be produced by applying an electric field. This technique is much more awkward to apply because of the need for electrodes, of complicated flow effects and of the screening of the field by the ions that may be found in the solvent of lyotropic LCPs. When the materials are too viscous to be aligned by a field, one has to use a rheological process to orient the mesophase. The most commonly used in the case of thermotropic LCPs is that of fiber drawing because it is very easy to achieve. It is however a very ill-defined process and sometimes gives rise to discrepancies in the literature. The use of fibers can also be deceptive because two different kinds of organization may be trapped in the skin and the core of the fiber which results in complicated X-ray scattering patterns. Shear alignment can also be obtained in a Couette shear cell. This cell is made of two concentric cylinders which rotate with respect to each other. The material is inserted in the gap between the two cylinders and sheared. Other rheological processes have been recently studied that closely mimic those taking place in the industry.

There are many kinds of X-ray scattering set-ups which can be used to probe the structures of the phases of liquid crystalline polymers. These set-ups range from simple cameras mounted on cheap fixed X-ray tubes to sophisticated and expensive diffractometers using a rotating anode X-ray generator or even synchrotron radiation. It is therefore very important to define carefully the type of desired information in order to choose the best suited apparatus to obtain it. A basic and cheap apparatus involves a fixed X-ray tube equipped with a Cu anode that delivers an X-ray beam of wavelength $\lambda = 1.54$ Å. This X-ray beam can be specifically filtered with a Ni foil and further defined with a pinhole collimator. The sample sits in an oven itself placed between the poles of a strong permanent magnet delivering at least 1 Tesla. The thermal stability of the oven needs not be better than 0.1 K unless phase transitions are specifically studied but temperatures as high as 300 °C are often needed for thermotropic main-chain LCPs. The detection can be made with a photographic plate at a distance of about 6 cm for thermotropic LCPs. This sample-detection distance will vary from a few cm to 50 cm for lyotropic LCPs depending on concentration. It is advisable to keep the sample and photographic plate inside an evacuated camera in order to get rid of air scattering. Such a simple set-up already allows phase identification, nematic order parameter evaluation, smectic period and tilt angle measurements which represents the information most frequently required. In a second step (Fig. 2), the use of a curved monochromator focussing the X-rays at the detection level and of imaging plates significantly improve this set-up and allows one to perform quantitative measurements of the scattering. Smectic reflection intensities can now be safely recorded and the diffuse scattering can also be studied.

More sophisticated experiments involve a rotating anode generator and a 4-circle diffractometer equipped with a very stable oven (0.001 K). Such experiments are meant to study the behaviour around phase transitions, the measurement of very large correlation lengths or the organization at interfaces. Finally, let us mention here the ever growing role of synchrotron radiation. The advent of the third generation synchrotron machines allows very specific applications. The first one makes use of the very small beams (1–10 µm) produced by microfocussing techniques and allows one to probe a liquid crystalline texture at the scale of a micron [10]. The second application is that of time-resolved studies. It is now possible to study the kinetics of alignment in an electric field for instance [11]. Very recently, industrial processing devices have been adapted to synchrotron beamlines and the structures of samples have been examined at the different stages of the process [12].

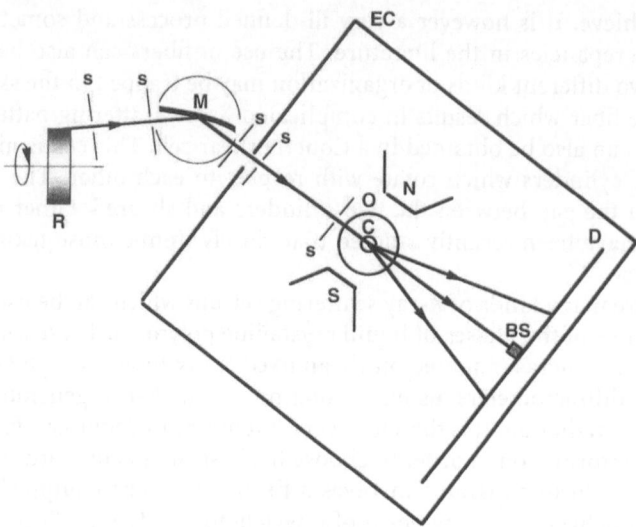

Fig. 2. Scheme of a typical X-ray scattering set-up. R: rotating Cu anode X-ray generator; s: slits; M: curved monochromator; EC: evacuated camera, C: capillary tube; O: oven; SN: permanent magnet; BS: beamstop; D: detection by photographic or imaging plate

3
Main-Chain Thermotropic Liquid Crystal Polymers

In this section we discuss the case of polymers made of rod-like mesogenic moieties linked together via flexible spacers. Linking the rod-like moieties together to create chains naturally strengthens their anisotropic character. Their interactions are therefore stronger and are responsible for the improvement of the mechanical properties which are comparable to those of steel. Therefore, these materials melt at relatively high temperatures (100–300 °C) and quite often directly to a nematic phase which is therefore widespread. In contrast, the fluid smectic phases, SmA and SmC, are not as frequent. Moreover, the SmC phase is as common as the SmA one. Besides, several studies have recently focussed on the crystallization kinetics of fibers using synchrotron radiation. In spite of their industrial relevance, they will not be described in this article.

3.1
The Measurement of the Nematic Order Parameter

The nematic phase differs from the usual isotropic liquid phase by the existence of a long-range orientational order [13]. The strength of this order may be quantified by the first non-zero moment $S = \frac{1}{2} \langle 3 \cos^2 \beta - 1 \rangle$ of the orientational distribution function $f(\beta)$. β is here the angle between the axis of the rod-like moiety with the average orientation direction called the director $\mathbf{n}$ (Fig. 3).

There are several ways of deriving the nematic order parameter S from the X-ray scattering pattern of an aligned sample of the nematic phase. We present

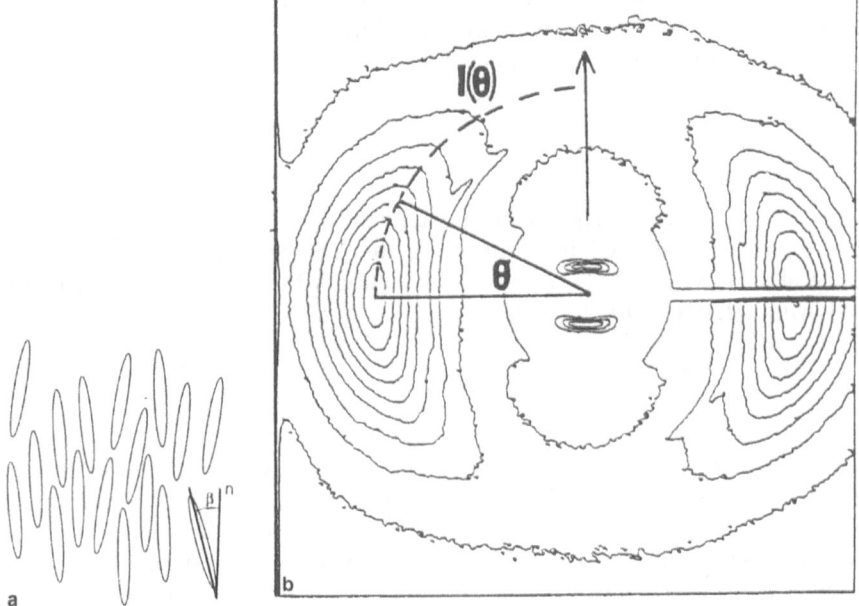

Fig. 3. a Schematic of the nematic phase. **b** Corresponding X-ray scattering pattern (From [14])

here a particularly simple approach first devised by Leadbetter and coworkers in 1979 [14], modified and thoroughly reviewed recently [15]. This approach is based on several assumptions.

1) The scattering moieties are uniform rod-like particles of uniaxial symmetry.
2) The so-called wide angle diffuse ring observable on the X-ray scattering patterns (Fig. 3b) arises purely from the lateral interferences among these moieties.
3) Therefore, a diffuse ring of scattering centered at the origin of reciprocal space is attached to each particle. Thus, the nematic order parameter which characterizes the distribution function of a *single* particle is derived from the interferences among a *cluster* of particles. This assumption is somewhat similar to a mean-field treatment and tends to overestimate S.
4) The form factor of the particle is not taken into account and the correlation length of the cluster is supposed to be infinite so that the scattering of this elementary cluster is reduced to a point. This assumption tends to underestimate S and more or less counterweighs the previous one.

From purely geometrical considerations, one can derive an integral relation between $f(\beta)$ and the scattered intensity $I(\theta)$ (Fig. 3b):

$$I(\theta) = \int_{\beta=0}^{\pi/2} \frac{f(\beta)\sin\beta}{\cos^2\theta\sqrt{\text{tg}^2\beta - \text{tg}^2\theta}}\,d\beta \tag{1}$$

This integral relation is very tedious to invert and the most practical approach there is simply to postulate a convenient form for the distribution function f(β). A very convenient, reasonable and widespread choice is the Maier-Saupe distribution function [13] of the form

$$f(\beta) = \tfrac{1}{Z}e^{m\cos^2\beta} \tag{2}$$

where Z is a normalization constant and m is the Maier-Saupe parameter which is closely related to S. Inserting this distribution function into the integral relation at Eq. (1) leads, after a straightforward calculation, to the following relatively simple formula:

$$I(\theta) = \frac{e^{m\cos^2\theta}}{\sqrt{m}Z\cos\theta}\frac{\sqrt{\pi}}{2}\operatorname{erf}\sqrt{m}\cos\theta \tag{3}$$

where erf is the error function. This simple form allows one to fit directly the experimental curve I(θ) with a small computer and obtain S. The advantage of this method over other ones is that it depends on a fit the quality of which can be judged. This procedure has been tested with a number of nematic compounds of low molar mass and PLCs as well [15]. In all cases, the description of the data was found to be very good or at least satisfactory except for the uncommon case of order parameters larger than 0.85. The accuracy of the method is about 5%. Figure 4 shows two examples of such fits for two members of a family of main-chain LCPs of general chemical formula:

The fairly good quality of the fits validates both Leadbetter's assumptions and the Maier-Saupe distribution function. However, the values of S obtained and even the quality of the fits obviously depend on the odd or even number of (CH_2) groups in the flexible spacer. This odd-even effect is widespread and well known in the field of main-chain LCPs and will be discussed later in this article. The nematic order parameter of main-chain LCPs may reach values as high as 0.85 which demonstrates the very high orientation of the nematic phase of these polymers. Such a large orientation is undoubtedly responsible for the good mechanical properties of this type of materials. The treatment described above therefore provides a very easy way of characterizing the orientational order of a nematic phase. It has also been tested for thermotropic side-chain LCPs and found to be satisfactory as well [15]. Unfortunately, it has not been used yet in the case of lyotropic LCPs except for some aqueous suspensions of mineral ribbons (Sect. 5) which are not quite typical of this family of materials.

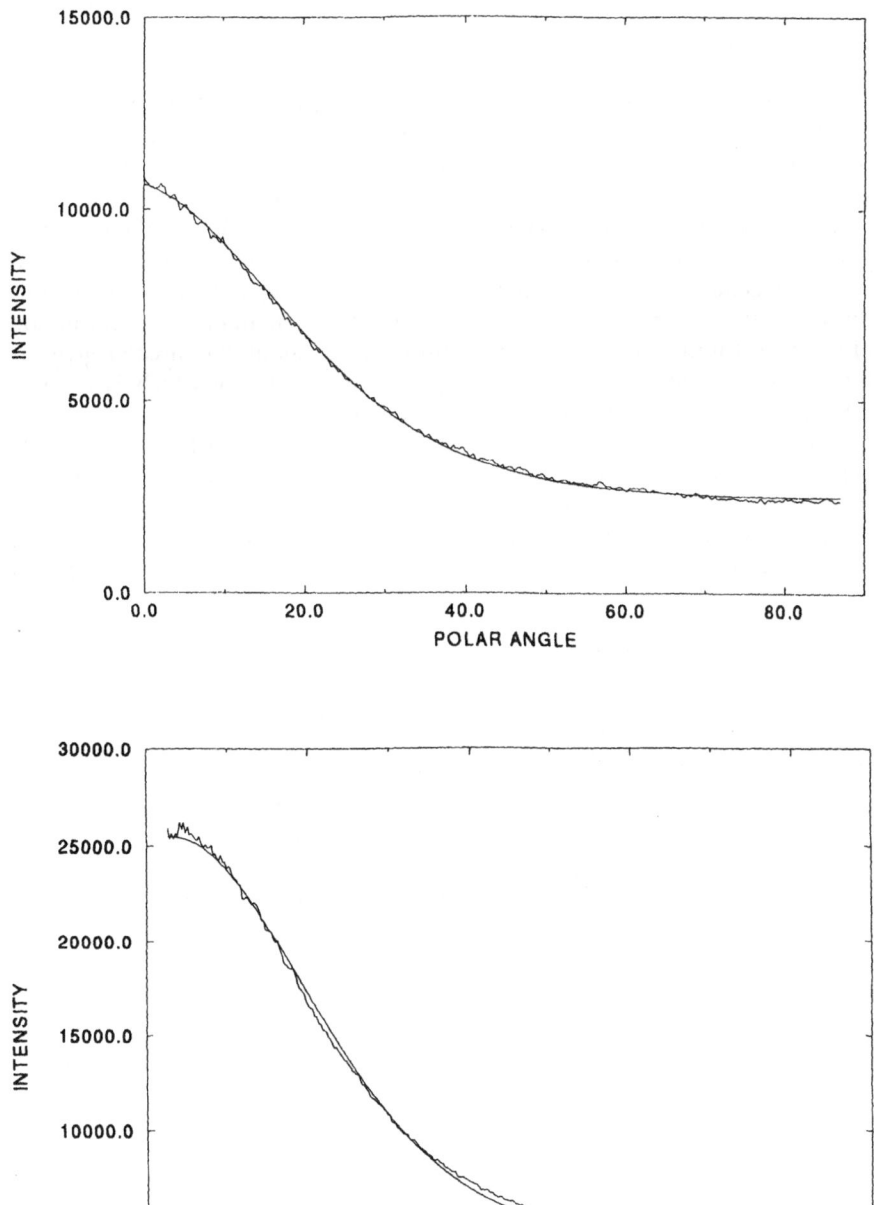

Fig. 4a,b. Fits of the scattered intensity inside the wide angle diffuse ring versus θ for two different main-chain LCPs: **a** n = 11, S = 0.65; **b** n = 10, S = 0.75

3.2
Chain-Like Correlations and the Odd-Even Effect

Overexposed X-ray scattering patterns of aligned samples of the nematic phase of main-chain LCPs often show a set of equidistant diffuse streaks perpendicular to the director [16]. This set of diffuse streaks represents the intersection with the Ewald sphere of a set of diffuse planes which is the Fourier transform of a modulated 1-dimensional object. Most of the time, this object is nothing more than the polymer chain itself with its alternation of rod-like mesogenic moieties and flexible spacers. Through its periodicity, this diffuse scattering provides us with a measure of the length of the conformation of the repeat unit. The width of the diffuse streaks along the director gives us information about the correlation length of this ordering which is a lower estimate of the persistence length of the polymer.

The conformation of the repeat unit of main-chain LCPs very strongly depends on the odd or even number of CH_2 groups in the flexible spacer. This is called the "odd-even effect" and is illustrated in Fig. 5.

In the schematic case of Fig. 5, the mesogenic cores are naturally parallel when the flexible spacer adopts an all-trans conformation if it has an even number of

a)

b)

Fig. 5a,b. Illustration of the odd-even effect: **a** the spacer has an even number of (CH_2) groups; **b** the spacer has an odd number of (CH_2) groups

(CH_2) groups. In contrast, when the spacer has an odd number of (CH_2) groups, it has to adopt some gauche conformations for the mesogenic cores to keep parallel. This decreases the degree of order of the nematic phase. Though this effect also exists for LMMLCs [17], it is very important for main-chain LCPs as it can even determine the type of phase, nematic or smectic, of the members of the homologous series. For instance, all even members of a series may have a nematic phase whereas all odd members have a smectic one [18].

Blackwell and coworkers [19] have extensively studied the case of liquid crystalline main-chain copolymers produced by the random polymerization of two or three different monomers. The X-ray scattering patterns of these materials usually display a set of non-periodic diffuse streaks. These diffuse streaks were analyzed by extending the model of disorder of Hosemann to take into account chemical disorder. The description of the data by this model gave the correlation length of the ordering, information on the conformations of the repeat units and proved the random sequence of the copolymers. Subsequent molecular modelling studies revealed the detailed conformation of the repeat units.

3.3
Hairpin Defects

The main molecular defects displayed by the nematic phase of main-chain polymers were predicted long ago by DeGennes and called "hairpins" [20]. These defects appear when a main-chain polymer suddenly performs a 180° rotation (Fig. 6).

If they exist, these defects should deeply affect the mechanical properties of these materials and therefore many teams have tried to prove their existence. However, these defects are difficult to study experimentally and it is only recently that Cotton and coworkers, in Saclay, France, have clearly proved their existence [16a,b, 21]. This was done by combining small angle neutron scattering (SANS) and X-ray scattering by single domains of the nematic phase. Mixtures of 50% hydrogenated and 50% deuterated polymers were carefully aligned in a magnetic field. The SANS signal due to the polymer conformation was found to be highly anisotropic and could be described by a model of oriented cylinders (Fig. 7). The length and diameter of these cylinders were obtained by a direct fit of the data. Then, the length of the conformation of the repeat unit was measured from the periodicity of the diffuse lines displayed by the X-ray scattering pattern. The polymerization degree of the polymer being known, it was possible to calculate the total contour length of the polymer. Finally, by comparing this contour length and that of the cylinder, the number of hairpins could be determined. As expected, this number increases with temperature and with polymerization degree.

3.4
Smectic Fluctuations

The nematic phase of main-chain polymers often shows smectic fluctuations, in particular of the smectic C type. This phenomenon is sometimes called

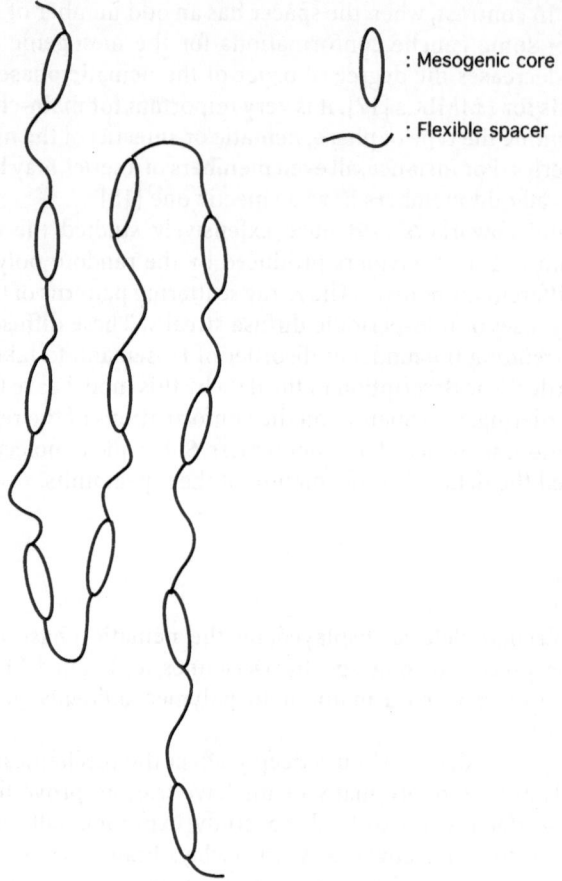

: Mesogenic core

: Flexible spacer

Fig. 6. Main-chain LCP showing two hairpin defects

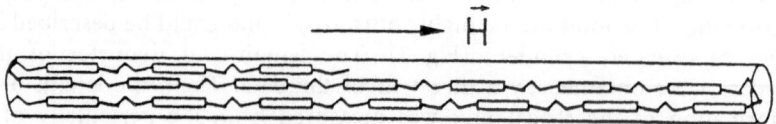

$\vec{H}$

Fig. 7. Conformation of a main chain LCP described as a cylinder. H is the magnetic field direction which aligns the nematic director

"cybotactic groups" in the literature. This name is very misleading however because it suggests that there are little clusters inside the phase in which the smectic ordering prevails, immersed in an ordinary nematic medium. Actually, the phase is uniform and all the rod-like moieties are equivalent. If one picks at random a moiety for the origin of coordinates, then the other moieties around show some smectic short-range order with a probability that decays with the distance from the origin. The distance over which this probability is reduced by a factor $1/e$ is called the smectic correlation length [13]. In the nematic phase, the smectic fluctuations have two correlation lengths $v_\parallel$ and $v_\perp$ along and perpendicular to the director, respectively. The smectic fluctuations are readily detected on the X-ray scattering patterns by the appearance of diffuse spots. When these diffuse spots are located in a direction parallel to the director, they point to the presence of smectic A fluctuations. When these diffuse spots appear at some angle with the director, they are due to smectic C fluctuations and the tilt angle can be directly measured on the pattern. This is illustrated in Fig. 8 which shows the X-ray scattering pattern in the nematic phase of an organometallic main-chain polymer [16c] of structure

2

The widths of the diffuse spots along the director and perpendicular to it are inversely proportional to the correlation lengths. For some reason, main-chain LCPs very often show SmC fluctuations in the nematic phase and when they crystallize, these diffuse spots usually condense into Bragg reflections.

4
Side-Chain Thermotropic Liquid Crystal Polymers

Side-chain LCPs are, in a way, more subtle than their main-chain counterparts because, thanks to the flexible spacers, the polymer main chain (called backbone) still retains many degrees of freedom. In fact, there is a true balance between the natural disorder of the backbones and the liquid-crystalline order of the rod-like moieties. For instance, in the SmA phase, it was recently demonstrated by SANS that the backbone performs a 2-dimensional random walk between the smectic layers [22]. Therefore, these materials have poor mechanical properties but display a rich variety of mesophases, even at room temperature. This makes them more suitable for applications in displays and electro-optic devices.

4.1
The Positional Long Range Order in the SmA Phase

The SmA phase differs from the nematic one by the onset of a 1-dimensional long-range positional order: the smectic layers [13]. This can be noticed by the appearance of the smectic reflections on the X-ray scattering patterns. Rigorously speaking, a 1-dimensional order cannot be

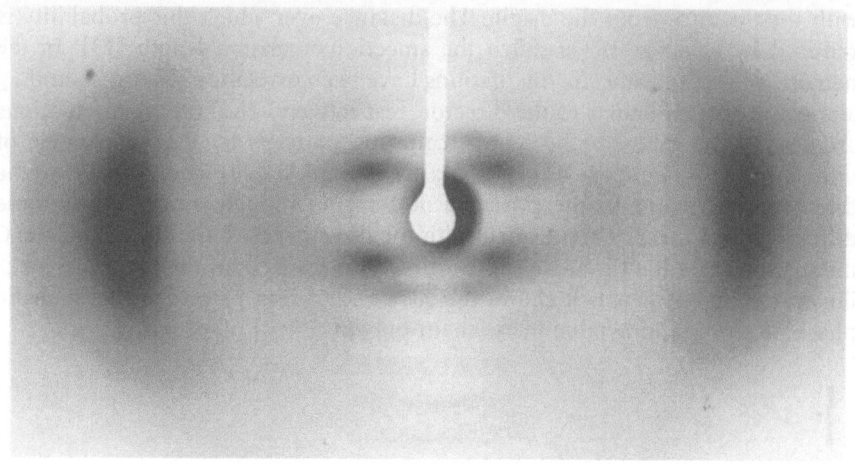

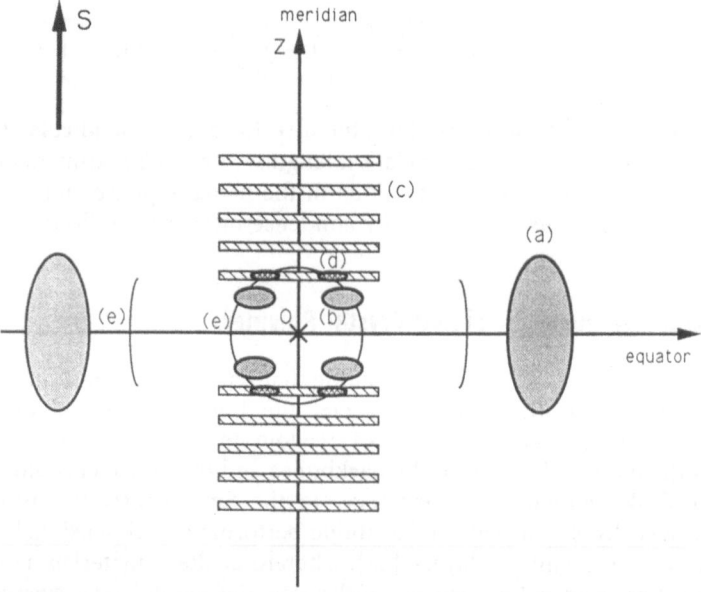

Fig. 8. a X-ray scattering pattern of a fiber of an organometallic main-chain LCP quenched from the nematic phase. **b** Schematic representation of a. (a) = wide angle diffuse ring; (b) = diffuse spots due to the SmC fluctuations; (c) = equidistant diffuse streaks; (d) = other weak diffuse spots; e = faint lines due to a slight crystallization

truely long-range. When using conventional apparatus, this effect can be safely neglected in most cases except perhaps in the vicinity of the phase transitions. However, by using sophisticated X-ray scattering set-ups, it can be shown that the smectic reflections are not resolution-limited which demonstrates that the smectic order only extends over a finite range [23].

In the special case of side-chain LCPs, it was found that the smectic order extends over about 1000 Å [24]. This value increases with temperature which was interpreted by the annealing of defects. The correlation length also increases with increasing spacer length which means that the smectic order develops more easily when the coupling between the mesogenic cores and the backbone gets weaker.

Moreover, the profiles of the smectic reflections can give very valuable information about the physical constants of the mesophase. Thus, Nachaliel et al. have obtained estimates of the elastic constants B for compression of the layers and K_1 for splay vs temperature [25]. Oddly enough, the critical exponent ϕ of the elastic constant B differs strongly from that of LMMLCs but agrees with theoretical predictions. DeJeu et al. have recently compared the reflection profiles of the SmA phase of side-chain LCPs and of side-chain liquid crystalline elastomers [26]. These latter materials are obtained by cross-linking the corresponding side-chain LCPs. Very interestingly, they have shown that, upon cross-linking, the reflection profiles become resolution limited which shows that the smectic order becomes truly long-range when 3-dimensional correlations are introduced.

From a different point of view, the most striking difference between the SmA phases of side-chain LCPs and those of LMMLCs is that the former usually show many orders of smectic reflections whereas the latter usually only show one [27, 28]. This observation will help us to answer the following basic question: how are the backbones of the side-chain LCPs affected by the smectic field (Fig. 9)? Do they keep a more or less random 3-dimensional conformation or are they strongly confined between adjacent sublayers of mesogenic cores?

It is actually possible to answer this question by the inspection of the projection $\rho(z)$ of the electron density along the normal to the layers which we call electron density profile in the following [28]. Indeed, the number of smectic reflections is related to $\rho(z)$. $\rho(z)$ can be expanded in a Fourier series and, since it is a centrosymmetrical function, only cosine terms are relevant:

$$\rho(z) = \sum_n a_n \cos \frac{n2\pi z}{d} \qquad (4)$$

where d is the smectic period and the a_n coefficients are real. The X-ray scattering experiment only provides the intensities of reflections which are proportional to the squared absolute values of the a_n coefficients so that the signs of the a_n coefficients remain unknown. This is the famous phase problem classical in crystallography but in a much simpler centrosymmetrical 1-dimensional version. Then, all the sign combinations must be generated and their number grows exponentially with the number of detected reflections. Fortunately, no more than five or six reflections are usually observed.

We shall illustrate this method by the example of a homologous series of polysiloxanes of the chemical formula:

Fig. 9a,b. Possible limit conformations for the backbones in the SmA phase: **a** microsegregated; **b** random. (The flexible spacers have been omitted for clarity)

 : Backbones

: Mesogenic cores

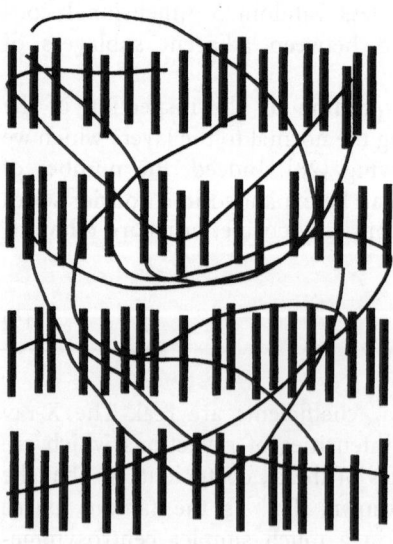

 : Backbones

: Mesogenic cores

$$(CH_3)_3\!-\!Si\!-\!(\!-\!O\!-\!\underset{\underset{\displaystyle (CH_2)_n\!-\!O\!-\!\text{⬡}\!-\!OCO\!-\!\text{⬡}\!-\!OC_mH_{2m+1}}{|}}{\overset{\overset{\displaystyle CH_3}{|}}{Si}}\!-\!)_{35}\!-\!O\!-\!Si\!-\!(CH_3)_3$$

3

Table 1. Amplitudes normalized to the first order reflection, of the different orders of reflection from the smectic layers for different $P_{n,m}$ polymers

	$P_{3,4}$	$P_{4,4}$	$P_{3,8}$	$P_{5,8}$
a_1	1	1	1	1
a_2	1.60	1.55	0.75	0.85
a_3	1.50	0.70	0.60	0.75
a_4	0.80	0	0.55	0.60
a_5	0	0	0	0.3

labelled $P_{n,m}$. Most of these polymers have a SmA phase and Table 1 shows the amplitudes of the different orders of reflection normalized to the amplitude of the first. Strikingly, for some members of this series, the second order of reflection is stronger than the first one. At this stage, it should be recalled that for usual LMMLCs, the amplitude of the second order is 10^{-3} that of the first. In that case, the electron density profile is a simple sinusoid of period d. The molecular organization in the SmA phase of these side-chain LCPs is therefore very different. Among all the sign combinations, it is possible to select the suitable one which represents $\rho(z)$ (Fig. 10).

This density profile shows a wide maximum in its middle (z = d/2) which represents the region of mesogenic cores. Another secondary maximum (z = 0) is observed in between the sublayers of mesogenic cores. This secondary maximum arises from the backbones which have a relatively large electron density because they have silicon atoms. Therefore, the backbones appear squeezed between the sublayers of mesogenic cores so that the structure of this SmA phase may be described in terms of "microphase separation". This is the situation depicted in Fig. 9a. Moreover, Fig. 10 also shows that adding aliphatic parts to the macromolecule helps the backbone to resist the confinement. Nevertheless, one should not conclude that all the SmA phases of LCPs can be described in this way. For instance, in the same homologous series, polymer $P_{4,1}$ shows only one order of smectic reflection so that $\rho(z)$ is sinusoidal and the backbones are not confined at all (Fig. 9b).

This method can be extended to neutron diffraction by considering two side-chain LCPs labelled PMA(H,D)OC$_4$H$_9$ of formula

$$-(\!-\!CR_2\!-\!\underset{\underset{\displaystyle CO_2\!-\!(CH_2)_6\!-\!O\!-\!\text{⬡}\!-\!CO_2\!-\!\text{⬡}\!-\!OC_4H_9}{|}}{\overset{\overset{\displaystyle CR_3}{|}}{C}}\!-\!)_m\!-$$

4

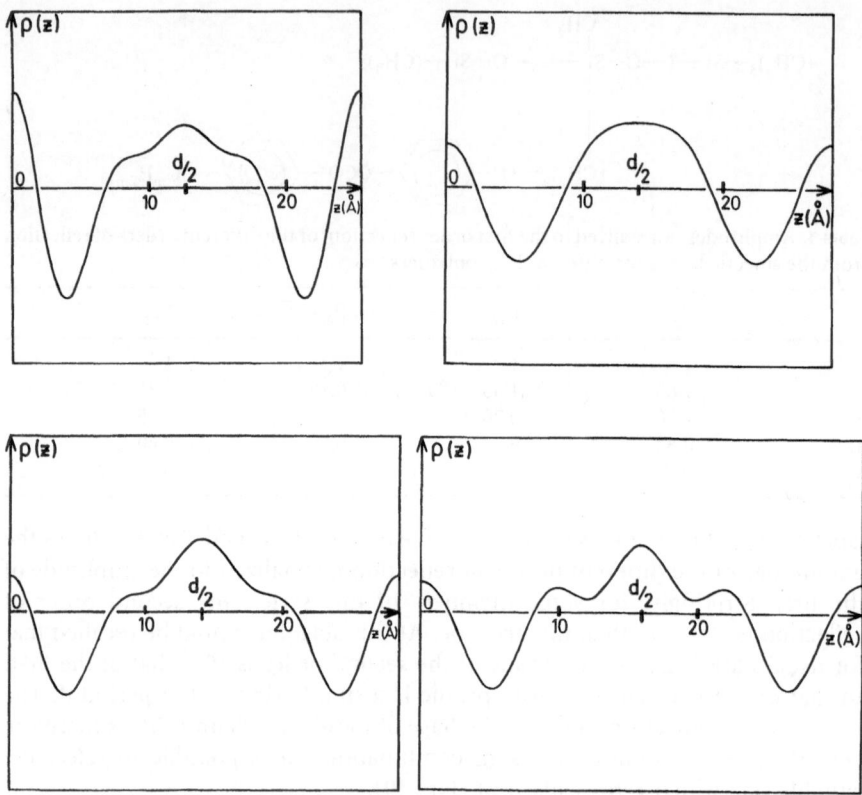

Fig. 10a–d. Electron density profiles of the $P_{n,m}$ polysiloxanes: **a** $P_{3,4}$; **b** $P_{4,4}$; **c** $P_{3,8}$; **d** $P_{5,8}$. The origin is chosen in between two successive sublayers of mesogenic cores

only differing by the fact that one is deuterated (R = D) on the backbone whereas the other one (R = H) is fully hydrogenated [29]. The intensities of the smectic reflections were recorded for both polymers. The difference between them directly gives, by Fourier transform, the profile of backbone units along the normal to the layers (Fig. 11).

This distribution again shows that the backbones are quite segregated in between adjacent layers of mesogenic cores and this segregation is less and less efficient as temperature increases.

4.2
Fluctuations in the SmA Phase

We have just seen that the backbones are very often confined in the SmA phase of side-chain LCPs. Therefore, they have lost many degrees of freedom and entropy. The backbones usually react to this confinement by inducing different kinds of fluctuations and disorder. This is obviously seen on overexposed X-

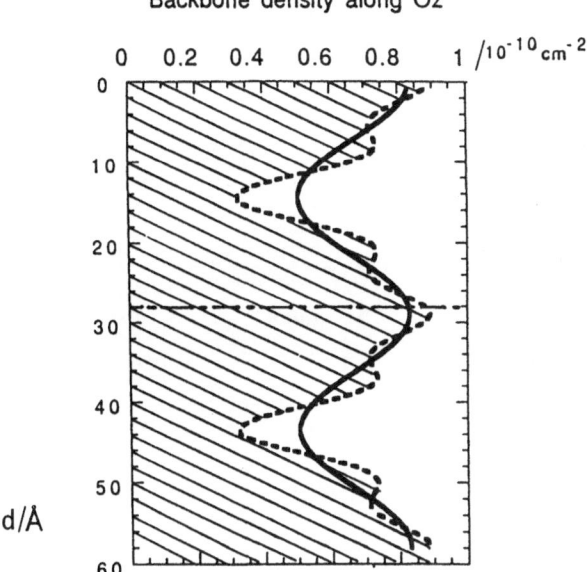

Fig. 11. Distribution of backbone units along the normal to the smectic layers. *Continuous line*: just below the N/SmA transition; *dashed line*: room temperature

ray scattering patterns (Fig. 12) which show that a very large part of the scattered intensity lies out of the smectic reflections and the wide angle diffuse rings [30]. This diffuse scattering in shape of diffuse spots and streaks is not usually detected on the X-ray scattering patterns of LMMLCs and directly reflects the influence of the backbones on the molecular organization of the SmA phase. In the following, we shall discuss two examples of fluctuations in order to illustrate how this kind of diffuse scattering can be dealt with.

Let us first consider the set of equidistant diffuse streaks (c) perpendicular to the director [1a,b, 30]. These streaks are very similar to those observed on the X-ray scattering patterns of the nematic phases of main-chain LCPs discussed in Sect. 3.2 but they must be interpreted differently because the SmA phase shows (quasi) long-range positional order. These diffuse streaks also correspond to the intersection with the Ewald sphere of a set of equidistant diffuse planes. But here this set represents the Fourier transform of uncorrelated rows of side-chains displaced along the director from their equilibrium position inside the layers (Fig. 13).

This is the classical description of an uncorrelated linear disorder and the results of this description are the following [9a]. Let us call d the smectic period, f the form factor of the side-chain, s the scattering vector (s = $2\sin\theta/\lambda$ where 2θ is the scattering angle and λ the wavelength), u the displacement of the side-chain away from its mean position in the layer and N the average number of side-chains in the rows. Furthermore, let us assume that the rows are completely uncorrelated. The scattered intensity is then given by:

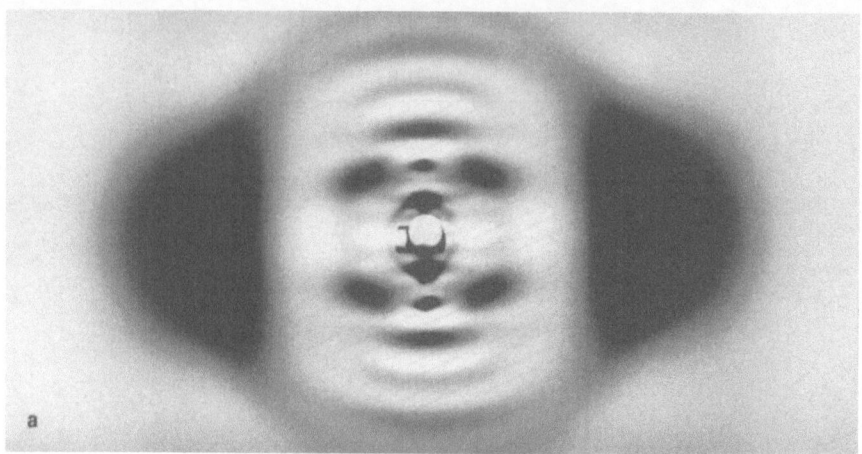

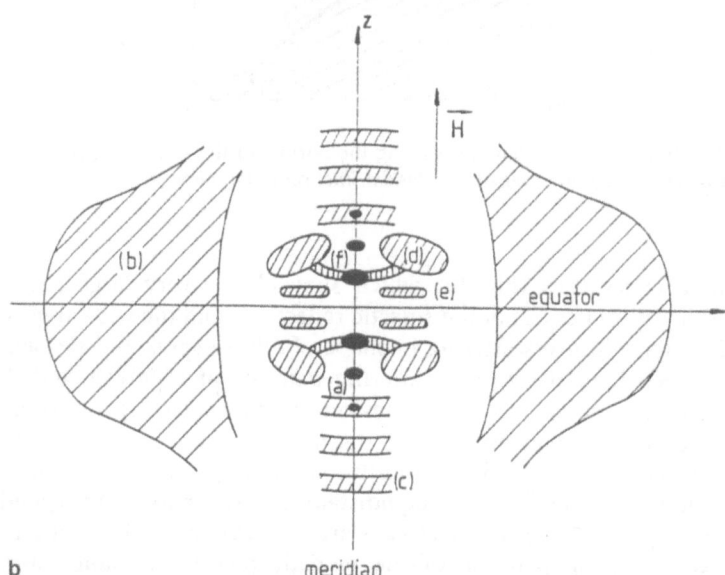

Fig. 12. a Overexposed X-ray scattering pattern of polymer $PMAOC_4H_9$ in the SmA phase. **b** Schematic representation of a. (a) = smectic reflections; (b) = wide angle diffuse ring; (c) = diffuse streaks; (d) = large diffuse spots due to the undulations; (e) = small diffuse spots due to some weak antiparallel ordering of the side chains; (f) = "moustaches" due to localized defects arising from layer crossings by the backbones. H is the magnetic field direction

$$I(s) = \frac{f^2 \sin^2 N\pi s.d}{\sin^2 \pi s.d} \sin^2 2\pi s.u \qquad (5)$$

The X-ray scattering is essentially localized in equidistant parallel planes of periodicity 1/d and perpendicular to the director (Fig. 14).

Their width along the director is 1/Nd, thus inversely proportional to the length of the row. No diffuse plane going through the origin can be found since **s.u** = 0 which is a typical feature of a purely displacive 1-dimensional disorder. The diffuse plane intensities are proportional to $\sin^2(2\pi \text{ s.u})$ and so the amplitude of the longitudinal displacement can also be estimated. Finally, in a given diffuse plane, the intensity is only described by the transverse form factor of the row which is that of the side-chains in most cases. For instance, the diffuse streaks (c) that can be seen in Fig. 12 can be interpreted in this way, leading to an average longitudinal displacement **u** of about 3–5 Å, a small fraction of the smectic period, and a longitudinal correlation length of about 100–200 Å, which represents a few layers. LMMLCs usually do not show such diffuse streaks. Their appearance is therefore a consequence of the polymeric nature of the compounds. Actually, each side-chain is chemically linked to a backbone so that when a side-chain fluctuates around its mean position in the smectic layer, this fluctuation is transmitted through the backbones to the other side-chains. This means that these correlations depend on the mechanical properties of the backbones sublayer. If this sublayer is very stiff, the correlations will extend over a long range. One can indeed notice that the diffuse streaks are stronger for polymers which show many smectic reflections.

Let us now turn to another type of fluctuations which very often affect the SmA phase of side-chain LCPs [1a,b, 30]. Figure 12 also shows some diffuse spots (d) located in the same reciprocal planes perpendicular to the director as the smectic reflections. Because of the uniaxial symmetry of the SmA phase, these diffuse spots actually represent the intersection of diffuse tori with the Ewald sphere. These diffuse tori arise from a transverse

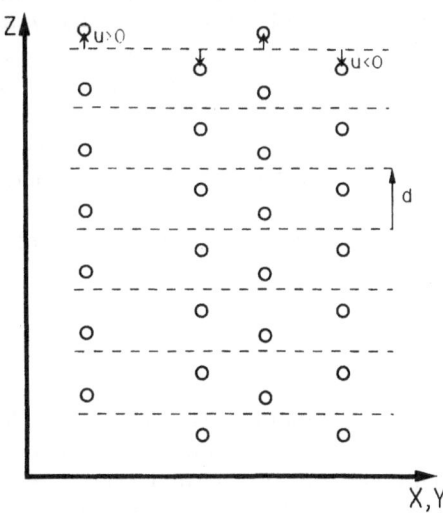

Fig. 13. Uncorrelated linear disorder. The *circles* represent the side chains centers of gravity. Four vertical uncorrelated rows have been represented

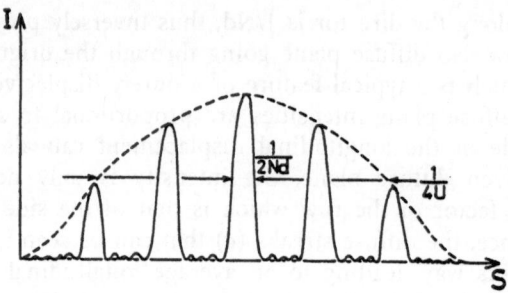

Fig. 14. X-ray scattered intensity in a direction parallel to the director. Each peak represents a diffuse plane. Their intensities are modulated (*dashed line*) by a factor $\sin^2(2\pi \mathbf{s} \cdot \mathbf{u})$ related to the displacement vector **u**

modulation of the smectic layers, of wave vector **a** and polarized along the director. The modulus of **a** (about 20 Å in the case of Fig. 12) is given by the inverse of the tori radius. Since the modulations of adjacent layers are in phase, we can consider a 2-dimensional local ordering based on the wave vector **a** and the smectic period **d**. The diffuse torus (d) may then be labelled "102". The fact that no such torus can be found in the equatorial plane shows that the modulation of the layers is purely displacive; thus the overall density remains uniform. Moreover, the fact that no other "hkl" diffuse spot other than "10l" can be detected in a given (00 l) reciprocal plane shows that the modulation is sinusoidal (Fig. 15).

Nevertheless, the "10l" diffuse spots being quite broad, this 2-dimensional ordering is only correlated over a short distance. Its correlation length is only about 50 Å. Besides, although the smectic period **d** is well defined, the modulation wave vector **a** is submitted to large fluctuations.

We need to explain why we only observe the "102" diffuse spot. We could have expected that the "101" diffuse spot should be stronger than the "102". This can be explained by considering that the undulations only affect a part of the layer so that the structure factor governing this modulation will be different from that governing the smectic reflection intensities. For simplicity, the electron density profile of the smectic layer may here be represented by step functions (Fig. 16).

Moreover, let us assume that the displacement follows the sinusoidal equation:

$$z = \frac{u}{2}\cos\frac{2\pi x}{a} \tag{6}$$

where u is the displacement amplitude. The Fourier transform of the sinusoid will then describe the displacement contribution to the intensity scattered along the $[h0s_z]$ reciprocal row as a term $J_h^2(\pi u s_z)$, where J_h is the Bessel function of order h and s_z is the z-component along the director, of the scattering vector **s**. Finally, the diffuse intensity $I_h(s_z)$ vs s_z along the reciprocal row $[h0s_z]$ will read as:

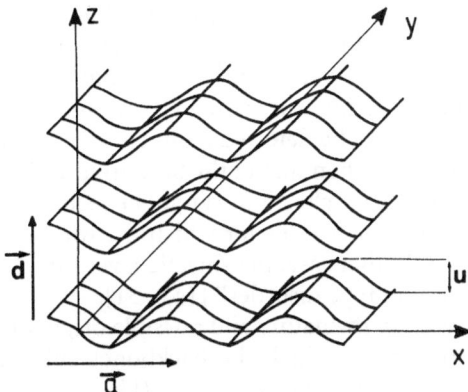

Fig. 15. Local undulations model for the backbones sublayer. The backbones perform a more or less two-dimensional walk on the undulating surface

$$I_h(s_z) = J_h^2(\pi u s_z) \left[\frac{np \sin \pi s_z p}{\pi s_z p} - \frac{np \sin \pi s_z (p+q)}{\pi s_z (p+q)} \right]^2 \qquad (7)$$

where the terms inside the brackets is the form factor of the profile shown in Fig. 16. The fit parameters are p, q and u and in the case of the diffuse spots seen in Fig. 12, we obtain $p \approx 6$ Å, $q \approx 4$ Å and $u \approx 10$ Å. These values are in good agreement with estimates made with molecular (Dreiding) stereomodels and with the values derived by inspecting the electron density profiles.

Now, the very weak intensity of the "103" diffuse spot and the absence of the higher order ones can be explained by the model of "second type disorder" developed by Guinier [9a]. In contrast with thermal fluctuations (which is a "first

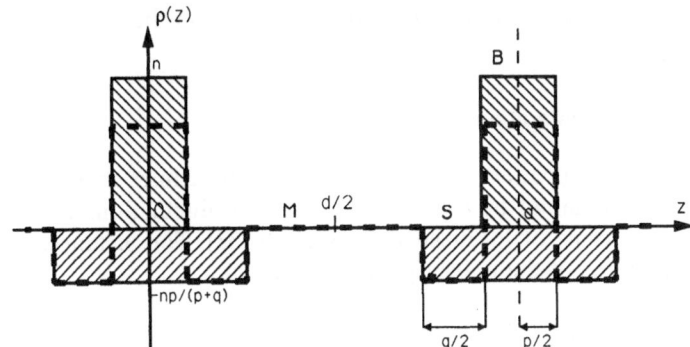

Fig. 16. Representation of the electron density in terms of step functions inside the smectic layer. X-rays being only sensitive to the contrast, the electronic density of the mesogenic cores has been arbitrarily assigned to zero. Since we do not make absolute intensity measurements and since we deal with a purely displacive disorder, the backbones and the spacers have been represented by two different step functions of equal *hatched areas*. The resulting electron density profile is shown in *thick broken line*. M, S and B stand for the mesogenic cores, spacers and backbones regions respectively

type disorder"), this type of disorder destroys the long-range correlations, and its main effect is to give a finite width to the Bragg reflections of the lattice it affects. In this frame, the width of a diffuse "hkl" spot increases with increasing hkl indices but its integrated intensity remains constant and is only governed by the form factor. However, to be detected a diffuse spot must have a maximum intensity appreciably larger than that of the background. This explains why the higher order "104" and "105" reflections cannot be seen. It should be noted here that there are other side-chain LCPs for which the whole "10l" row of diffuse spots can be observed. In that case, the smectic layer undulates as a whole and there is only little "second type disorder" [1a,b, 31].

We now comment upon the nature and possible origin of these undulations. They create small regions in which the backbones sublayer is curved. This sublayer is a local array of mounds and wells (Fig. 17).

In these domains, the spacers should assume a splay-like configuration within which all the side-chains probably point in the same direction. In the case of $PMAOC_4H_9$, considering circular mounds and wells of diameter ≈ 10 Å, the area encompassed by each mound or well will be roughly 100 Å^2 so that about five mesogenic cores are needed to build up one domain. For this particular LCP, the detailed conformation of the backbones has been obtained from SANS experiments. The backbones perform an almost 2-dimensional random walk with a gyration radius in the layer plane of $R_\perp \approx 100$ Å. This shows that a single backbone runs through a fairly large number of these mounds and wells. A possible explanation of such undulations is that the packing density of the mesogenic cores within a smectic layer may be different from that of the backbones segments. This would create an intrinsic frustration depending on the details of the random walk performed by the backbones. The modulation would then appear to relieve this frustration in a similar way as that of the "Skoulios pinches" [32] in discotic columnar phases. This hypothesis should be checked by studying the influence on the layer undulations of the mesogenic core grafting density on the backbone.

4.3
Scattering by Edge-Dislocations in the SmA Phase

The X-ray scattering patterns of side-chain LCPs sometimes show some diffuse scattering which cannot be explained in terms of fluctuations but has to be interpreted in terms of localized defects [1a,b, 33]. For instance, the X-ray scattering patterns of the polyacrylates of formula

in the SmA phase shows a diffuse scattering in the shape of "butterfly wings" which probably arises from edge-dislocations (Fig. 18).

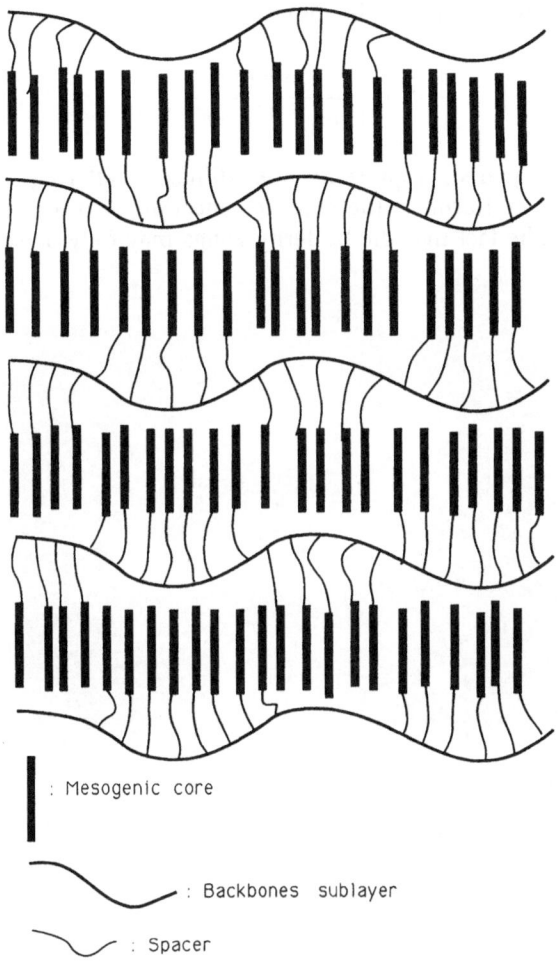

: Mesogenic core

: Backbones sublayer

: Spacer

Fig. 17. 2-dimensional cut of the undulation local order schematic picture. Only a part of the smectic layer undulates, i.e. the backbones and the spacers. The layers undulations induce mounds and wells regions where the spacers adopt a splay-like configuration

The displacement field of an edge-dislocation in the SmA phase has been calculated by De Gennes in the framework of the elastic continuum theory [34]. This field depends on a typical length $\lambda = (K_1/B)^{1/2}$ where K_1 and B represent the elastic constants of the SmA phase for splay deformation and compression of the layers respectively. The distorted region is essentially located inside the parabola (P) of equation: $z = \lambda^{-1}x^2$ (Fig. 19).

The elastic deformation field may be expressed in the form:

$$u_n(x) = \frac{d}{4} + \frac{d}{4\pi} \int \frac{dq}{iq} e^{iqx} e^{-nd\lambda q^2} \tag{8}$$

and its Fourier transform will be described by

$$A'\left(s_x, \frac{p}{d} + \delta s_z\right) \approx \frac{pF\left(\frac{p}{d}\right)}{s_x} \frac{4\pi^2 \lambda d s_x^2}{4\pi^2 (\delta s_z)^2 d^2 + (4\pi^2 \lambda d s_x^2)^2} \tag{9}$$

which, when squared, represents the diffuse scattering intensity at a wavevector $(s_x, \delta s_z)$ around the smectic reflection of order p and structure factor $F(p/d)$. The fact that the butterfly wings may only be seen around the

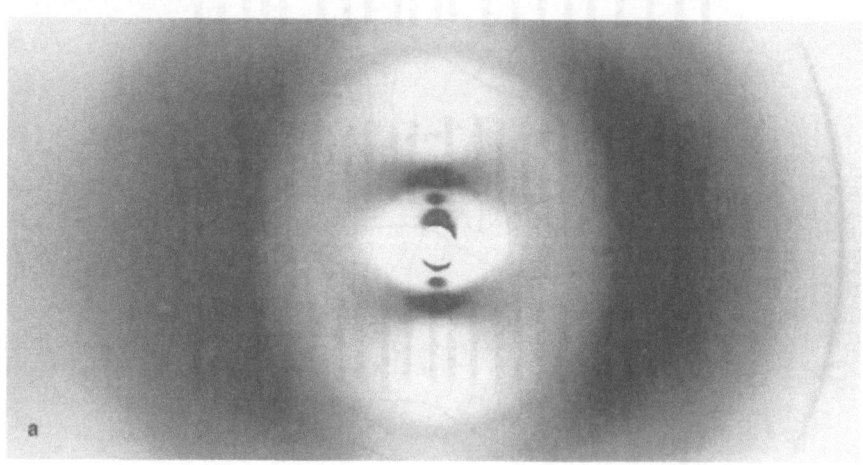

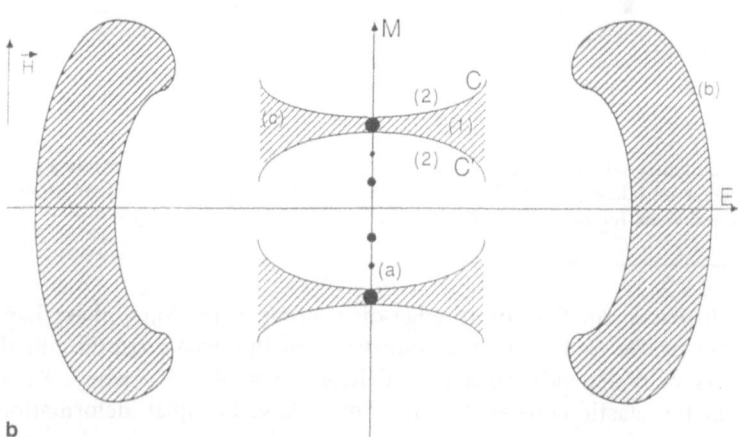

Fig. 18. a Overexposed X-ray scattering pattern of the cyanosubstituted P_8 polyacrylate in the SmA phase. **b** Schematic representation of a. (a) = smectic reflections; (b) = wide angle diffuse ring; (c) = diffuse streaks called "butterfly wings". Parabolae (C) and (C') are essentially contrast lines which separate region (1) of stronger scattering from the background (2). M and E stand for meridian and equator, respectively. H is the magnetic field direction

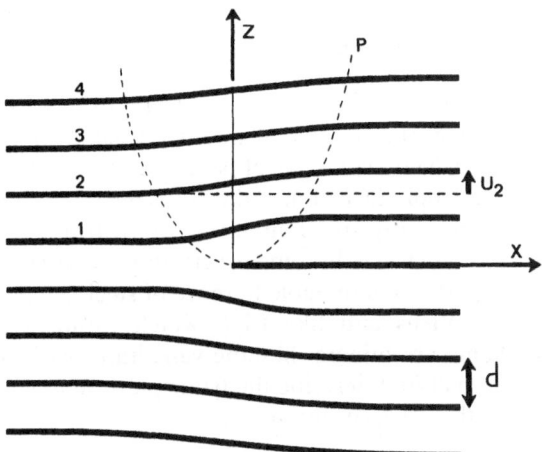

Fig. 19. Scheme of an edge dislocation in a smectic A phase. d is the smectic period and u_2 the displacement of layer #2

third order smectic reflections comes partly from the influence of the prefactor p in the formula of the scattered amplitude and partly because the diffuse scattering follows the peculiar structure factor F(p/d) of the smectic layers. The isointensity lines and the curves (C) and (C′) can also be calculated and compared to the experiment. The only fit parameter is λ and the value obtained is $\lambda \approx 1$ Å, a value significantly less than that of the SmA phase of LMMLCs which is usually around 10 Å [13]. The values of K_1 for several side-chain polymers have already been measured and were always found to be of the same order of magnitude as those of LMMLCs [35]. This suggests that this low value of λ is due to the fact that the elastic constant B for compression of the layers should be about 100 times larger for the SmA phase of side-chain LCPs than that of LMMLCs. Indeed, the high resolution X-ray study of Davidov and coworkers shows that B may be extrapolated to large values away from the N/SmA phase transition [25]. The same conclusion can be reached from the large number of observable reflections of polymeric SmA phases and also from the reflectivity experiments of Shashidhar and coworkers [36]. It should also be recalled here that the smectic layer fluctuation modes cannot usually be detected in the X-ray scattering patterns of the SmA phase of side-chain LCPs. This suggests a priori different orders of magnitude for B. The relatively large value of the elastic constant B may be explained by the intrinsic rigidity of the backbones sublayer and also by the fact that the permeation effect in side-chain LCPs also involves the diffusion of the backbone. Finally, the ratio of the diffuse scattering to the smectic reflection intensities was estimated and leads to an edge-dislocation density of $\approx 10^8$ cm^{-2}. Thus, fairly large dislocation densities are needed in order to observe their contribution to the X-ray scattering patterns.

5
Lyotropic Liquid Crystal Polymers

In this section, we consider the case of solutions of rigid or semi-flexible polymers which display one or several liquid crystalline phases in a given range of concentration. The main control parameter is not the temperature as is the case for thermotropic LCPs but rather the concentration of polymer in the solvent. There are many different kinds of lyotropic LCPs. Some are synthetic like Kevlar which has become a very important structural material with mechanical properties comparable to those of steel. Some are natural like the Tobacco Mosaic Virus and like DNA which shows a nematic and a hexagonal phase. Some are mineral like the vanadium pentoxide ribbons. In the next section, we shall first describe the lyotropic system which is probably best known, namely the tobacco mosaic virus.

5.1
The Example of the Tobacco Mosaic Virus

The tobacco mosaic virus (TMV) is one of the best documented examples of lyotropic LCPs [37]. In fact, this particle is known down to atomic resolution. The TMV is a rigid rod of length L = 3000 Å and diameter D = 180 Å. It is a highly charged particle with a linear charge density of 1–2 e^-/Å at pH 7. Aqueous suspensions of TMV show an isotropic phase, a nematic phase, a smectic A phase and a crystalline phase in well defined ranges of concentration and ionic strength. There have also been reports of a hexagonal mesophase with a hexagonal packing of the particles in the plane perpendicular to the director but only short range correlations along the director. The main advantage of the TMV is that all particles are rigorously identical in mass, length, diameter and charge density. Therefore, the suspensions of TMV constitute a good model system to test the theories of the nematic ordering such as the Onsager theory which will be briefly reviewed in the next section. In particular, the mesophases are unsensitive to temperature which suggests that the system may be described by hard core theories.

Another crucial advantage of the suspensions of TMV is that they can be oriented in a magnetic field. Meyer and coworkers could therefore obtain highly oriented X-ray scattering patterns of suspensions of TMV [38]. They measured the nematic order parameter S by a different method from that described in Sect. 3.1 because, in this case, interparticle interference effects only appear at small angles and independent particle scattering can be easily measured at larger angles. Very large order parameters (0.77 < S < 0.95) were obtained in good agreement with a modified Onsager theory. Aligned samples of TMV suspensions were also studied by small angle neutron scattering [39]. A small angle peak at q = 0.14 nm^{-1} = $2\pi/450$ $Å^{-1}$ was detected and is due to interparticle interferences. Altogether, as expected, the nematic phase of TMV only shows short-range positional order. Moreover, the correlation length of this positional order remains the same on both sides of the N/I phase transition. A striking feature of the scattering by the TMV is that all molecular details are mostly irrelevant. The scattering patterns may be described by models only involving uniform cylindrical particles. This feature is actually typical of most lyotropic LCPs.

The TMV suspensions show a second order N/SmA phase transition that can be studied by light scattering [40] rather than X-ray scattering because the length of the particles is closer to the wavelength of visible light. A selective Bragg reflection of light was observed which demonstrated the existence of a layered structure of periodicity comparable to the virus length. Moreover, smectic fluctuations were also detected in the nematic phase at concentrations close to that of the N/SmA phase transition. The same experiment was also reported using suspensions of the fd virus, a similar but more flexible linear virus [41]. The main effect of the increased flexibility is to make the phase transition first order. The light scattering could be described by the usual theories of the scattering by smectics in good agreement with the intuitive image of linear particles stacked in register to form layers. Moreover, five orders of smectic reflections could be recorded in the case of the fd virus which shows that it forms a highly layered phase not unlike the micro-segregated organization of the SmA phase of side-chain LCPs.

5.2
Theoretical Predictions

We shall first quickly describe the Onsager theory because it is one of the most important statistical physics theories of the nematic ordering though it does not aim at predicting precisely the molecular organization of the nematic phase. In 1949, Onsager derived a microscopic theory to describe the nematic ordering of the suspensions of TMV [42]. The building blocks are considered to be hard rigid rods of very large aspect ratio. The free energy is only determined by entropy. In this frame, the nematic ordering results from the competition between two types of entropy. The first one is the orientational entropy of the rods which is maximum when the rods are oriented at random. The second one is the translational entropy which is maximum for the smallest free volume per rod. The central quantity in this theory is the excluded volume $b = \pi L^2 D/4$ which is related to the second virial coefficient. Note that the excluded volume is larger than the particle volume by a factor L/D. In the limit of $L/D \gg 1$, Onsager predicted that the N/I phase transition should be first order with coexistence concentrations $bc_i = 3.3$ and $bc_n = 4.2$ where c_i and c_n are the number densities of the isotropic and nematic phases respectively. A very large jump of the nematic order parameter $S \approx 0.80$ was also predicted at the transition. Moreover, the transition is temperature independent since hard-rod fluids are athermal systems. Onsager also adapted his model to the case of electrically charged objects. He showed that the model could still be used provided that the bare particle diameter be replaced by an effective diameter obtained by taking into account the Debye screening length. Moreover, Onsager also treated the case of the nematic ordering of suspensions of hard disks in the same seminal paper.

Since then, a large body of work, both theoretical and experimental, has dealt with the Onsager model and the N/I phase transition of hard rods. This subject has been recently reviewed by Vroege and Lekkerkerker [43]. A major improvement was brought by Khokhlov and Semenov who considered the effects of rod flexibility [44].

More recently, Selinger and Bruinsma [45] have developed a theory of the hexagonal and nematic phases of chains. This theory describes not only the phase transitions but also the correlation functions and it predicts the main features of the X-ray scattering patterns. Several types of systems are addressed ranging from thermotropic main-chain LCPs to lyotropic LCPs and even worm-like micelles and columns of discotic liquid-crystals. The main predictions of this theory are the following. In the nematic phase, in contrast with LMMLCs, the small angle scattering should be very anisotropic and should show a peculiar butterfly pattern. In particular, a line of extinction of the scattering should appear along the director ($q_\perp = 0$). However, the presence of a large defect concentration in the nematic phase [45c] should alter this pattern back to a merely anisotropic scattering not unlike that expected for LMMLCs. Moreover, at a fixed finite $q_\perp$ and small q_z, the lineshape should be Lorentzian. Although true Bragg reflections are predicted for the hexagonal phase, these reflections should nevertheless be surrounded by a large thermal diffuse scattering extended along the director. The small angle scattering should also display a butterfly pattern. In principle, it should be possible to extract the values of the elastic constants from the quantitative study of the X-ray scattering patterns. However, to the best of our knowledge, such studies are still extremely uncommon.

5.3
Orientation Under Flow of Lyotropic LCPs

Lyotropic LCPs are often used as structural materials and are therefore processed in industrial facilities. These processes usually involve flow so that it is extremely important to understand the rheological properties of these materials. These properties are of course related to the molecular organization which can be probed by X-ray scattering. Unfortunately, the flow behavior of LCPs is very complicated both from the theoretical and the experimental points of view. Therefore, we shall only briefly review this topic.

A number of studies have appeared in the literature which describe in-situ X-ray scattering by mesophases of LCPs submitted to various kinds of flows. Although the three major classes of LCPs have been investigated under flow conditions, the class of lyotropic LCPs is probably the best documented. Depending on the nature of the mesophase, the relative values of its different viscosities, the concentration and the shear rate, different situations are expected. An intuitive limit case is the flow aligning regime found at high shear rates in which the director lies close to the direction of flow. In contrast, many other situations called tumbling, wagging, log-rolling, etc. were also predicted [46] and experimentally found at lower shear rates. One has to pay particular attention to the initial state of the sample. Whether it is unoriented or a single domain will determine the "cascade" of regimes that will be observed. Most experimental studies have dealt with initially unoriented samples.

Mitchell and coworkers have recently investigated the solutions of hydroxypropylcellulose (HPC) subjected to shear flow in both steady and transient states [47]. They observed a transition from a low orientation at low shear rates to a rather large orientation (S ≈ 0.5) at higher shear rates. A first transition takes place at shear rates in the range 0.3–1 s^{-1} and was interpreted as follows: at low shear rates, the director describes an orbit that lies out of the shearing plane whereas, at higher shear rates, this orbit is confined within the shearing plane. The transition to the flow-aligning regime occurs at still higher shear rates (≈3–10 s^{-1}) and only if the concentration is large enough (>50%). The same authors also examined the relaxation phenomena observed upon cessation of shear. Different behaviors were detected depending on concentration and shear history. Relaxation from the various tumbling regimes induces a single exponential decay of the orientation until an isotropic state is recovered. In contrast, relaxation from the flow-aligning regime induces a more complicated decay of the orientation and the isotropic state is not fully recovered. In both cases, topological defects obviously play a major role in the steady state and relaxation regimes. Indeed, half-strength disclination loops lying in the shearing plane have been recently observed [48]. A similar X-ray scattering study of the orientation of aramid solutions by shear flow has also been performed by Picken et al. [49].

Hongladarom et al. have studied the molecular orientation under shear of HPC and poly(benzylglutamate) (PBG) by birefringence, X-ray and neutron scattering [50]. All three techniques show that the PBG solutions differ from the HPC ones by the fact that a moderate orientation already appears at very low shear rates. Moreover, PBG solutions only display a single transition from the tumbling to the flow-aligning regimes. Relaxation experiments also show that the orientation of the PBG solutions does not decrease upon cessation of the shear. It may actually even increase under certain conditions.

The orientation of several thermotropic main-chain LCPs under flow has also been studied by X-ray scattering both in the nematic and SmA phases. Windle and Romo-Uribe have observed a transition from the so-called "log-rolling" regime to the flow-aligning one in copolyesters of the Vectra type [51]. Alt et al. also detected two different kinds of alignment for the nematic phase of another main-chain LCP [52]. The same polymer in its SmA phase aligns with the layer normal either along the vorticity direction or, more surprisingly, along the velocity direction. The latter orientation occurred at small strain amplitude and moderate to low frequency. This odd behavior contrasts with both theoretical predictions and with the behavior of the SmA phases of LMMLCs and surfactants [53].

There are very few structural studies of thermotropic side-chain polymers under shear flow. Recent work [54] describes the neutron scattering by the liquid crystalline polymethacrylate labelled PMAOC$_4$H$_9$ (Sect. 4.1) in the SmA phase. Typical liquid crystal behavior was observed in which the layers orient parallel to the shearing plates. The backbones are then elongated along the velocity direction and still confined between sublayers of mesogenic cores.

5.4
Mineral Lyotropic LCPs

The above theories (Sect. 5.2) of the nematic ordering in LCPs do not specify in any way the mineral or organic nature of the polymers. However, there are now hundreds of organic LCPs whereas only very few mineral LCPs have been reported in the literature so far. In the following, we shall describe two mineral systems that display a nematic phase.

The first one belongs to the family of the Chevrel clusters which is well known in the field of solid-state chemistry because it provided one of the first examples of superconducting non-metal compounds. The Chevrel phases have the general formula MMo_6X_8 where M = Cu, Pb, Sn, Ag etc. and X is a chalcogenide S, Se or Te and they are based on the $[Mo_6X_8]$ octahedral cluster core [55]. The $[Mo_6X_8]$ motif may be regarded as dimerized $[Mo_3X_3]$ units capped by two chalcogenides. These $[Mo_3X_3]$ units may also polymerize to form the infinite 1-dimensional polymer $([Mo_3X_3^-])_\infty$ (Fig. 20) [56].

$LiMo_3Se_3$ is one of the members of this last family of mineral polymers and thus the objects described here are 1-dimensional chains of transition metal clusters.

In the crystalline phase, these chains pack in a hexagonal lattice in which the Li^+ cations are located between the chains to ensure electrical neutrality. These polymers have a conducting core made of Mo and an insulating envelope made of Se, so that they have been compared to molecular electrical wires. These wires bear a nominal charge of ≈ 0.5 $e^-/Å$. Tarascon and coworkers have shown that this material could be dispersed in very polar solvents to yield a suspension of discrete chains [57].

These suspensions show typical threaded textures (Fig. 21) which immediately demonstrates their liquid-crystalline nature [58]. Further experiments have shown that this mesophase is a nematic one. The X-ray scattering pattern of a partially oriented sample of the nematic phase revealed a small angle diffuse ring which is due to the lateral interferences among discrete chains. The scattering patterns of these suspensions are actually very similar to those of the TMV. The position in reciprocal space of the diffuse ring showed that the average distance between chains in the plane perpendicular to the director is ≈ 70 Å. From the concentration of the suspension, it is then possible to calculate an effective radius of ≈ 5 Å for the scattering object. This value compares very well with the Van der Waals radius (R = 5.00 Å) of the discrete chain. In addition, a sharp wide angle diffraction line can be detected on the scattering patterns. This reflection comes from the periodicity of 2.24 Å along the chain axis and its sharpness sets a lower limit of ≈ 1000 Å for the persistence length of the chains. Unfortunately, these compounds are slightly sensitive to the oxygen of the atmosphere with which they react on a time scale of hours and therefore need to be handled in an oxygen free atmosphere.

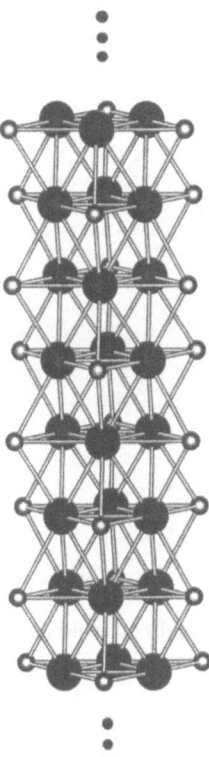

Fig. 20. Molecular structure of a $[Mo_3Se_3^-]_\infty$ wire (*large rings*: Mo, *small rings*: Se) (courtesy JC Gabriel)

Aqueous suspensions of vanadium oxide, V_2O_5, constitute a second example of mineral lyotropic nematic phase [59]. These suspensions are made of V_2O_5 ribbon particles dispersed in water. These ribbons are 10 Å thick, several hundred Ångströms wide and several thousand Ångströms long, the latter two dimensions being subject to some polydispersity. The ribbons also bear a rather large electrical charge density of $\approx 1e^-/Å$.

Figure 22 shows the phase diagram of these suspensions. The suspensions are isotropic at low concentrations. The nematic phase is found at higher concentration and is separated from the isotropic liquid by a biphasic (N/I) region. Moreover, a sol-gel transition also takes place in the middle of the nematic phase. Nematic sols display typical Schlieren textures (Fig. 23) whereas samples of nematic gels sometimes show banded textures if they were submitted to flow on filling the optical capillary tubes.

The nematic sols are also readily aligned by moderate magnetic fields (≈ 0.5 T) so that oriented X-ray scattering patterns (Fig. 24) can be obtained which are very similar to those of the TMV suspensions.

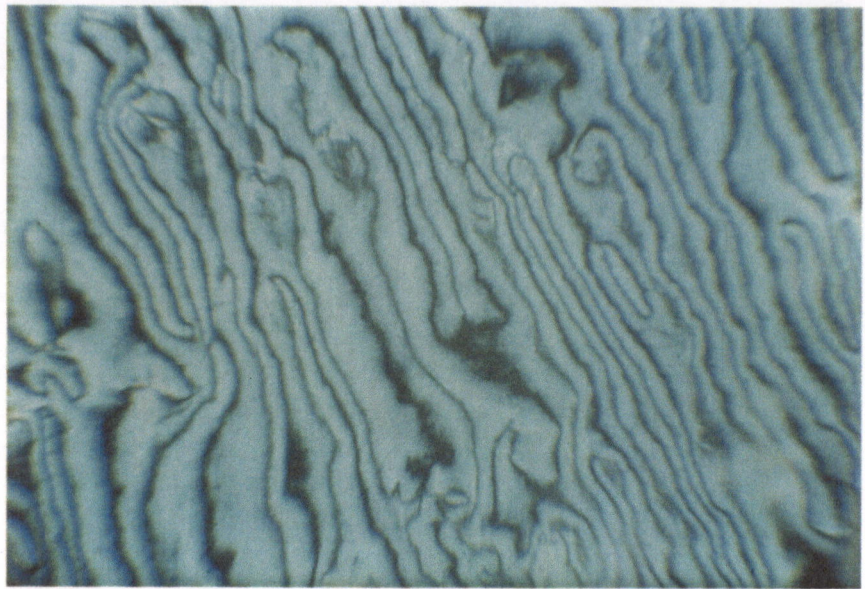

Fig. 21. Threaded texture of a 10^{-1} mol/l suspension of $LiMo_3Se_3$ in NMF in the nematic phase (courtesy JC Gabriel)

There again, one observes a small angle diffuse ring which arises from lateral interferences among ribbons. The position of this diffuse ring is inversely proportional to the average distance between ribbons which varies with the concentration. At low concentrations, this distance follows

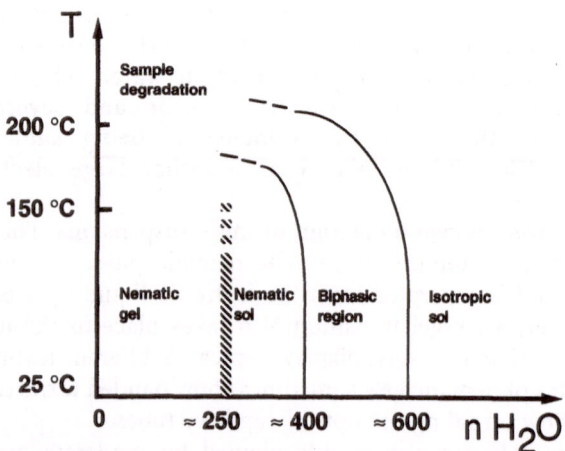

Fig. 22. Phase diagram of (V_2O_5, n H_2O) suspensions vs n and temperature for samples held in sealed flat capillary tubes. (This phase diagram is not isobaric)

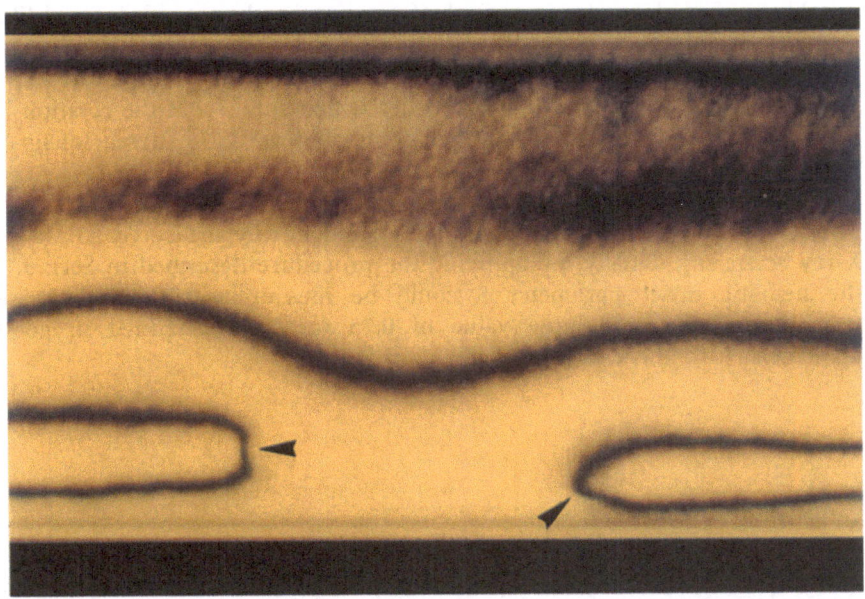

Fig. 23. A (V_2O_5, 400 H_2O) sample of the nematic sol showing a typical Schlieren texture

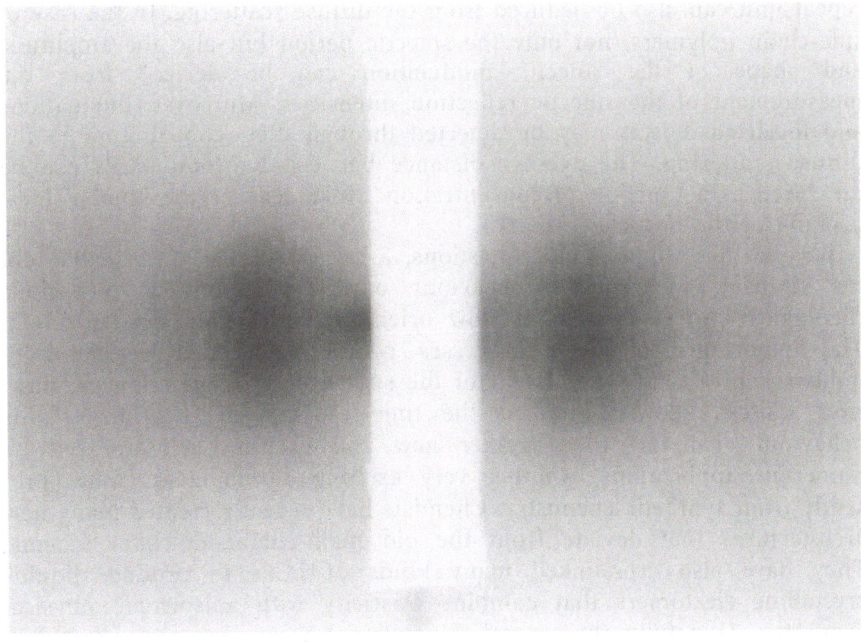

Fig. 24. Small angle X-ray scattering pattern of a (V_2O_5, 400 H_2O) sample of nematic sol aligned by a 1 Tesla magnetic field

roughly a $q^{-0.5}$ dependence typical of a nematic phase made of rod-like particles which shows that the ribbons rotate about their long axis. However, at large concentrations this distance shows a crossover to a q^{-1} dependence indicative of a lamellar order which shows that the rotation is now severely hindered. From another point of view, the scattered intensity beyond the diffuse ring is driven by the form factor of the particle and was used to confirm the ribbons transverse dimensions mentioned above. The length, being in the micron range, could not be measured so far by X-ray scattering. Finally, by applying the procedure described in Sect. 3.1, the nematic order parameter S could be measured at the N/I phase transition. The rather large value of 0.75 was thus obtained in good agreement with the Onsager model.

6
Conclusion

Through the study of the different topics considered in this article, it was shown how X-ray scattering is a useful tool to characterize the most salient features of the mesophases of LCPs. For instance, a simple procedure can be used to measure the nematic order parameter and it is so far valid for all kinds of LCPs based on rod-like moieties. In the case of main-chain polymers, useful information about the conformation of the repeat unit can also be deduced from the diffuse scattering. In the case of side-chain polymers, not only the smectic period but also the amplitude and shape of the smectic modulation can be derived from the measurement of the smectic reflection intensities. Moreover, fluctuations and localized defects may be detected through their contribution to the diffuse scattering. The average distance between lyotropic LCPs can be measured as a function of concentration which tells us the kind of local packing of the particles.

Beyond these fundamental questions, X-ray scattering is a valuable tool for studying in situ the behaviour of LCPs submitted to various rheological processes such as flow orientation and fiber crystallization. The importance of these processes comes from their relevance to industrial processes. The advent of the so-called 3rd generation synchrotron sources allows nowadays the time-resolved investigation of this behaviour. This is still a rather new and exciting field spurred by numerous applications. Another very exciting development comes presently from synthetic chemistry. Chemists have recently created many new architectures that deviate from the old main-chain/side-chain scheme. They have also crosslinked many kinds of LCPs to produce liquid-crystalline elastomers that combine elasticity with anisotropic physical properties. Hopefully, the general principles of X-ray scattering, illustrated through the examples described above, will prove helpful to navigate through this new wealth of fascinating materials.

7
References

1. (a) Davidson P, Levelut AM (1992) Liquid Crystals 11:469; (b) Davidson P (1996) Progress in Polymer Science 21: 893; (c) Tsukruk VV, Bliznyuk VN (1997) Progress in Polymer Science 22: 1089
2. (a) Azaroff LV (1987) Mol Cryst Liq Cryst 145: 31; (b) Chin HH, Azaroff LV, Saini AR (1989) J of Pol Science Part B, Pol Physics 27: 1993; (c) Chin HH, Azaroff LV, Saini AR (1989) J of Pol Science Part B, Pol Physics 27: 2001; (d) Noël C, Navard P (1991) Progress in Polymer Science 19: 55
3. DuPré DB (1982) Techniques for the evaluation of material constants in lyotropic systems and the study of pretransitional phenomena in polymer liquid crystals. In: Ciferri A, Krigbaum WR, Meyer RB (eds) Polymer liquid crystals. Academic Press, New York
4. Dobb MG, McIntyre JE (1984) Advances in Polymer Science 60: 61.
5. Livolant F, Leforestier A (1996) Progress in Polymer Science 21: 1115
6. (a) Economy J, Goranov K (1994) Advances in Polymer Science 117: 221; (b) Preston J (1978) Synthesis and properties of rod-like condensation polymers. In: Blumstein A (ed) Liquid crystalline order in polymers. Academic Press, New York
7. (a) Finkelmann H, Rehage G (1984) Advances in Polymer Science 60: 99; (b) Shibaev VP, Plate NA (1984) Advances in Polymer Science 60: 173
8. (a) Noirez L, Keller P, Cotton JP (1995) Liquid Crystals 18: 129; (b) Cotton JP, Hardouin F (1997) Progress in Polymer Science 22: 795
9. (a) Guinier A (1994) X-ray diffraction in crystals, imperfect crystals and amorphous bodies. Dover Publications, New York; (b) Warren BE (1990) X-ray diffraction. Dover Publications, New York; (c) James RW (1962) The optical principles of the diffraction of X-rays. G.Bell and Sons, London; (d) Amoros JL, Amoros M (1968) Molecular crystals: their transforms and diffuse scattering, Wiley, New York
10. Dreher S, Zachmann HG, Riekel C, Engstrom P (1995) Macromolecules 28: 7071
11. (a) McNamee SG, Bunning TJ, Patnaik SS, McHugh CM, Ober CK, Adams WW (1995) Liquid Crystals 18: 787; (b) McNamee SG, Bunning TJ, McHugh CM, Ober CK, Adams WW (1994) Liquid Crystals 17: 179
12. (a) Butler MF, Donald AM, Ryan AJ (1998) Polymer 39: 39; (b) Butler MF, Donald AM, Ryan AJ (1998) Polymer 38: 5521; (c) Pople JA, Hamley IW, Fairclough JPA, Ryan AJ, Komanschek BU, Gleeson AJ, Yu GE, Booth C (1997) Macromolecules 30: 5721; (d) Hsiao BS, Barton R, Quintana J (1996) J of Applied Polymer Science 62: 2061
13. De Gennes PG (1974) The physics of liquid crystals. Clarendon Press, Oxford
14. (a) Leadbetter AJ, Norris EK (1979) Molecular Physics 38: 669; (b) Leadbetter AJ, Wrighton PG (1979) Journal de Physique Colloques 40: C3; (c) Leadbetter AJ (1979) Structural studies of nematic, smectic A and smectic C phases. In: Luckhurst GR, Gray GW (eds) The molecular physics of liquid crystals. Academic Press, New York
15. Davidson P, Petermann D, Levelut AM (1995) Journal de Physique II 5: 113
16. (a) Li MH, Brulet A, Davidson P, Keller P, Cotton JP (1993) Phys Rev Lett 70: 2297; (b) Li MH, Brulet A, Cotton JP, Davidson P, Strazielle C, Keller P (1994) Journal de Physique II 4: 1843; (c) Alonso PJ, Puertolas JA, Davidson P, Martinez B, Martinez JI, Oriol L, Serrano JL (1993) Macromolecules 26: 4304
17. Gray GW (1967) The influence of molecular structure on liquid crystalline properties. In: Brown GH, Dienes GJ, Labes MM (eds) Liquid crystals. Gordon and Breach, New York
18. (a) Samulski ET, DuPré DB (1979) Advances in Liquid Crystals 4: 121; (b) Lenz RW (1985) J of Polymer Science Polym Symp 72: 1; (c) Strzelecki L, Liebert L (1988) European Polymer Journal 17: 1271
19. (a) Ishaq M, Blackwell J, Chvalun SN, Kricheldorf HR (1996) Polymer 37: 477; (b) Wu TM, Blackwell J, Chvalun SN (1995) Macromolecules 28: 7349; (c) Ishaq M, Blackwell J, Chvalun SN (1996) Polymer 37: 1765

20. DeGennes PG (1982) Mechanical properties of nematic polymers. In: Ciferri A, Krigbaum WR, Meyer RB (eds) Polymer liquid crystals. Academic Press, New York
21. Hardouin F, Sigaud G, Achard MF, Brulet A, Cotton JP, Yoon DY, Percec V, Kawasumi M (1995) Macromolecules 28: 5427
22. Noirez L, Boeffel C, Daoud Aladine C (1998) Phys Rev Lett 80: 1453
23. (a) Als-Nielsen J, Litster JD, Birgeneau RJ, Kaplan M, Safinya CR, Lindegaard-Andersen A, Mathiesen M (1980) Phys Rev B 22: 312; (b) Safinya CR, Roux D, Smith GS, Sinha SK, Dimon P, Clark NA, Bellocq AM (1986) Phys Rev Lett 57: 2718
24. (a) Tsukruk VV, Shilov VV, Lipatov YuS (1985) Acta Polymerica 36: 403; (b) Tsukruk VV, Shilov VV, Lokhonia OA, Lipatov YuS (1987) Soviet Phys Crystallography 32: 88, (c) Tsukruk VV, Shilov VV (1990) Polymer 31: 1793; (d) Nachaliel E, Keller EN, Davidov D, Zimmermann H, Deutsch M (1987) Phys Rev Lett 58: 896; (e) Keller EN, Halfon R, Nachaliel E, Davidov D, Zimmermann H (1988) Phys Rev Lett 61: 1206
25. Nachaliel E, Keller EN, Davidov D, Boeffel C (1991) Phys Rev A 43: 2897
26. Wong GCL, DeJeu WH, Shao H, Liang KS, Zentel R (1997) Nature 389: 576
27. (a) Tsukruk VV, Shilov VV, Lipatov YuS (1985) Soviet Phys Crystallography 30: 115; (b) Tsukruk VV, Shilov VV, Lipatov YuS (1986) Macromolecules 19: 1308; (c) Tsukruk VV, Shilov VV, Lokhonia OA, Lipatov YuS (1987) Soviet Phys Crystallography 32: 88; (d) Tsukruk VV, Shilov VV (1990) Polymer 31: 1793; (e) Gudkov VA (1984) Soviet Phys Crystallography 29: 316; (f) Gudkov VA (1987) Soviet Phys Crystallography 31: 686
28. (a) Davidson P, Strzelecki L (1988) Liquid Crystals 3: 1583; (b) Davidson P, Levelut AM, Achard MF, Hardouin F (1989) Liquid Crystals 4: 561; (c) Davidson P, Levelut AM (1992) Liquid Crystals 11: 469
29. Noirez L, Davidson P, Schwarz W, Pepy G (1994) Liquid Crystals 16: 1081
30. Davidson P, Keller P, Levelut AM (1985) J de Physique 46: 939
31. Davidson P, Levelut AM (1989) J de Physique 50: 2415
32. DeGennes PG (1983) J de Physique Lettres 44: L657
33. Davidson P, Pansu B, Levelut AM, Strzelecki L (1991) J de Physique II 1: 61
34. De Gennes PG (1972) Comptes Rendus de l'Academie des Sciences B275: 939
35. (a) Fabre P, Casagrande C, Veyssie M, Finkelmann H (1984) Phys Rev Lett 23: 372; (b) Rupp W, Grossmann HP, Stoll B (1988) Liquid Crystals 3: 583
36. (a) Geer RE, Shashidhar R, Thibodeaux AF, Duran RS (1993) Phys Rev Lett 71: 1391; (b) Geer RE, Shashidhar R (1995) Phys Rev E 51: R8; (c) Geer RE, Qadri SB, Shashidhar R, Thibodeaux AF, Duran RS Phys Rev E 52: 671
37. Fraden S (1995) Phase transitions in colloidal suspensions of virus particles. In: Baus M, Rull LF, Ryckaert JP (eds) Observation, prediction and simulation of phase transitions of complex fluids. Kluwer Academic Publishers, Dordrecht Germany
38. Oldenburg R, Wen X, Meyer RB, Caspar DLD (1988) Phys Rev Lett 61: 1851
39. Maier EE, Krause R, Deggelmann M, Hagenbuchle M, Weber R, Fraden S (1992) Macromolecules 25: 1125
40. Wen X, Meyer RB, Caspar DLD (1989) Phys Rev Lett 63: 2760
41. Dogic Z, Fraden S (1997) Phys Rev Lett 78: 2417
42. Onsager L (1949) Annals of the New York Academy of Sciences 51: 627
43. Vroege GJ, Lekkerkerker HNW (1992) Reports on Progress in Physics 55: 1241 and references cited therein
44. Khokhlov AR, Semenov AN (1981) Physica A 108: 546
45. (a) Selinger JV, Bruinsma RF (1991) Phys Rev A 43: 2910; b) Selinger JV, Bruinsma RF (1991) Phys Rev A 43: 2922; (c) Selinger JV, Bruinsma RF (1992) J de Physique II 2: 1215
46. (a) Doi M (1981) J of Polymer Science, Polymer Physics 19: 229; (b) Larson RG, Ottinger HC (1991) Macromolecules 24: 6270
47. (a) Keates P, Mitchell GR, Peuvrel-Disdier E, Navard P (1993) Polymer 34: 1317; (b) Keates P, Mitchell GR, Peuvrel-Disdier E, Navard P (1996) Polymer 37: 893
48. De'Nève T, Navard P, Kleman M (1995) Macromolecules 28: 1541
49. Picken SJ, Aerts J, Visser R, Northolt MG (1990) Macromolecules 23: 3849

50. Hongladarom K, Ugaz VM, Cinader VK, Burghardt WR, Quintana JP, Hsiao BS, Dadmun MD, Hamilton WA, Butler PD (1996) Macromolecules 29: 5346

51. (a) Romo-Uribe A, Windle AH (1993) Macromolecules 26: 7100; (b) Romo-Uribe A, Windle AH (1996) Macromolecules 29: 6246

52. Alt DJ, Hudson SD, Garay RO, Fujishiro K (1995) Macromolecules 28: 1575

53. (a) Bruinsma RF, Safinya CR (1991) Phys Rev A 43: 5377; (b) Safinya CR, Sirota EB, Plano RJ (1991) Phys Rev Lett 66: 1986; (c) Safinya CR, Sirota EB, Bruinsma RF, Jeppesen C, Plano RJ, Wenzel LJ (1993) Science 261: 588

54. (a) Noirez L, Lapp A (1997) Phys Rev Lett 78: 70; (b) Noirez L, Lapp A (1996) Phys Rev E 53: 6115

55. (a) Chevrel R, Sergent M, Prigent J (1971) J of Solid State Chem 3: 515; (b) Hughbanks T, Hoffmann R (1982) J Am Chem Soc 105: 1150

56. (a) Potel M, Chevrel R, Sergent M (1980) Acta Cryst B 36: 1545; (b) Tarascon JM, Hull GW, DiSalvo FJ (1984) Materials Research Bulletin 19: 915

57. Tarascon JM, DiSalvo FJ, Chen CH, Carroll PJ, Walsh M, Rupp L (1985) J Solid State Chem 58: 290

58. (a) Davidson P, Gabriel JC, Levelut AM, Batail P (1993) Europhysics Letters 21: 317; (b) Davidson P, Gabriel JC, Levelut AM, Batail P (1993) Advanced Materials 5: 665

59. (a) Zocher H (1925) Zeitschrift Anorg Allg Chem 147: 91; (b) Davidson P, Garreau A, Livage J (1994) Liquid Crystals 16: 905; (c) Davidson P, Bougaux C, Schoutteten L, Sergot P, Williams C, Livage J (1995) J de Physique II 5: 1577; (d) Commeinhes X, Davidson P, Bourgaux C, Livage J (1997) Advanced Materials 9: 900

Columnar Order in Thermotropic Mesophases

Daniel Guillon

Institut de Physique et Chimie des Matériaux de Strasbourg, Groupe des Matériaux
Organiques, 23, rue du Loess, 67037 Strasbourg Cedex, France
E-mail: guillon@michelangelo.u-strasbg.fr

Thermotropic columnar mesophases were observed with soap molecules a long time ago.
Since then, they have been widely investigated thanks to the synthesis of disk-shaped
molecules. In such mesophases, the molecules are packed one upon another to form
columns. These then organize according to a two-dimensional hexagonal lattice. In recent
years, a variety of compounds with quite different molecular architectures which also exhibit
columnar mesophases have been reported. A review of the most recent significant systems is
presented. It concerns discotic molecules, polycatenar molecules, side-chain polymers
without mesogenic moieties, dendrimers, pseudo-polymeric chains of metal soaps, self-
assembled systems driven by specific interactions such as hydrogen bonding or dipolar
interactions, and polar columnar phases. Typical examples are presented, with emphasis
given to the relationships between the columnar structure and the molecular architecture,
along with the specific interactions responsible for the establishment of this type of
supramolecular organization.

Keywords: Columnar mesophase, Structure, Interactions, Molecular architecture, Discotic

Structure and Bonding, Vol. 95
© Springer Verlag Berlin Heidelberg 1999

1
Introduction

In smectic mesophases, the rod-like shaped molecules retain one preferred orientation and are localized within parallel and equidistant layers resulting in a long range positional order. These stratified structures may exhibit different molecular organizations depending on the degree of order of the molecular mass centres inside the layers and the tilt of molecules with respect to the normal to the layers or on the stacking period compared with molecular lengths [1]. Besides smectic mesophases, columnar mesophases generally present a more pronounced order, since they result from the self-assembly of columns of indefinite length according to a two-dimensional lattice [2]. The symmetry of the latter is found to be hexagonal, rectangular or oblique (Fig. 1) [3].

The formation of the columns can result from different molecular or macromolecular species and/or from different types of interactions, as we will see in detail throughout this chapter. In all cases however, one column has in general a central rigid core surrounded by a flexible part directly attached to it (for example aliphatic chains) or free to move around it (organic solvent). This ensures the sliding of columns with respect to adjacent ones, and thus the liquid crystal nature of the two-dimensional phases encountered. One typical example is the case of thermotropic columnar mesophases obtained from pure metal soaps [4]. The molecules of these materials associate through their polar heads in indefinite columns, which are laterally assembled according to different two-dimensional lattices. In these structures, the aliphatic chains are quite disorganized, in a quasi liquid state, and uniformly fill the space available between the polar heads forming the core of the columns. Another typical example is the case of polypeptides [5]. Here, the columns are composed of individual molecules. Taking into account that their helical conformation is stabilized by intramolecular hydrogen bondings, the molecules of polypeptide do indeed have the shape of long rods with a diameter of 20 Å and a relatively strong rigidity corresponding to a persistance length of about 300 Å. In the presence of an appropriate solvent, and in concentrated solutions, these columns keep away from each other and organize themselves into a lyotropic columnar mesophase with hexagonal symmetry. The same type of lyotropic mesophase is encountered in concentrated solutions of nucleid acids (such as DNA) which have a double helix conformation within a rigid rod-like shape [6].

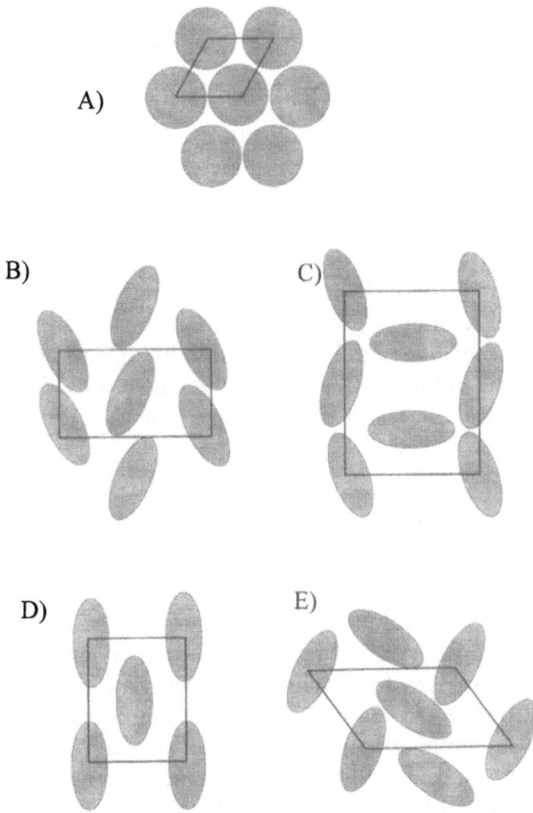

Fig. 1a–e. Different types of lattices found in columnar mesophases. Two-dimensional lattice with **a)** hexagonal symmetry, **b)** rectangular symmetry (P2₁/a), **c)** rectangular symmetry (P2/a), **d)** centred rectangular symmetry (C2/m), **e)** oblique (P1). (From Levelut AM (1983) J Chimie Physique 80:149)

In this chapter, the columnar mesophases obtained during recent years from various systems will be analyzed from their structural point of view. These systems are not limited to the well known disk-shaped molecules [7]. Depending on the molecular structure and the intermolecular interactions involved, they can be side-chain polymers without a mesogenic moiety, dendrimers, pseudo-polymeric chains of metal soaps, polycatenar molecules, or other systems driven by specific interactions such as hydrogen bonding or dipolar interactions. The next section describes briefly the determination of the key structural parameters needed for a good understanding of the molecular arrangement within the columnar mesophases. Then, typical examples of the systems listed above will be analyzed in detail. The emphasis will be given on the relationships between the columnar structure and the molecular architecture, along with the specific interactions responsible for the establishment of this type of organization.

2
Structural Parameters of the Columnar Mesophases

From a geometrical point of view, the columnar structure is mainly characterized by two parameters: the cross sectional area of the column, S (see Fig. 1), and the periodicity along the columnar axis, h (see Fig. 2).

As a rule, the two-dimensional ordering of the columns is developed over large distances and hence is easy to determine experimentally by small angle X-ray diffraction. On the other hand, the internal structure of the columns can be more difficult to handle. It is especially the case for tilted intracolumnar arrangements, and for non-oriented samples, because of the high viscosity of columnar mesophases in general. The periodicity, h, can be deduced from the value of the specific volume, v, as it can be measured by dilatometry [8], and from the cross-sectional area of the columns, S, as determined by small angle X-ray reflections through the relation:

$$Sh = vM/N_A \tag{1}$$

in which M is the molecular weight of the repeat unit and N_A the Avogadro's number [9].

When the molecular volume cannot be measured (mainly because the quantity of the sample is too small), an alternative method using only the lateral packing parameter of the columnar order, can be applied to determine h. It requires a systematic X-ray study of the columnar lateral packing as a function of the length of the peripheral chains. With the reasonable assumption of the additivity law of partial molar volumes, the volume of one repeat unit in the columns may indeed be taken as increasing linearly with

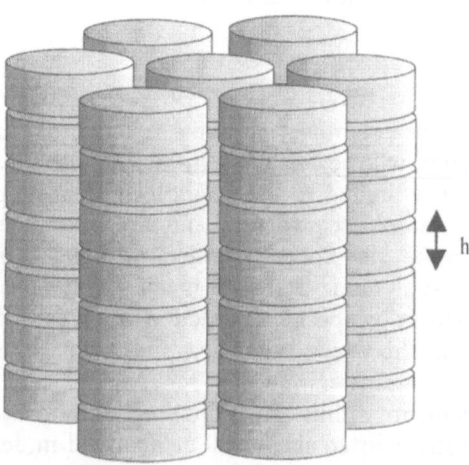

Fig. 2. Schematic of a two-dimensional arrangement of columns. *h* is the stacking period along the columnar axis

the number of methylene groups, n, in the peripheral aliphatic chains. If h is constant as a function of n, then it can be deduced from the linear variation of S with n as shown in the following equation:

$$Sh = V_0 + Z \cdot n \cdot V_m \tag{2}$$

where V_m is the volume of one methylene group, V_0 the volume of one repeat unit deprived of its lateral methylene groups and z the number of peripheral chains. This method has been found to be successful in many studies, as we will see in more detail in the following sections.

3
Molecular Architecture and Columnar Order

3.1
Discotic Molecules

Among the large variety of flat discotic molecules which have been synthesized in recent years, substituted phthlacyanine molecules represent one large class of materials designed for obtaining self-assembled supramolecular wires [10]. Substituted with eight rather long peripheral chains, it is well known that phthalocyanines produce mesophases with columnar symmetry [11]. Upon heating, the alkyl chains become disordered at the transition from the crystal to the liquid-crystalline state, and the system adopts a structure corresponding to the stacking of the aromatic cores in columns laterally assembled according, in general, to a two-dimensional hexagonal lattice. The interest in phthalocyanine molecules lies in several characteristic features. First, their flat aromatic core is quite large and offers the possibility for extended π-electron delocalization; second, strong van der Waals interactions between the flat cores enhance the thermal stability of the mesophases; finally, there is the possibility to include a variety of metal atoms in the central cavity of the molecules, in an attempt to produce one-dimensional metallic systems with potentially interesting conducting properties.

The nature and the length of the peripheral aliphatic chains play an important role in the intra- as well as in the intercolumnar ordering of these phthalocyanine derivatives (Fig. 3). Using Eq. 2, the stacking period of the molecules along the columnar axis was calculated from the variation of S as a function of the length of the terminal chains. It was found to be equal to 3.5 Å for the alkoxy and 4.9 Å for the alkyl and methylenoxyalkyl derivatives. This is interpreted in terms of two stacking modes: in the alkoxy series the molecules are oriented upright with respect to the columnar axis, while in the alkyl and methylenoxyalkyl series the molecules are tilted by an angle of 46°. In both cases, the aromatic cores were found to be in close contact with each other. The distinction between upright and tilted stacks of molecules is in agreement with observations of aromatic compounds showing that oxygen atoms directly attached to a planar aromatic system exert significant effect in steering from a tilted to an upright stack structure through weak hydrogen bonding [12]. The

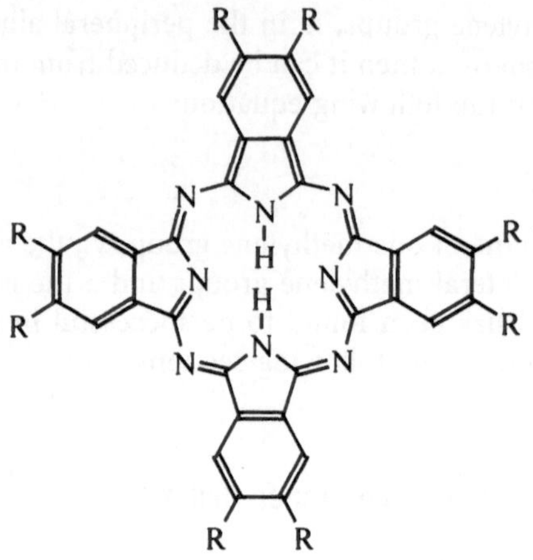

$$R = -O(CH_2)_nCH_3, \quad -CH_2O(CH_2)_nCH_3, \quad -(CH_2)_nCH_3$$

Fig. 3. Chemical structure of substituted phthalocyanine molecules with some examples of peripheral chains

distinction may also be due to steric effects. Indeed, the presence of an oxygen atom directly attached to the core acts as a rotula and let the peripheral chains to point in different directions of space, in order to adapt their conformations to neighbouring chains. On the contrary, when the chains are directly attached to the core, they are more constrained and the cores have to tilt in order for the chains to find enough freedom (Fig. 4).

Regarding the lateral arrangement of the columns, which is achieved according to a two-dimensional hexagonal lattice, three packing modes were found to be related to the shape of the columnar cores and their azimuthal orientation about their axis. With alkoxy peripheral chains, the aromatic cores are fairly circular in shape and therefore in agreement with a pure two-dimensional hexagonal lattice (case A in Fig. 1). With methylenoxyalkyl and alkyl peripheral chains, the aromatic columnar cores are elliptical in shape (due to the tilt with respect to the columnar axis) and their orientation with respect to the hexagonal lattice becomes a matter of great importance. Thus, for short side chains, the cross-sectional anisotropy of the columns is relatively pronounced, their azimuthal orientational ordering is well developed, and their packing results in a herringbone configuration (case B in Fig. 1). For long side chains, the cross-sectional anisotropy of the columns is strongly reduced and their hexagonal packing is achieved with complete rotational disorder about the columnar axis. Finally, for intermediate chain lengths, the columnar packing develops in a herringbone configuration at low temperatures and in a rotationally disordered configuration at high temperatures.

A great number of phthalocyanine derivatives with various peripheral moieties and other discotic molecules has been studied so far, mainly for their

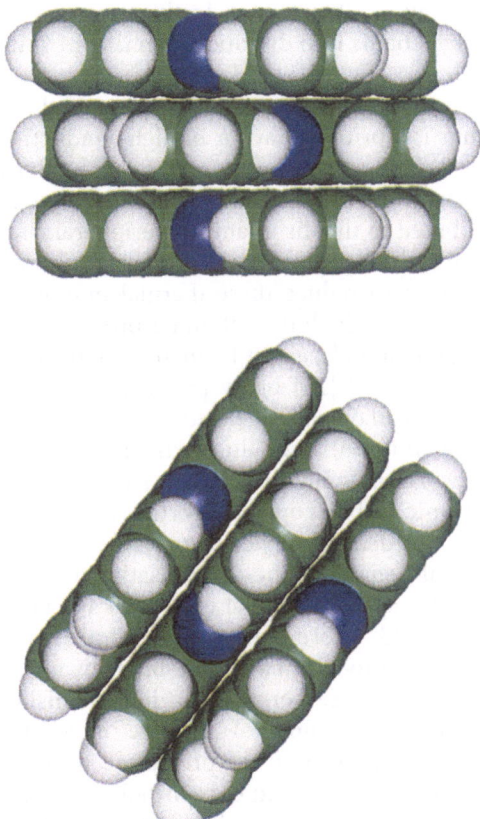

Fig. 4. Upright and tilted stacking arrangement of the phthalocyanine cores with respect to the columnar axis in the mesophase

photoconductive properties [13]. As a matter of fact, the columnar structure of flat discotic conjugated molecules in the mesophase is well suited for electronic transport parallel to the columnar axis [14]. The conducting chains are formed by the spine of the columnar structure surrounded by an insulating region of hydrocarbon chains. To tailor the mesophase behaviour of such materials, substituted lutetium phthalocyanine derivatives [15], triphenylene derivatives [16], naphthalocyanine derivatives [17], metal-free and metal-containing phthalocyanines substituted with long-chain amide groups [18] or containing a single poly(oxyethylene) side chain [19], porphyrazine derivatives [20] have been considered. Among them, some indeed exhibit very high values for the one-dimensional charge carrier mobility. Other efforts have been undertaken to increase the phase stability and the temperature range of the columnar mesophase. This can be achieved with large discotic molecules, since usually the larger the central part of the molecules and the greater the delocalization, the more the columnar phase is stable. For example, hexabenzocoronene derivatives can exhibit a columnar phase within a temperature range of about 300 °C [21]. In the same way, the introduction of semi-fluorinated alkyl chains into discotic systems contributes also to broaden the mesophase [22].

However, if the π-π overlap occurring in most of the discotic molecules stacked into columns is a priori in favour of high conductive properties, many studies performed so far seem to indicate that conductivity is greater in the crystalline phase than in the columnar mesophase of these materials [23]. This behaviour was attributed to the disorganized peripheral chains inducing a less efficient face to face ordering, and also to the molecular rotation around the columnar axis. In order to reduce these thermal motions, it was decided to prepare glassy materials in which the columnar order can be frozen in [24], or high molecular weight materials, either by increasing the molecular complexity (star-like discotic molecules) [25] or by preparing discotic liquid crystalline polymers [26] and macroscopically ordered discotic columnar networks [27] which can induce, in addition, a certain mechanical stability to the columnar assembly [28]. It is worth noticing that, in order to obtain molecular materials with extended electronic and magnetic properties, core-core interactions could also be enhanced if the density of peripheral side-chains is increased. This favours a better lateral register of the cores relative to each other [29].

In order to improve also the physical properties of columnar structures, it was envisaged to build more ordered structures by processing the corresponding materials into thin Langmuir-Blodgett films [30]. As a matter of fact, due to their ability to self-organize in highly anisotropic structures at the air/water interface, thin films of discotic molecules may present a considerable interest for applications as pressure sensors [31] or anisotropic conductors [32] for example. The studies reported so far in the literature show that discotic molecules can adopt two different orientations at the air/water interface, depending on their molecular architecture [33]. When the attractive forces between the mesogenic moieties and the liquid phase are strong enough, the central parts of the molecules hold face-on to the interface, whereas if the π-π interactions between adjacent molecules are predominant, then the central parts of the molecules stand edge-on to the interface. For example, melamine derivatives form Langmuir monolayers at the air/water interface due to their amphiphilic nature, with the polar central molecular parts lying flat on the water, whereas the lateral alkyl chains are oriented perpendicular to the water surface [34]. On the contrary, porphyrin molecules, which have a natural tendency to self-organize in columnar mesophases in the bulk, organize themselves also in a columnar structure within Langmuir films where the edge-on arrangement is favoured. In this case, the columnar stacking arrangements are parallel to the water surface [35]. More generally, discotic compounds with a polar central core and relatively long hydrocarbon tails adopt the face-on arrangement, whereas discotic compounds with hydrophobic cores and short tails adopt the edge-on one (Fig. 5).

Ordered thin layers can also been obtained from dilute solutions of discotic compounds onto gold or silicon surfaces. For example, triphenylene derivatives with a single long ω-alkylthiol chain have been shown to self-assemble into monolayers, in which the two types of orientation, face-on and

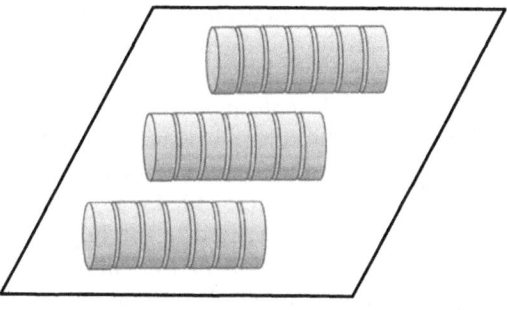

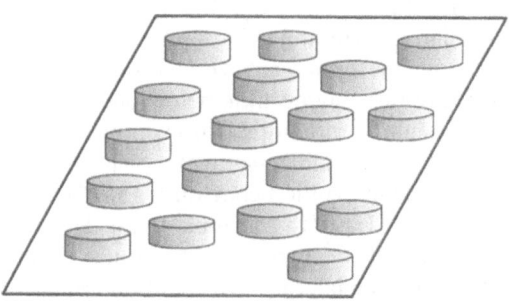

Fig. 5. Schematic of side-on and edge-on arrangement of mesogenic discotic molecules at the air/water interface in Langmuir films

edge-on, can be achieved depending of the molecular structure. In both cases, well ordered surface superstructures in the form of parallel columns on a gold surface can be obtained [36]. In the same way, ordered monomolecular films from discotic donor-acceptor twins can be formed on silicon surfaces. In these latter, the columns are also lying parallel to the solid surface, the discoid molecules standing edge-on on the substrate [37]. Finally, the spin-coating technique has recently been used with success to produce thin films of discotic molecules where the columns are oriented parallel to the substrate [38]. Such a process seems to be quite efficient in light-emitting devices, where fast responses could be expected from the large carrier mobility along the columnar axis.

3.2
Polycatenar Calamitic Molecules

If it is straightforward to understand why columnar mesophases are easily obtained with disk-like shaped molecules (the columns are formed by the piling of the flat cores of the molecules, the molten aliphatic chains filling the space between the columnar cores), it may appear rather surprising, at first sight, to observe columnar mesophases with polycatenar calamitic molecules, for example phasmidic or biforked molecules. These latter are composed of long rigid aromatic cores (containing in general five rings) with two or three paraffinic chains at both ends [39–42]. They are one of the several examples of liquid-crystalline compounds with molecular shapes deviating from the usual rod or disc-like ones. Biforked molecules (two chains at both ends) give rise to very interesting polymorphisms which may include nematic, lamellar, cubic and columnar liquid-crystalline phases [39, 40, 43–45]. The general molecular architecture of these materials is represented in the Fig. 6.

Some general trends for the formation of the columnar mesophase have been deduced [46] from the large number of compounds synthesized since the discovery of the first phasmidic mesomorphic material [39]. Columnar mesophases are always formed when the number of terminal chains attached in the meta position (N_m) is larger than the number of chains attached in the para position (N_p). When $N_m = N_p$, there is a competition between the occurrence of lamellar and columnar mesophases, the longest chains being in favour of the columnar mesophases. Several studies have also been published in order to understand the polymorphic behaviour as a function of molecular parameters of polycatenar compounds. For example, the introduction into the centre of the rigid core of flexible segments such as oxymethylene (CH_2O-) groups destroys the mesomorphic behaviour [45, 47], whereas the introduction of two methylene (CH_2-CH_2-) groups at the last junction between the rings very much favours the occurrence of columnar mesophases [47]. On the other hand, the presence of a lateral substituent on one of the rings composing the rigid core suppresses the occurrence of columnar mesophases in favour of smectic and/or nematic phases [48], strongly suggesting that the internal packing of polycatenar molecules in the columnar cores is mainly due to efficient van der Waals interactions between the rigid cores. Moreover, the ratio between the number, n_c, of carbon atoms of the paraffinic moiety of a molecule and the number of phenyl rings, n_ϕ, forming the core has been invoked also to predict the nature of the mesophase that this type of

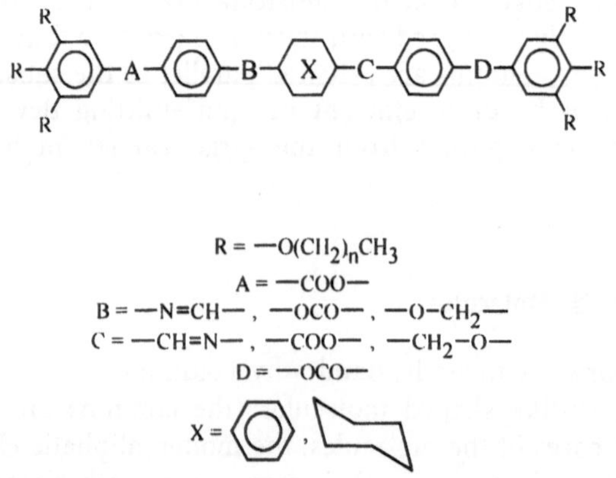

$$R = -O(CH_2)_nCH_3$$
$$A = -COO-$$
$$B = -N=CH-,\ -OCO-,\ -O-CH_2-$$
$$C = -CH=N-,\ -COO-,\ -CH_2-O-$$
$$D = -OCO-$$

$$X = $$

Fig. 6. General molecular architecture of calamitic polycatenar molecules

compound might exhibit [49]. For example, this ratio, n_c/n_ϕ, should be larger than 9 so that columnar mesophases could be observed.

The general description of the molecular packing in the columnar phase of polycatenar molecules is based on a model previously proposed for a hexacatenar compound [50]. This model considers the transversal section of the columns to be constituted by several calamitic rigid cores disposed side by side and surrounded by the corresponding disorganized aliphatic end chains. The number of molecules in a slice of a column depends upon the number and the length of the terminal aliphatic chains; three molecules are found for six-chain phasmids, four to five molecules for biforked mesogens.

Most of the X-ray diffraction patterns obtained in the columnar mesophase of these compounds are characteristic of a hexagonal columnar structure. With the assumption that the spacing corresponding to the lateral inter-molecular interaction is identical to the mean intracolumnar periodicity, h, the average number, n_b, of calamitic rigid cores associated in a columnar core can be calculated from the ratio between the volume of the hexagonal cell and that of the molecular volume, by using the following relation developed in Ref. 50:

$$n_b = \frac{\frac{\sqrt{3}}{2}a^2 h}{V_m} \tag{3}$$

where a is the distance between the axes of two neighbouring columns, and V_m the molecular volume.

For example, for the biforked compound in Fig. 7, the values of n_b obtained are close to 4.5 and are weakly temperature dependent. This number ($\cong 4.5$ molecules in each elementary cluster of the column) would correspond to an average paraffinic crown formed by 18 alkyl chains. Now, considering a disc of diameter of 30 Å equal to the length of the rigid core, its circumference corresponds precisely to 18 times the average distance between molten aliphatic chains (about 5 Å). Thus, the circumference of the columnar rigid core seems to be an essential parameter to control with respect with the number of terminal chains in each molecule in order to fully embed the core in

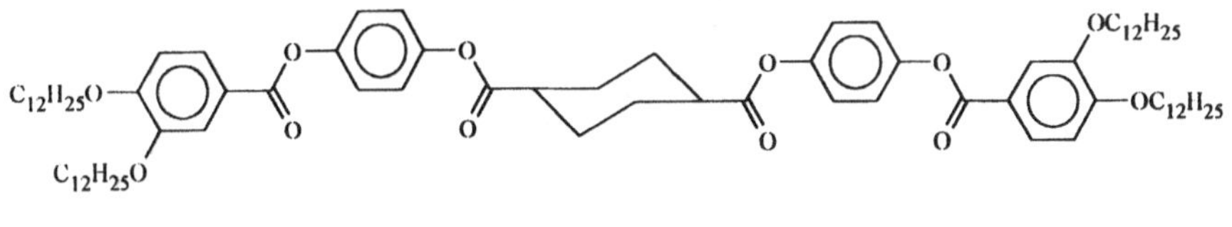

$$K \xrightarrow{\quad} S_C \xleftrightarrow{\quad} \phi_h \xleftrightarrow{\quad} I$$
$$\quad 113°C \qquad 149°C \qquad 156°C$$

Fig. 7. Chemical structure of a biforked compound and its corresponding thermotropic behaviour. K is the crystalline phase, S_C the lamellar smectic C phase, ϕ_h the columnar mesophase and I the isotropic phase

a paraffinic matrix and then to be able to produce columnar mesophases with calamitic polycatenar compounds. This result is also in perfect agreement with other data reported for hexacatenar and pentacatenar compounds containing central rigid cores of same length; the values of n_b found for these compounds being respectively 3 and 3.6, correspond in both cases to 18 chains around the central core [41]. From the general agreement between all the data for the columnar mesophase of different polycatenar materials, it seems reasonable to consider, for a more general description of columnar mesophases obtained with this type of materials, that the columnar cores have a diameter corresponding to the length of the rigid core of the molecules. The clusters of rigid cores pile one over another and the aliphatic end chains forming a liquid matrix around. Figure 8 illustrates how the clusters pile one over the other to form a cylinder with a roughly circular section.

The same type of columnar structure is also proposed for other polycatenar materials, such as tetrone derivatives [51], phasmidic macrocyclic liquid crystals [52], silver (I) complexes [53], or 2,2'-bipyridine derivatives [54].

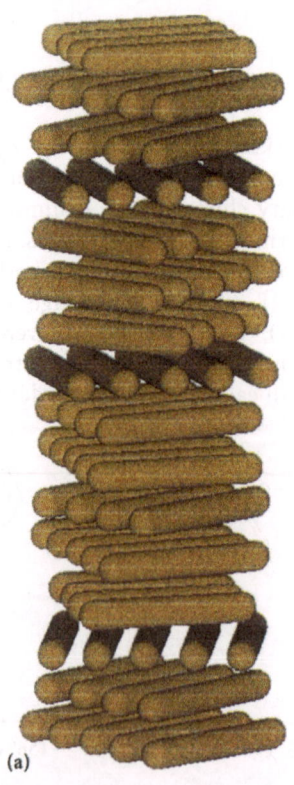

(a)

Fig. 8a, b. Simulation of one columnar core formed by the piling of clusters of molecules. **a** side view, **b** top view

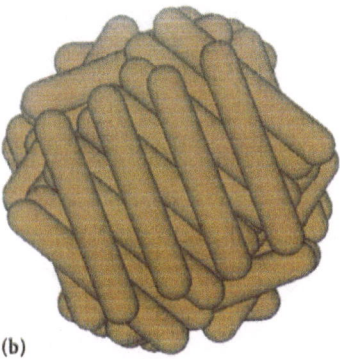

(b)

Fig. 8. (Cont).

At this point, it is interesting to consider how such biforked materials can undergo a change from a lamellar to a columnar mesophase. Based on dilatometry and X-ray diffraction experiments, a model for this phase transition has been proposed recently [55]. In the smectic C phase, the sublayer formed by the rigid cores of the molecules undergoes undulations, in order for the amphipathic character of the molecules [56] to be best satisfied, and thus to keep the aliphatic chains enough away from the rigid cores' sublayer, when the sublayer thickness is at a maximum. The difference in the transverse areas of both molecular moieties is compensated, when the sublayer thickness is at a minimum, by the correlated shift in the positions of the rigid cores. With increasing temperature, the volume of the alkyl chains and therefore the amplitude of the layer undulations increase. At the smectic to columnar phase transition, the amplitude of these undulations reaches the whole thickness of the rigid cores' sublayer. Above the transition, the rigid cores' sublayers have broken into columns, separated by the aliphatic chains forming a continuous medium (Fig. 9).

Even if they are not, strictly speaking, polycatenar molecules, double swallow-tailed compounds with long branches can be considered as a special case of tetracatenar mesogens, where the branches are not directly attached to the terminal rings [57], as shown in Fig. 10.

Such molecules exhibit columnar mesophases and smectic C phases. In contrast to the corresponding tetracatenar compounds described above, the columnar phase is, in most of these swallow-tailed compounds, the low temperature phase with respect to the smectic C phase [58]. This inversion of thermotropic sequence with respect to the biforked compounds described just above has not found any explanation in relation with molecular parameters. The symmetry of the columnar two-dimensional lattice is found to be hexagonal, oblique and even rectangular [59] in some cases. For the latter case, the model proposed involves several molecules packed side by side to form the lattice unit, the different moieties being packed alternately in order to form a two-dimensional centred cell as shown in Fig. 11.

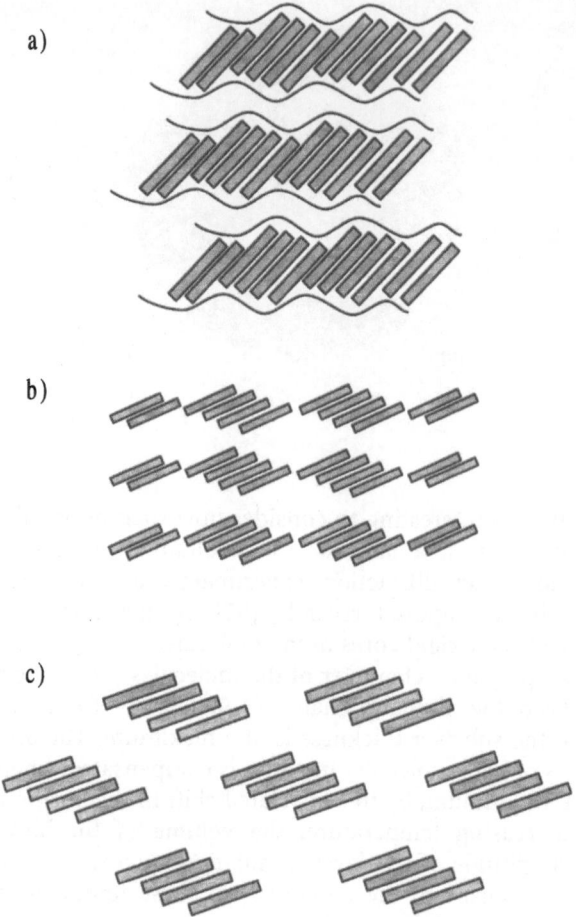

Fig. 9a–c. Structural models of **a** the smectic C phase, with undulations of the hard cores' sublayer, **b** the smectic C to columnar phase transition, **c** the columnar phase

Notice that this structure has already been reported to describe the ribbon structure observed with sodium soaps [60], with the S_A antiphase [61], and for the two-dimensional smectic structure of dissymmetrical and highly polar mesogens [62].

Very recently, it has been shown that the introduction of only one sufficiently large hydrophilic group into a lateral position of conventional rod-like molecules provides another way for obtaining materials exhibiting columnar mesophases [63]. In this case, the ribbons are formed by a parallel arrangement of the rigid cores of the molecules which are separated by the hydrophilic domains of the lateral groups. The alkyl chains are molten and fill up the space between the ribbons. As for the polycatenar compounds described above, segregation effects between the different constituent parts of the molecules are responsible for the formation of columnar mesophases.

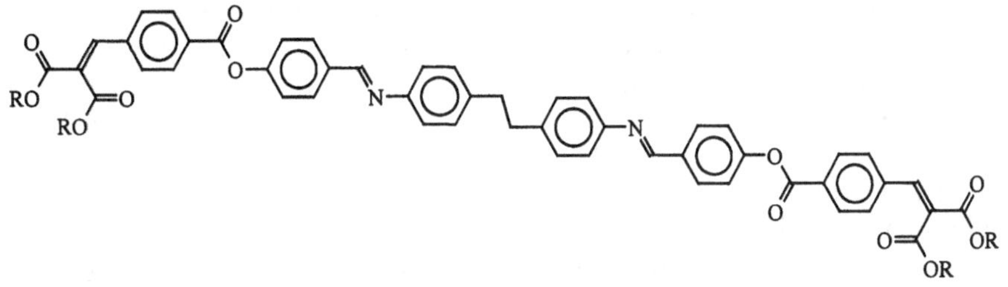

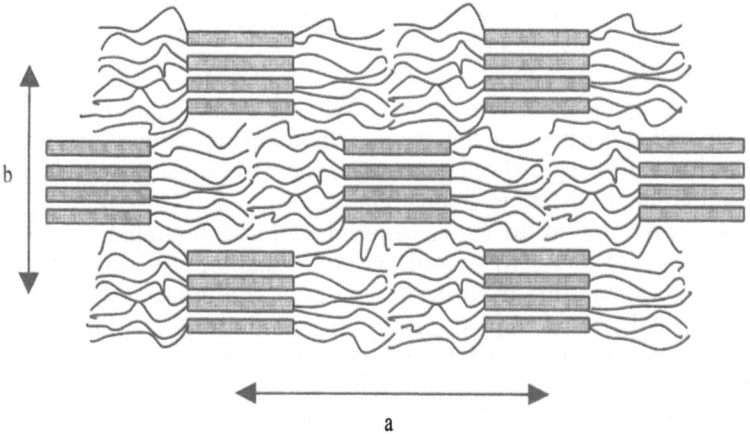

$$R = C_nH_{2n+1}$$

Fig. 10. Typical example of the double-swallow-tailed molecules

Fig. 11. Model proposed for the rectangular columnar phase of bis-swallow-tailed molecules

However, the intermolecular interactions responsible for the segregation are different in both types of compounds.

Another example where the amphipathic feature of the molecules plays an important role is the case of some thermotropic ionic calamitic liquid crystals such as ditholium salts [64]. In spite of their rod-like shape, these compounds exhibit columnar mesophases with supramolecular organization similar to that of amphiphilic molecules. In this case, it is clear that the driving force for the supramolecular organization is the amphipathic character of the molecules [56].

3.3
Side-Chain Polymers

Side-chain liquid crystalline polymers containing disk-like pendant mesogenic groups are well known to exhibit thermotropic columnar mesophases [65], the structure of which is identical to their corresponding low molecular weight disk-like counterparts. Rigid chain polymers, such as nucleic acids and polypeptides [5b, 66], are also well known to produce lyotropic columnar mesophases with appropriate solvents. The poly(di-n-alkylsilanes) and poly(di-n-alkylsiloxanes), Fig. 12, with only alkyl side groups and deprived of any mesogen [67] are examples of polymers showing thermotropic columnar mesophases which need to be analyzed in more detail.

Polysilanes have a chain backbone consisting only of silicon atoms and characterized by an electron delocalization along the σ-bonds [68]. These polymers have attracted a high interest since they can be used for microlithography [69], for nonlinear optical devices [70] or for photoconductivity [71]. The poly(di-n-alkylsilanes) undergo a first order transition (from 40 to 80 °C depending upon the alkyl chain length) from a low temperature crystalline to a high temperature columnar mesomorphic phase. The columnar nature of the mesophases was established by using polarizing optical microscopy and X-ray diffraction [72]. Their structure consists of a two-dimensional hexagonal packing of elongated silicon backbones surrounded by aliphatic chains spreading outwards in a disordered conformation. A systematic study as a function of the alkyl side chain length has shown that the squared distance between the axes of neighbouring columns, deduced from the experimental hexagonal cell parameters increases linearly with the number of carbon atoms in the alkyl chains, Fig. 13.

The interpretation of the above experimental data (extrapolated value to p = 0 and slope of $D^2 = f(p)$) indicates first that on and after the fourth methylene group starting from the silicon atom, the molar volume of all the disordered methylene groups is exactly the same. In other words, there is a great compactness of the packing near the silicon backbone. Second, it indicates that the mean stacking period of silicon atoms along the columnar axis is a constant for the whole series of poly(di-n-alkylsilanes) with p ranging from 4 to 14. The value of this stacking period (see Eq. 2, Sect. 2) is thus found to be 1.68 Å. Such a low value for the silicon period along the columnar axis is by far too small to be compatible with an all-trans conformation for the silane

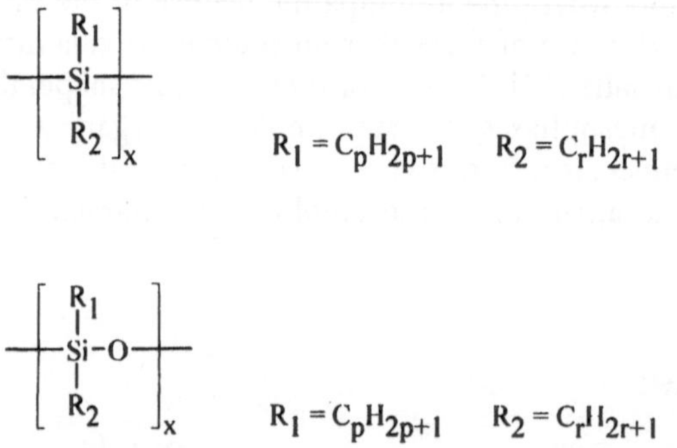

$$\left[\begin{array}{c} R_1 \\ | \\ -Si- \\ | \\ R_2 \end{array} \right]_x \qquad R_1 = C_pH_{2p+1} \qquad R_2 = C_rH_{2r+1}$$

$$\left[\begin{array}{c} R_1 \\ | \\ -Si-O- \\ | \\ R_2 \end{array} \right]_x \qquad R_1 = C_pH_{2p+1} \qquad R_2 = C_rH_{2r+1}$$

Fig. 12. Chemical structure of poly(di-n-alkylsilanes) and poly(di-n-alkylsiloxanes)

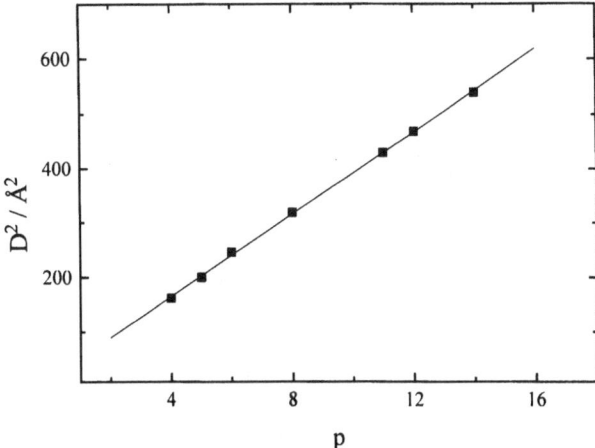

Fig. 13. Square of the intercolumnar distance as a function of the number, p, of methylene groups in the alkyl side chains for the columnar mesophase of poly(di-*n*-alkylsilanes) at 100 °C

backbone. It rather indicates an almost all-gauche conformation [73], or at least a coiled structure involving a significant number of gauche conformers. This period is also independent of temperature. This means that the conformation of the polymer backbone is not affected by the bulkiness and the thermal agitation of the pendant alkyl groups, having presumably reached an enhanced stability upon the transition into the columnar liquid crystalline phase, Fig. 14.

Since the films of symmetrically-substituted poly(di-*n*-alkylsilanes) are often brittle and tend to crack, it was thought necessary to introduce asymmetry into the side chains [74]. The resulting material can be processed into thin films which are then more flexible and less susceptible to producing cracks. As for symmetric polysilanes, the asymmetric ones exhibit hexagonal columnar mesophases at temperatures above 45 °C [74, 75]. On the contrary, the introduction of a methylene group between two silicon atoms in the backbone lead to materials showing high glass transition temperatures and therefore having a low backbone flexibility ; these poly(silylene-methylene)s symmetrically substituted with alkyl side groups do not exhibit any stable hexagonal mesophase [76].

From the above results, it is clear that the presence of classical mesogens is not necessary for linear chains like poly(di-*n*-alkylsilane)s or poly(di-*n*-alkoxyphos-phazene)s [77] to show a mesomorphic behaviour. This concept has also been applied to the case of poly(di-*n*-alkylsiloxane)s. For alkyl chains containing between 2 and 6 carbon atoms, the materials were shown to form a columnar mesophase, in which the molecules are positionally and orientationally ordered in a two-dimensional hexagonal lattice, but along the chain no long range correlations exists [78]. The most famous member of this series of polysiloxanes is polydiethylsiloxane which exhibits at room temperature a

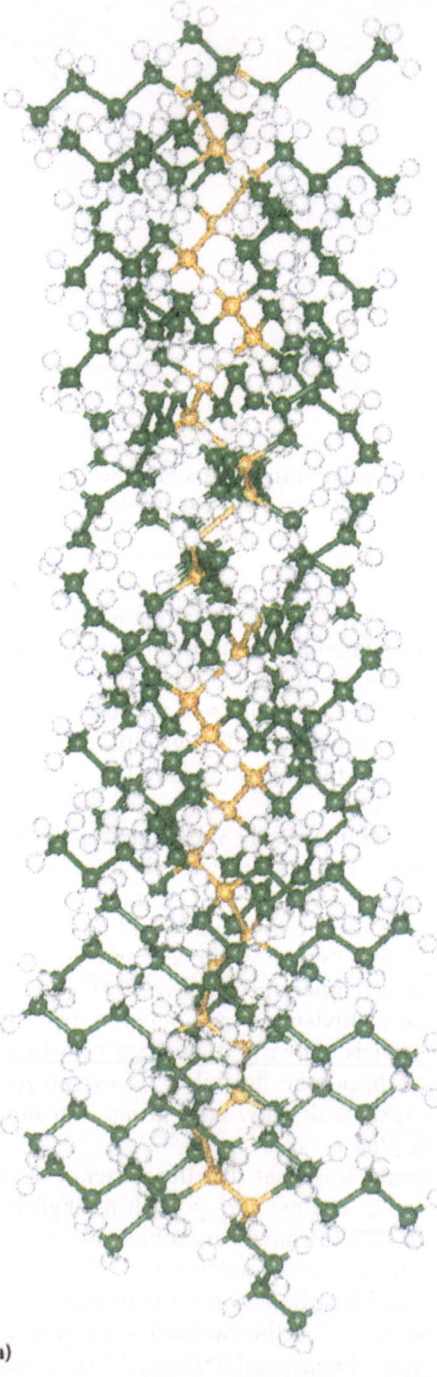

(a)

Fig. 14a, b. Simulation of one column formed by a poly(di-*n*-butylsilane). Silicon and carbon atoms are yellow and green respectively. **a** side view, **b** top view

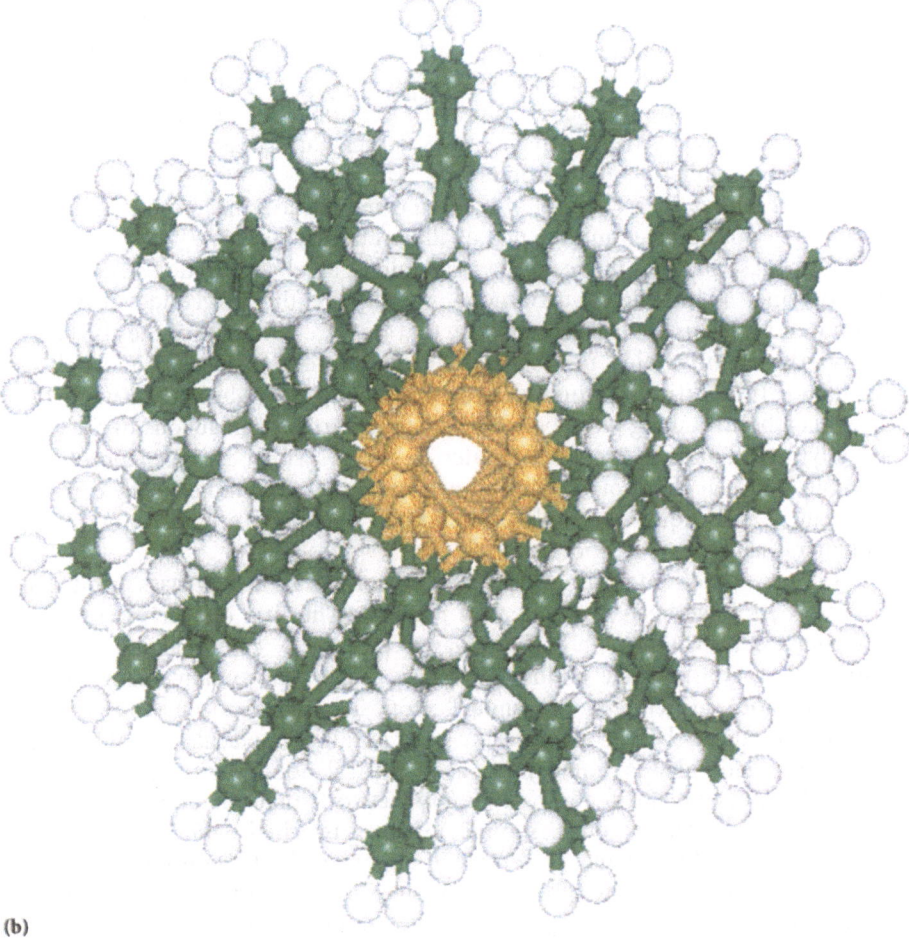

(b)

Fig. 14. (Cont).

columnar mesophase, the temperature range of which was found to be strongly molecular weight dependent [79]. Elongation of the alkyl side groups from ethyl to n-hexyl improved the stability of the hexagonal columnar mesophase, which is reflected in an increase in the mesophase temperature domain from $\Delta T = 43$ °C for poly(diethylsiloxane) to more than 310 °C for poly(di-*n*-hexylsiloxane) [80]. In contrast, poly(di-*n*-alkylsiloxane)s with *n*-alkyl side groups containing 7 to 10 carbon atoms show no hexagonal columnar mesophase [81]. Thus, the stabilization of the columnar state by long alkyl side groups seems to be overruled at a certain alkyl length of the substituents. As the alkyl lateral chains become longer, the packing and the motion of the main chain and of the side groups become more and more decoupled. This results in the possibility for the long alkyl side groups to crystallize without ordering of the polymer backbone. In order to study the role of the introduction of irregularities within the molecular architecture upon the the packing efficiency of the polymer molecules, random copoly(di-*n*-alkylsiloxane/di-*n*-hexylsiloxne) have been investigated [82]. All the copoly(di-*n*-pentylsiloxane/di-*n*-hexylsiloxane)s exhibit also a columnar mesophase. But, as soon as the difference in alkyl side group length is larger than one carbon atom, no columnar mesophase could be obtained.

More recently, the formation of columnar mesophases has been found with *N*-acylated poly(ethylenimine) such as illustrated in Fig. 15.

In the mesophase of the resulting polyamides, a helically folded structure of the polymer main chain surrounded radially by the alkyl side chains is proposed [83]. It is believed in this case that the direct linkage of the alkyl side groups by a stiff amide may contribute to enforce an overall rigid helical structure. In addition the helical structure should be stabilized through the hydrogen bonds taking place between the amide groups, analogous to the case of poly(L-glutamate)s [84]. These materials are described as systems bridging the gap between thermotropic columnar mesophases obtained with discotic molecules and liquid crystalline columnar structures formed by linear polymers just described above.

Two new other interesting examples of columnar mesophases produced by polymers have been reported very recently. The first one involves rigid-rod polyesters with flexible side chains [85]. The columnar mesophase formed is a consequence of the microsegregation of aromatic main chains and aliphatic side chains. The model proposed for the occurrence of the columnar order is based on a specific association of molecules. In this case, the molten side chains would form a cylindrical domain and the aromatic main chains would form the honeycomb network that surrounds the side-chain domains. The second example involves telechelic alternating multi-component semifluorinated polyethylene oligomers [86]. Here the columnar order is induced by the presence of several perfluorinated and perhydrogenated segments which prevents the ability of the fluorinated and hydrogenated blocks to form individual layers as for diblock low molecular weight compounds [87].

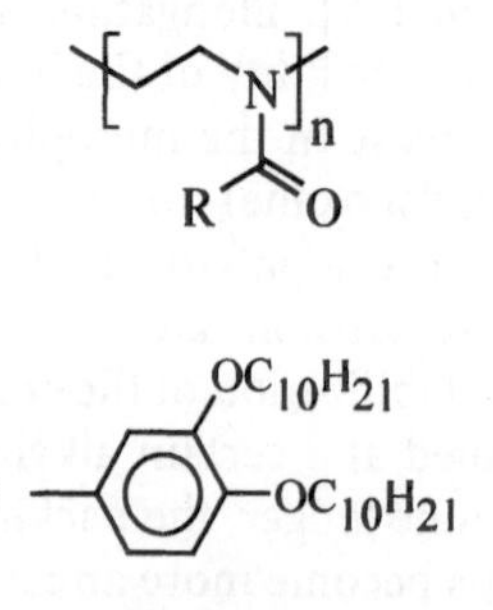

Fig. 15. Molecular structure of oligo and polyamides

3.4
Dendrimers

Dendritic molecules, commonly called dendrimers, represent a new class of polymeric materials with tree-like architectures which can be synthesized with broad or uniform molecular weight distributions [88]. They are hyper-branched structures of oligomers and polymers containing a branching point in each structural unit. They are considered to be promising new materials for many applications, such as phase transfer catalysis, recyclable extracting agents [89], or regulators for drug release [90]. The morphology of branched dendrimers evolves from an open, extended form to a more globular shape, as the number of generations increases. Most of these high generation dendrimers do not crystallize, and some of them can exhibit thermotropic liquid crystalline behaviour [91] when appropriately designed. In particular, it was shown that the fluorophobic effect increases the stability of the self-assembly of semifluorinated tapered monodendrons into supramolecular columnar dendrimers which organize themselves according to a hexagonal columnar mesophase [92]. It is worth noticing that in this case, the corresponding dendrimers do not correspond strictly to the definition given above, since they do not present any obvious tree-like architecture.

In the structural model of this columnar mesophase, the ether crowns are placed side-by-side in the centre of the column with the melted alkyl tails surrounded radially at its periphery, the perfluorinated segments being segregated from the perhydrogenated and aromatic parts of the columns. Several wedge-shaped monodendrons (five in the case of the monodendron represented in Fig. 16) form the elementary slice of the column. Moreover, the columns of these supramolecular semifluorinated systems spontaneously orient homeotopically on untreated glass slides, thus forming monodomain liquid crystals with the columns oriented with their long axes perpendicular to the glass surface. Finally, it is also important to stress that the self-assembly in

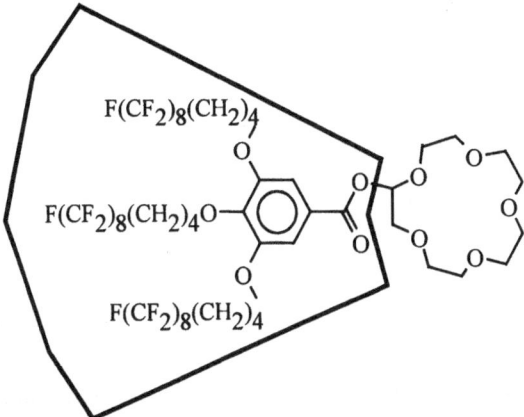

Fig. 16. Example of a tapered block used for self-assembly into a supramolecular dendrimer

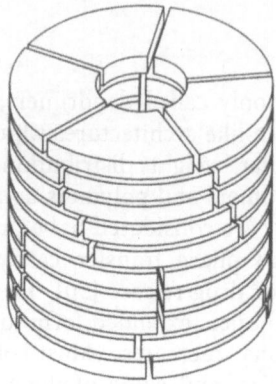

Fig. 17. Self-assembly of tapered blocks into a supramolecular dendrimer (from ref. 91 and 92)

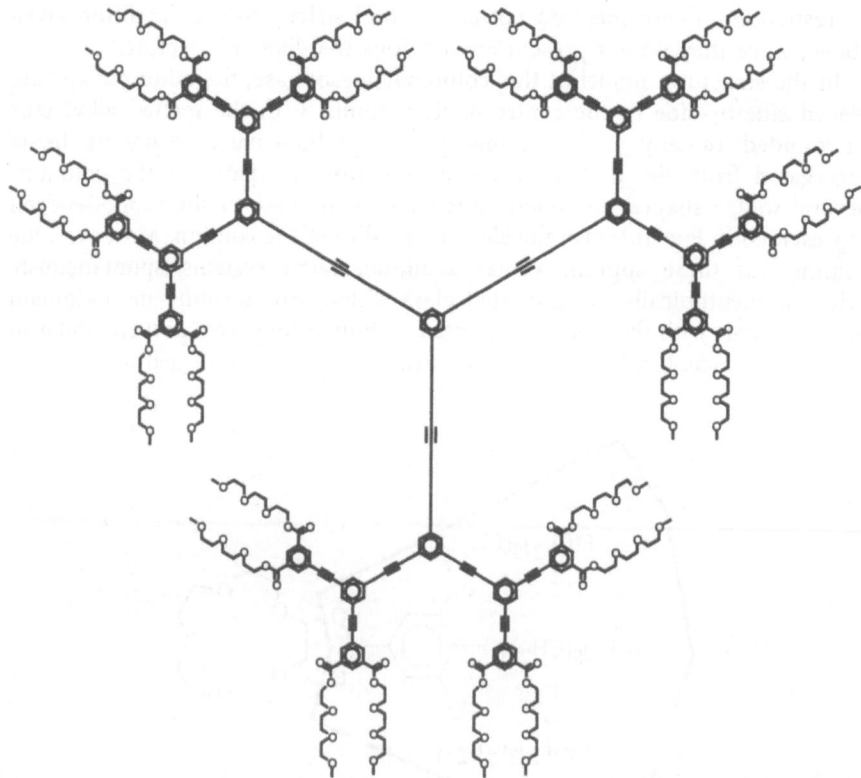

Fig. 18. Dendrimer of third generation from a phenylacetylene tridendron (from ref. 93)

these systems occurs also through the fluorophobic effect which is in fact an extension of the more general hydrophobic effect; in other words, this suggests that the hydrophobic effect is also able to produce such an organization which is based on repulsive rather than attractice forces, Fig. 17.

Liquid crystalline dendrimers characterized by a well-defined molecular weight and exhibiting thermotropic columnar mesophases have been reported very recently. An example is the dendrimers based on the stiff, triconnected phenylacetylene monomer [93] is shown in Fig. 18.

For this particular case, the mesophase behaviour is displayed over a wide temperature range, about 250 °C. A further example is given by the carbosilane dendrimers with perfluoroalkyl end groups [94]. For these dendrimers, the generation-dependent thermotropic phase behaviour was

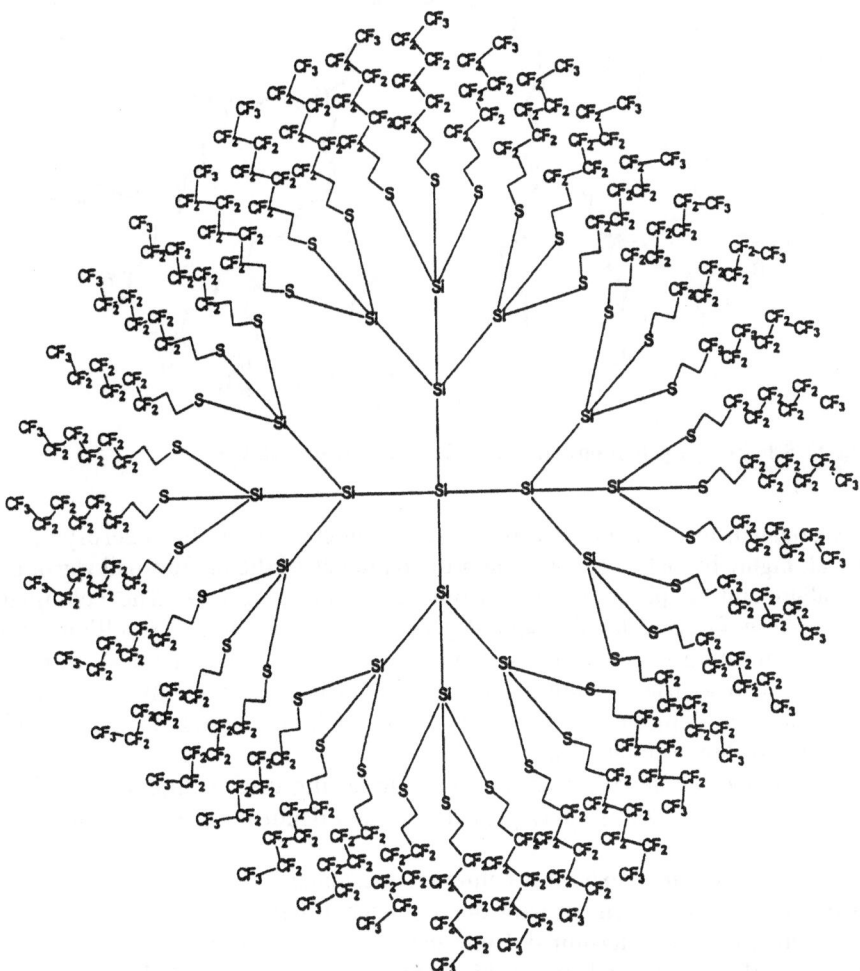

Fig. 19. Carbosilane dendrimer of the third generation containing 36 end groups (from ref. 94)

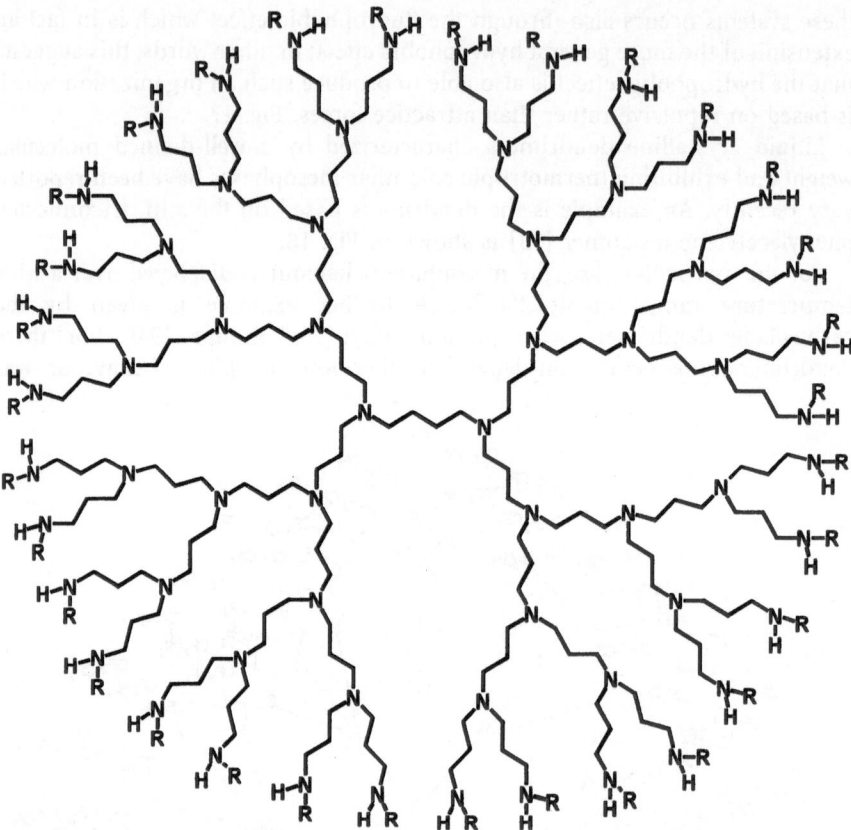

Fig. 20. Scheme of a poly(propyleneimine) dendrimer (from ref. 95)

investigated, in particular because of the conflict between the spherosymmetrical, highly branched topology and the tendency of the perfluoroalkylgroups to align and to phase separate from the carbosilane core. The columnar mesophase observed for the highest generation is described as resulting from the organization of the molecules into rows of stacked dendrimers, with the perfluorinated surface of the constituting dendrons pointing outward, such as schematically depicted in the Fig. 19. These columns (rows) then assemble into a two-dimensional lattice.

The last example concerns the very interesting case of liquid crystalline dendrimers starting from poly(propyleneimine) dendrimers [95], as shown in Fig. 20:

These compounds exhibit a columnar mesophase, the lattice constants of which are not in agreement with the diameter of flat, two-dimensional, disk-like molecules as determined from molecular simulations. This is due of course to the extremely dense packing of the aliphatic chains in the outer shell of the dendrimer. Therefore, it is proposed that the columnar structure of the liquid crystalline dendrimers should result from the piling of three dimensional cylindrical entities. These should be constituted by a polar core and an apolar shell as the cylinder casing. The driving force to build such a structure would be the microphase separation between polar and apolar regions along the hexagonal columnar lattice. It is interesting to notice here that the next generation of the dendrimers shown in Fig. 20 do not present any mesomorphic behaviour. Presumably because of the too high density of peripheral alkyl chains, the dendrimers transform into a spheroid structure with a close apolar shell, and the cylindrical domains cannot exist any longer.

3.5
Polymeric Chains of Metallomesogens

Recently, there has been great interest in combining the anisotropy and fluidity of liquid crystalline phases with the properties derived from the presence of metals [96]. Among the various metallomesogens which have been synthesized and studied, binuclear carboxylates of divalent transition metals, of general formula $M_2(RCO_2)_4$ are now well known to produce columnar hexagonal mesophases at ca. 100 °C. This represents another example of molecular engineering to tailor materials for their one-dimensional magnetic or electric properties. Interestingly, for M = Cu(II), Rh(II), Ru(II) and Mo(II), the structure of the mesophase is independent of the nature of the metal [97]. All the binuclear complexes of this family exhibit the so-called "lantern structure" in the crystalline phase where two metal atoms are bridged by four bidentate carboxylate groups [98]. Still existing as binuclear units in the mesophase as proved by Exafs investigations [99], the polar cores of the soap molecules are stacked to form columns of indefinite length, surrounded by the paraffinic chains with disordered conformations. The dimeric species are coordinated in axial positions by oxygen atoms of neighbouring molecules, thus giving rise to polymeric chains. Oriented parallel to each other, the columns are assembled following a two-dimensional hexagonal lattice, Fig. 21.

The transition from the crystal to the columnar mesophase is characterized by a change in the repeat distance of the binuclear cores along the pseudo-polymeric axis. In the crystalline phase, these cores are all oriented in the same direction with a repeat distance of 5.2 Å; in the columnar mesophase, the polar cores are perpendicular to the columnar axis and superposed in a four-fold helicoidal fashion, at least on a local scale, with a repeat distance of 4.7 Å.

X-ray diffraction experiments performed under pressure show that the area of the two-dimensional lattice decreases with increasing pressure and, at sufficiently high pressure, the columnar mesophase transforms into a crystalline lamellar phase. By combining Pressure-Volume-Temperature measurements and X-ray diffraction measurements as a function of pressure, it is shown that the stacking period of the binuclear cores increases as a function of increasing pressure [100]. Contrary to the lateral distance between columns which decreases upon compression, the stacking period increases leading to a stretching of the columnar cores. Starting from 4.65 Å at room pressure, with the polar cores oriented perpendicular to the columnar axes, the stacking period increases to about 4.8 Å at a pressure of 800 bar, thus

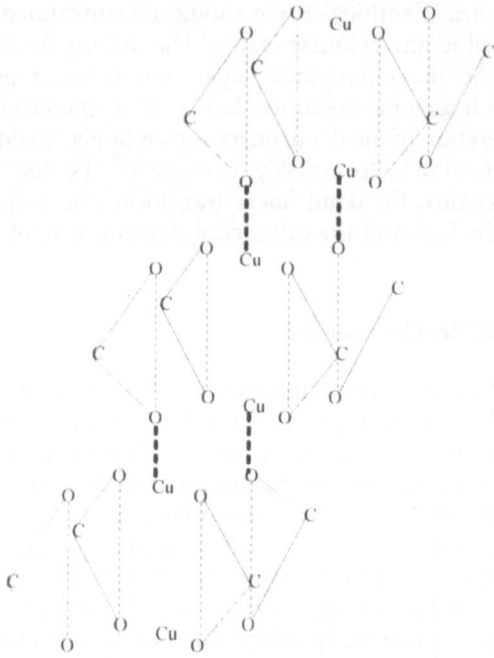

Fig. 21. Pseudopolymeric chain of binuclear transition metal complexes in the columnar mesophase

covering a significant proportion expansion to the period (5.2 Å) characteristic of the crystal. This stretching of the columns occurs concurrently with a tilting of the molecules and thus the action of pressure contributes to bring the stacking mode of the polar heads nearer to that of the crystal. This is achieved through a modification of the sequence of the oxygen atoms involved in the axial ligation, which progressively opens up the peripheral hollows between the polar cores (see Fig. 22b), by tilting the molecular polar heads with respect to the columnar axis.

Mixed divalent ruthenium (II,III) tetracarboxylates represent another interesting case where the nature of the counteranion X, present for the electroneutrality of the material, may have an important role on the mesomorphic properties. As a matter of fact, the nature of X strongly affects the appearance of the liquid crystalline order [101]. The chloro complexes are not mesomorphic, whereas the complexes with carboxylate (X = RCO_2) or dodecyl sulphate (X = DOS) anions exhibit a columnar hexagonal mesophase above ca. 150 °C. The columnar surface is about three times larger than the calculated one for the divalent Ru(II,II) analogues corresponding to the same chain length. The molecular arrangement in the mesophase has been understood from the crystalline structure of the Ru_2(propionate)$_5$ analogue [102], where the binuclear complexes are bridged through the axial propionate counteranion to give alternating cation-anion chains. In the mesophase, the

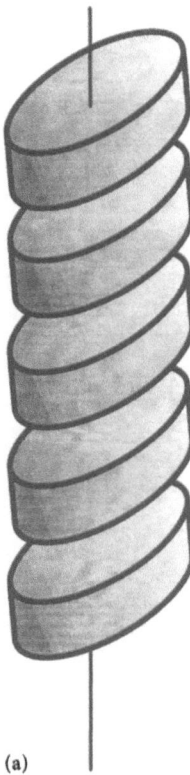

(a)

Fig. 22 . Schematic representation of the stacking of the transition metal soap polar heads in **a** the crystalline phase and in **b** the columnar hexagonal phase

alternating chains – dimer-carboxylate – are retained in the mesophase. Each column of the hexagonal array is made up of four molecular chains entangled and axially shifted by half a dimer, as schematically represented in Fig. 23.

This reflects that one of the key parameters, which has to be tuned in order to stabilize the mesomorphic behaviour of these compounds, is an efficient filling of the space around a rigid skeleton of stacked polar cores within a columnar structure.

A recent strategy has been explored to obtain self-organized, oriented materials with high refractive index and birefringence. It is based on the principle of photopolymerizing in-situ reactive monomers based on disubstituted carboxylates and disubstituted benzoates of Mg(II) and of Zn(II) [103]. The hexagonal columnar structure of the mesomorphic monomers was retained in the polymers, thus giving a material with an anisotropic structure locally. This is an example of processing polymeric materials containing a high density of various metal centres with potentially interesting physical properties, thanks to the columnar mesophase exhibited by the corresponding monomers.

(b)

Fig. 22. (Cont).

4
Columnar Order Through Specific Intermolecular Interactions

4.1
Hydrogen Bonding Interactions

The occurrence of columnar mesophases can also be due to the existence of specific intermolecular interactions between identical or different individual molecules, such as intermolecular hydrogen bonding [104]. One of the first sytems studied was the di-isobutyl-silanediol presenting a columnar mesophase where the basic unit is a dimer in which two molecules are held together by conventional hydrogen bonds [105]. Then, supramolecular cylinders have been obtained by mixing tartaric acid and a complementary unit as uracil or diacylamino-pyridine [106]. Another preliminary example was the case of two chain diols compounds which can self organize via hydrogen bonding, six single molecules arranging into a disk-like plate,

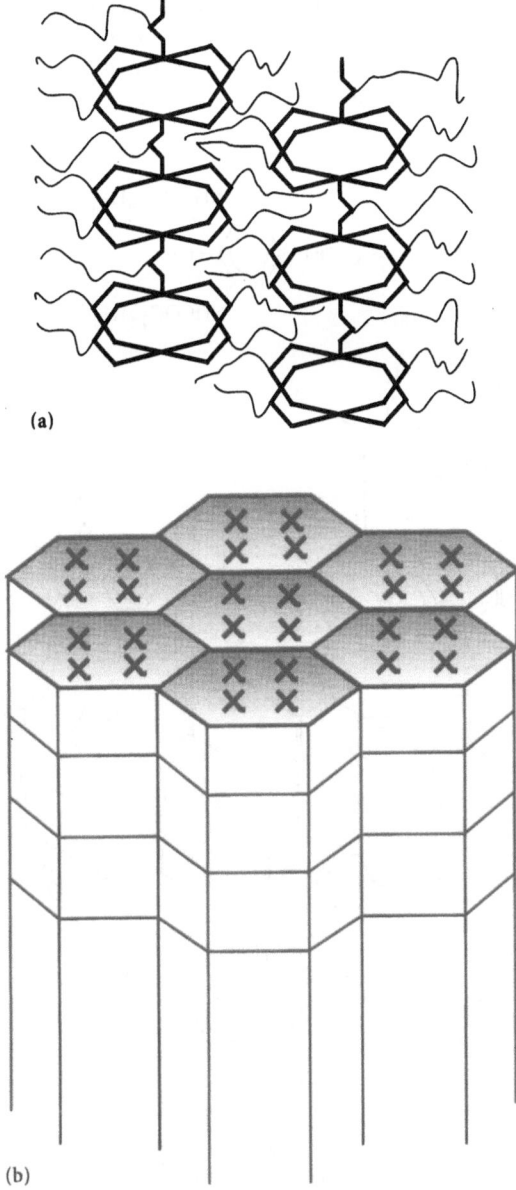

Fig. 23a, b. Schematic representation of the structural model proposed for the thermotropic mesophase of mixed-valent diruthenium pentacarboxylates. **a** Two polymeric chains shifted by half a dimer, **b** columnar hexagonal mesophase built up by four polymeric chains

resulting in the formation of a columnar mesophase [107]. Following these first examples, a similar supramolecular organization has also been found for inositol liquid crystals [108], benzamides [109], mesogenic low molecular and

polymeric hydrazides [110] or for diacylaminobenzenes [111]. Let us consider
more particularly the case of these latter diamide derivatives surrounded by
three long alkyl chains, which exhibit a two-dimensional columnar mesophase
and a nematic phase as a function of temperature (see Fig. 24).

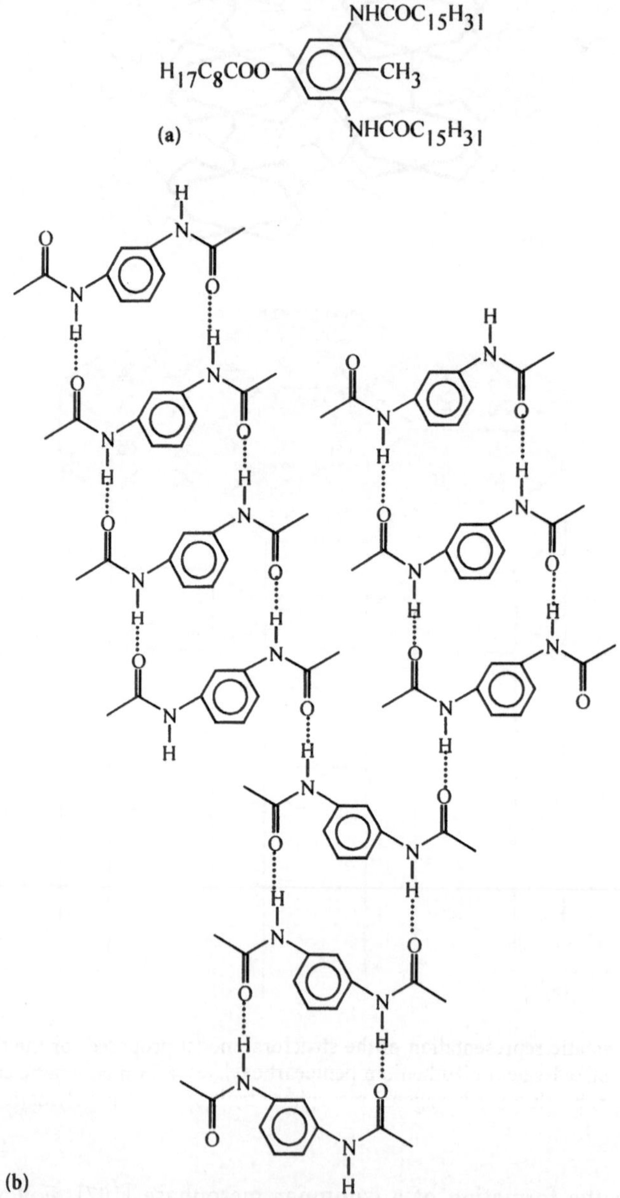

Fig. 24. a Chemical structure of the trisubstituted triamide, **b** Schematic representation of the hydrogen bonds array in the columnar structure (from ref. 112)

The molecules are linked together by hydrogen bonds, and form some kind of cross-linked polymers. If the major part of the molecules constitute quasi infinite simple wires, the nodes correspond to a small number of molecules which are bound to three different neighbours. A precise study performed by X-ray diffraction and dilatometry has demonstrated that the hexagonal columnar phase can be described as an array of columns containing defects which evolve as a function of temperature [112].

Polycatenar liquid crystals can be formed from small building units connected by hydrogen bonds, for example through the interaction of pyridines and carboxylic acids [113]. Three chain benzamides form also hexacatenar compounds where the central part is the hydrogen-bonded cyclic dimer[109]. These systems exhibit spontaneously columnar mesophases as the polycatenar compounds described in Sect. 3.2 do. Hydrogen bonding is also invoked to understand the self-assembly of twin tapered bisamides into supramolecular columns which arrange themselves according to a two-dimensional hexagonal lattice. In the model of supramolecular organization proposed, the melted aliphatic chains occupy uniformly the space around the hydrogen-bonded amide channel [114]. The hydrogen bonds should be directed vertically along the column axis and the column should be interlinked by a hydrogen-bonded network.

One of the most recent examples reported in the literature is quite representative of the building game made possible with complementary chemical units in supramolecular chemistry [115]. It concerns equimolar mixtures of melamine and disubstituted-benzoic acid shown in Fig. 25.

Molecular recognition between both different units leads to the formation of disk-shaped supramolecular entities, which then can pile up one over the other (like discotic molecules, see Sect. 3.1) to form one column. The columns organize themselves according to a two-dimensional lattice to produce a columnar mesophase.

4.2
Charge Transfer Interaction

The induction of various columnar mesophases can be realized through the doping of low molecular weight electron rich molecules with electron acceptors as schematically represented in Fig. 26.

One of the first studies involved non-mesomorphic triphenylene derivatives mixed with 2,4,7-trinitrofluorenone (TNF) [116] shown in Fig. 27.

The resulting mixtures revealed the existence of columnar mesophases, with a pronounced intracolumnar order, the stacking periodicity being in the range from 3.40 to 3.45 Å. The same type of ordered columnar mesophase has been induced by mixing discotic nematic compounds with TNF. In this case, the interactions between the disks are too small in themselves to favour a columnar mesophase, but the incorporation of electron acceptor molecules such as TNF contributes to enhance the intermolecular interactions and to generate an ordered columnar mesophase. The stabilization of the columnar structure in these binary systems has been found for other types of molecules

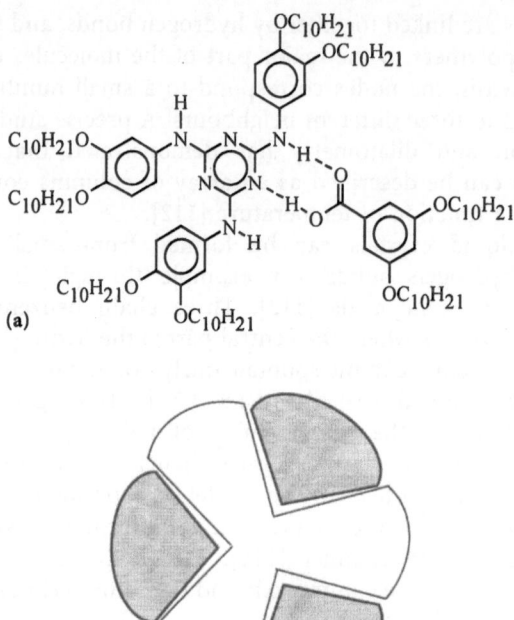

(a)

(b)

Fig. 25. a Chemical structure of the melamine and of the disubstituted benzoic acid used in equimolar mixture, **b** Schematic of the molecular recognition between the two chemical units to form an elementary disk-shaped entity (from ref. 115)

[117], and revealed to be a general characteristic due to an alternant stacking of donor and acceptor molecules [118]. As judiciously mentioned by the authors of reference 118, ordered columnar mesophase simply means that the columns are more ordered than in the so-called "disordered" mesophases. As a matter of fact, from the width of the X-ray reflection relative to the stacking periodicity along the columnar axis, the corresponding correlation length is only of the order of 20 molecules. So, it is better to invoke a fairly regular stacking within the columns.

In summary, addition of the electron acceptor TNF to rich electron compounds generally results in the formation of ordered columnar mesophases in wide temperature ranges. There is a regular alternated stacking within the columns, and in addition some azimuthal correlations between the donor and acceptor molecules within the plane perpendicular to the columnar axis [118]. In these charge transfer (CT) induced mesophases, the donor/acceptor interaction is predominant compared to the donor/donor and acceptor/acceptor ones, leading to an alternate stacking and to an enhanced stability and intracolumnar ordering. This CT concept could be quite useful for the molecular engineering of columnar systems from disk-like electron rich molecules and poor electron ones. It could be possible to use these mixtures to tune the mesomorphic properties, and in particular the temperature mesomorphic range.

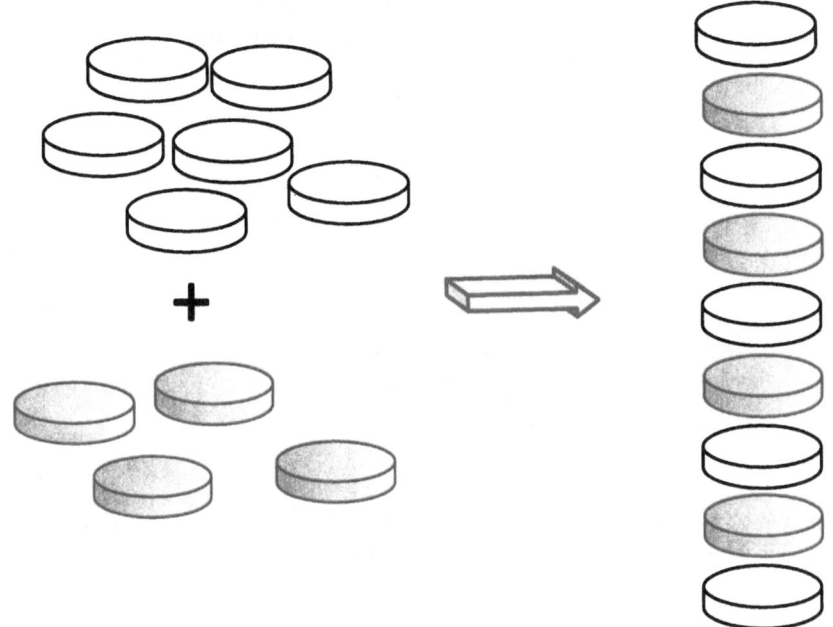

Fig. 26. Schematic of the induction of ordered columnar mesophases through charge transfer interaction

4.3
Dipolar Interactions

In order to design molecular materials with cooperative properties, it is necessary to try to control the intermolecular interactions, and for example the dipolar interactions. This has been achieved for some liquid crystalline materials exhibiting a columnar order. A recent example is the case of non-discoid molecules such as unsymmetrical copper bis-β-diketonates [119]. The latter present columnar mesophases with no order along the columnar axis. Even if one individual complex does not have a disk-shaped architecture, dimers obtained by a 90° rotation between two complexes can present indeed a disk-shaped structure, which is able to produce a columnar mesophase. Moreover, the presence of electron-withdrawing substituents in the molecule strongly stabilizes the mesophase through intermolecular dipolar interactions. This behaviour corroborates previous ones obtained with non-discoid molecules exhibiting columnar mesophases, which had been shown to be stabilized by a specific dipolar organization [120].

4.4
Molecular Recognition

Molecular recognition has been claimed to induce self-assembly into supramolecular architectures displaying columnar hexagonal mesophases

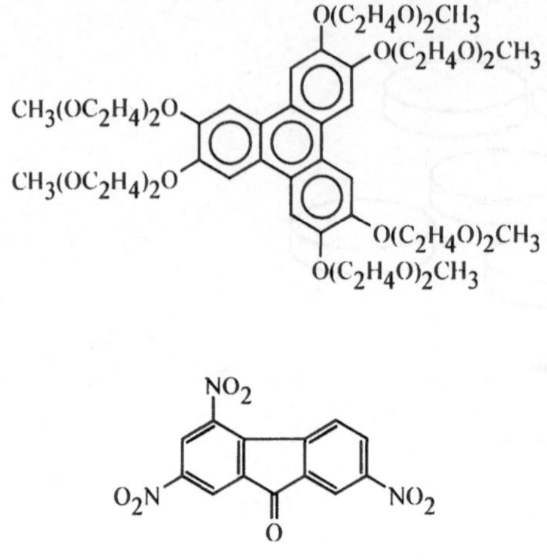

Fig. 27. Chemical structures of non mesomorphic discoid compound and TNF molecule used for induction of columnar mesophase through charge transfer interaction

according to principles which are similar to those of the self-assembly occurring in tobacco mosaïc virus [121]. This was particularly the case of taper-shaped molecules [122]. The driving force for such a cylindrical self-assembly is described as being the result of a subtle balance between what are called "exo" and "endo" recognition processes. The tapered shape of the molecules and the hydrophobic character of some constituent part act as the *exo*-recognition process, whereas dipolar and ionic interactions with the presence of crown ethers containing a well suited cavity for the binding of charged substrates, can act as *endo*-recognition process (see also Sect. 3.4).

5
Polar Columnar Mesophases

5.1
Columnar Mesophases with Axial Polarity

The possibility of building up polar molecular materials is of particular importance nowadays, since it offers great potential for tuning specific physical properties such as ferroelectricity for example. However, it is generally difficult to generate spontaneous stable polar assemblies, because the materials usually adopt antiferroelectric structures in order to avoid internal electric fields. To overcome this difficulty, hexagonal columnar super-structures with axial polarity are of particular interest, since the hexagonal symmetry cannot accomodate bulk antiferroelectric order. Several attempts to induce this type of ferroelectricity base on bowl-shaped and pyramidic molecules [123, 124] as well as metallomesogenic vanadyl complexes [125] have been reported, Fig. 28.

The vanadyl group is attractive since it exhibits a square pyramidal structure and has a large directional dipole associated with the V=O bond. The axial polarity should result from the head-to-tail alignment of these groups in a linear chain structure: $-V=O-V=O-$. Even if these materials have the molecular requirements necessary to yield columnar mesophases with an axial polarity, unfortunately it was not possible to define the existence of a spontaneous polarization. However, a very recent paper [126] seems to indicate that ferroelectric columnar mesophases with the polarization parallel

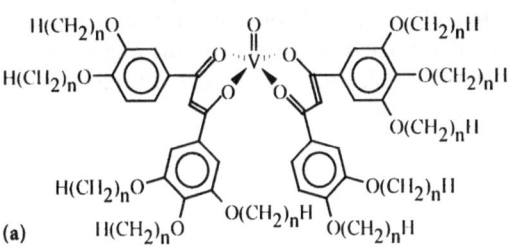

(a)

(b)

Fig. 28. a Chemical structure of a vanadyl complexe used to produce hexagonal columnar mesophase with polar order. **b** Schematic of the intramolecular arrangement within one polarized column

to the optical axis are observed with pyramidic molecules subsituted with six chiral chains. In this case, the switching time would be in the range of seconds and the polarization about 10 nC/cm^2.

It is worth recalling in this section that polar columns located at the nodes of a two-dimensional lattice can be oriented all parallel or alternant if the lattice is tetragonal [127]. In this case, ferroelectric or antiferrolectric mesophases can be obtained, depending on the specific interactions between columns. On the contrary, if the mesophase has the hexagonal symmetry, only a ferroelectric arrangement of the columns is possible. In this context, recent Monte Carlo simulations [128] of discotic mesogens with axial dipole have shown that nematic and hexagonal columnar mesophases can be generated. In the columnar mesophase, each column contains aligned dipolar domains, but no column is fully polarized, and the dipolar domains are paired with some others of neighbouring columns.

5.2
Chiral Columnar Mesophases

Similarly to the molecular engineering of calamitic molecules to produce ferroelectric smectic C* phases [129], disk-like molecules with chiral peripheral chains tilted with respect to the columnar axis were predicted to lead to ferroelectric columnar mesophases [130]. Indeed, as it is the case with all flat disk-shaped mesogenic molecules, the tilt is mainly associated with the flat rigid aromatic cores of the molecules, the side-chains being in a disordered state around the columnar core. Thus, the nearest part of the chains from the cores makes an angle with the plane of the tilted aromatic part of the molecules. If the chiral centre and the dipole moment are located close to the core, then each column possesses a non-zero time averaged dipole moment, and therefore a spontaneous polarization. For reasons of symmetry, this polarization must be, on average, perpendicular to both the columnar axis and to the tilt direction; in other words, the polarization is parallel to the axis about which the disk-shaped molecules rotate when they tilt as shown in Fig. 29.

The first report on ferroelectric switching in columnar mesophases appeared in 1992 [131]. Since then, a few other examples of compounds exhibiting switchable columnar mesophases have been reported [127, 132–134]. If the measured spontaneous polarizations are of the same order as those of lamellar smectic C* phases, the switching times on the contrary are much slower (of the order of ms compared to μs in case of smectic C*). This is probably due to the high viscosity of the material related to the two-dimensionnal lattice of the columns, to the relatively large size of the molecules, and to the elliptic shape of the columns caused by the tilt of the aromatic cores with respect to the columnar axis. Thus, columns cannot rotate freely from one state to the other, or cannot rearrange themselves internally, depending whether the switching mechanism involves rotation of the whole columns or a reversing of the tilt without molecular rotation. An attempt to

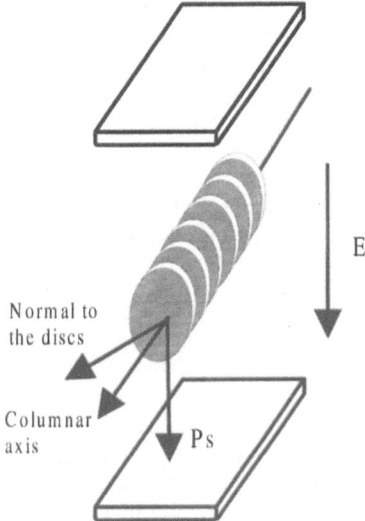

Fig. 29. Molecular arrangement within a column of chiral disk-shaped molecules

understand the switching mechanism was reported from very recent structural studies on such chiral columnar mesophases. It was possible to establish relations between the structural features and the electro-optical behaviour [135]. In particular this latter can be linked to the different orientational ordering possibilities according to a triangular lattice of columns.

The interest in these materials is not purely academic, although their switching rates are slow and the orientation in the electro-optical cells, obtained by shearing, difficult to obtain in an industrial development. However, switchable ferroelectric columnar liquid crystals possess several advantages. First, they are probably resistant from the mechanical point of view, due to a certain stabilization induced by the two-dimensional arrangement of the columns. Second, the tilt angle of the aromatic cores does not vary significantly with temperature (as far as we can state from the few known examples of such ferroelectric columnar mesophases). Third, as the electro-optical properties seem to depend on the electric field strength [136], it is possible to imagine multistable switching devices by setting the tilt angle of the flat aromatic cores at different values.

7
Conclusion

Even if the list of supramolecular systems described in this chapter is not completely exhaustive (nearly 400 papers dealing with columnar mesophases appeared in the last ten years), it is obvious that a considerable progress has now been accomplished in the construction of supramolecular columnar mesophases. As discussed in detail in the previous sections, such a construction can involve specific molecular architectures and/or specific intermolecularly forces. There are now enough molecular and theoretical tools to better control the organization of these systems better, and further steps are conceivable to tune the properties of these materials finely, in order to develop novel molecular devices.

Acknowledgments. The author thanks C. Cruz, S. Méry and A. Skoulios for many helpful discussions, and B. Heinrich for his help during the preparation of this manuscript.

7
References

1. Gray GW, Goodby JWG (1984) Smectic Liquid Crystals, Leonard Hill, Glasgow
2. Chandrasekhar S (1992) Liquid Crystals, Cambridge University Press
3. a) Levelut AM (1983) J Chim Phys 80: 149 b) Leadbetter AJ, Gray GW (ed) (1987) Thermotropic Liquid Crystals, John Wiley, Chichester, p 1–27
4. a) Skoulios A, Luzzati V (1959) Nature 183: 1310 b) Skoulios A, Luzzati V (1961) Acta Cryst 14:278 c) Gallot Y, Skoulios A (1962) Acta Cryst 15: 826
5. a) Robinson C (1958) Disc Far Soc 25: 29 b) Saludjan P, Luzzati V (1967) Poly-α-amonoacids, Dekker, New York, p 157
6. a) Luzzati V, Nicolaïeff (1959) J Mol Biol 1: 127 b) Robinson C (1961) Tetrahedron 13: 219
7. Chandrasekhar S (1993) Liq Cryst 14: 3
8. Guillon D, Skoulios A (1977) Mol Cryst Liq Cryst 39: 139
9. Weber P, Guillon D, Skoulios A (1991) Liq Cryst 9: 369
10. van Nostrum CF (1996) Adv Mater 8: 1027
11. a) Piechocki C, Simon J, Skoulios A, Guillon D, Weber P (1982) J Am Chem Soc 104: 5245 b) Guillon D, Skoulios A, Weber P (1985) Mol Cryst Liq Cryst 130: 223 c) Hanack M, Beck A, Lehmann H (1987) Synthesis Commun 8: 703 d) Sauer T, Wegner G (1988) Mol Cryst Liq Cryst 162: 97 e) Ohta K, Jacquemin L, Sirlin C, Bosio L, Simon J (1988) New J Chem 12: 751 f) van der Pol JF, Neeleman E, Zwikker JW, Nolte RJM, Drenth W, Aerts J, Visser R, Picken SJ (1989) Liq Cryst 6: 577
12. Desiraju GR, Rahda Kishan KV (1989) J Am Chem Soc 111: 4838
13. a) Boden N, Bushby RJ, Clements J, Movaghar B, Donovan KJ, Kreouzis T (1995) Phys Rev E 52: 13274 b) Reichert A, Ringsdorf H, Schuhmacher P (1996) Compr Supramol Chem 9: 313
14. Boden N, Bushby RJ, Clements J (1993) J Chem Phys 98: 5920
15. a) van de Craats AM, Warman JM, Hasebe H, Naito R, Ohta K (1997) J Phys Chem 101: 9224 b) Toupance T, Bassoul P, Mineau L, Simon J (1996) J Phys Chem 100: 11704
16. a) Kumar S, Schumacher P, Henderson P, Ringsdorf H (1996) Mol Cryst Liq Cryst 288: 211 b) Rego JA, Kumar S, Dmochowski IJ, Ringsdorf H (1996) Chem Commun 1031 c) Haarer D, Adam D, Simmerer J, Closs F, Funhoff D, Häussling L, Siegensmeyer K, Ringsdorf H, Shumacher P (1994) Mol Cryst Liq Cryst 252: 155 d) Adam D, Schumacher P, Simmerer J, Häussling L, Siemensmeyer K, Etzbach KH, Ringsdorf H, Haarer D (1994) Nature 371: 141
17. Dennis KP, Yee-O Y, Wai KC, Sze-Chit Y (1997) Tetrahedron Letters 38: 6701
18. Duro JA, de la Torre G, Barbera J, Serrano JL, Torres T (1996) Chem Mater 8: 1061
19. Clarkson GJ, Cook A, McKeown NB, Treacher KE, Ali-Adib Z (1996) Macromolecules 29: 913
20. van Nostrum CF, Nolte RJM (1994) Macromol Symp 77: 267

21. Herwig P, Kayser CW, Müllen K, Spiess HW (1996) Adv Mater 8: 510
22. Dahn U, Erdelen C, Ringsdorf H, Festag R, Wendorff JH, Heiney PA, Maliszewsky NC (1995) Liq Cryst 19: 759
23. Schouten PG, Warman JM, de Haas MP, van Nostrum CF, Gelinck GH, Nolte RJ, Copyn MJ, Zwikker LW, Engel MK, Hanack M, Chang YH, Ford WT (1994) J Am Chem Soc 116: 6880
24. Treacher KE, Clarkson GJ, McKeown NB (1995) Liq Cryst 19: 887
25. Plesnivy T, Ringsdorf H, Schumacher P, Nütz, Diele S (1995) Liq Cryst 18: 185
26. a) Boden N, Bushby RJ, Cammidge AN (1995) J Am Chem Soc 117: 924 b) Sauer T (1993) Macromolecules 26: 2057 c) van der Pol JF, Zwikker JW, Warman MP (1990) Recl Trav Chim Pays-Bas 109: 208 d) Sielcken OE, van de Kuil LA, Drenth W, Schoonman J, Nolte RJM (1990) J Am Chem Soc 112: 3086
27. Favre-Nicolin CD, Lub J, van der Sluis P (1996) Adv Mater 8: 1005
28. Disch S, Finkelmann H, Ringsdorf H, Schumacher P (1995) Macromolecules 28: 2424
29. Zheng H, Lai CK, Swager TM (1995) Chem Mater 7: 2067
30. a) Karthaus O, Ringsdorf H, Tsukruk VV, Wendorff JH (1992) Langmuir 8: 2279 b) Vandevyver M, Albouy PA, Mingotaud C, Perez J, Barraud A, Karthaus O, Ringsdorf H (1993) Langmuir 9: 1561
31. Jin S, Tiefel TH, Wolfe R, Sherwood RC (1992) Science 446: 356
32. Vincent PS, Barlow WS (1980) Thin Solid Films 71: 305
33. Maliszewskyl NC, Heiney PA, Josefowicz JY, Plesnivy T, Ringsdorf H, Schuhmacher P (1995) Langmuir 11: 1666
34. Goldmann D, Janietz D, Festag R, Schmidt C, Wendorff JH (1996) Liq Cryst 21: 619
35. Sartori E, Fontana MP, Dalcanale E, Costa M (1996) Mol Cryst Liq Cryst 290: 31
36. Schönherr H, Kremer FJB, Kumar S, Rego JA, Wolf H, Ringsdorf H, Jaschke M, Butt HJ, Bamberg E (1996) J Am Chem Soc 118: 13051
37. Tsukruk VV, Reneker DH, Bengs H, Ringsdorf H (1993) Langmuir 9: 2141
38. Christ T, Glüsen B, Greiner A, Kettner A, Sander R, Stümpflen V, Tsukruk V, Wendorff JH (1997) Adv Mater 9: 48
39. Malthête J, Levelut AM, Nguyen HT (1985) J de Phys Lett 46: L-875
40. a) Malthête J, Collet A (1985) Nouv J Chim 9: 151 b) Zimmermann H, Poupko R, Luz Z, Billard J (1985) Z Naturforsch 40a: 149 c) Lin Lei (1987) Mol Cryst Liq Cryst 146: 41 d) Dalcanale E, du Vosel A, Levelut AM, Malthête J (1991) Liq Cryst 10: 185 e) Diele S, Ziebarth K, Pelzl G, Demus D, Weissflog W (1990) Liq Cryst 8: 211
41. Levelut AM, Malthête J, Destrade C, Nguyen HT (1987) Liq Cryst 2: 877
42. Nguyen HT, Destrade C, Levelut AM, Malthête J (1986) J de Phys 47: 553
43. Destrade C, Nguyen HT, Roubineau A, Levelut AM (1988) Mol Cryst Liq Cryst 159: 163
44. Nguyen HT, Destrade C, Malthête (1990) Liq Cryst 8: 797
45. Destrade C, Nguyen HT, Alstermark C, Lindsten G, Nilsson M, Otterholm B (1990) Mol Cryst Liq Cryst 180B: 265
46. Nguyen HT, Destrade C, Malthête J (1997) Adv Mater 9: 375
47. Malthête J, Nguyen HT, Destrade C (1993) Liq Cryst 13: 171
48. Nguyen HT, Destrade C, Mathête J (1998) Handbook of Liquid Crystals, Ed VCH, Vol 2B p 865
49. Fang Y, Levelut AM, Destrade C (1990) Liq Cryst 7: 265
50. Guillon D, Skoulios A, Malthête J (1987) Europhys Lett 3: 67
51. Nütz, Diele S, Pezlz G, Ringsdorf H, Paulus W, Willson G (1995) Liq Cryst 18: 699
52. Tuffin RP, Toyne KJ, Goodby JW (1996) J Mater Chem 6: 1271
53. Donnio B, Heinrich B, Gulik-Krzywicki T, Delacroix H, Guillon D, Bruce DW (1997) Chem Mater 9: 2951
54. Rowe KE, Bruce DW (1998) J Mater Chem 8: 331
55. Guillon D, Heinrich B, Ribeiro A, Cruz C, Nguyen HT (1998) Mol Cryst Liq Cryst in press
56. Skoulios, Guillon D (1988) Mol Cryst Liq Cryst 165: 317
57. Weissflog W, Letko I, Diele S, Pelzl G (1996) Adv Mater 8: 76

58. Weissflog W, Rogunova M, Letko I, Diele S, Pelzl G (1996) Liq Cryst 21: 13
59. Diele S, Zierbarth K, Pelzl G, Demus D, Weissflog W (1990) Liq Cryst 8: 211
60. Skoulios A, Luzzati (1961) Acta Cryst 14: 278
61. Hardouin F, Levelut AM, Achard MF, Sigaud G (1983) J Chim Phys 80: 53
62. Guillon D, Skoulios A (1984) J Physique 45: 607
63. Hildebrandt F, Schröter JA, Tschierske C, Festag R, Wittenberg M, Wendorff JH (1997) Adv Mater 9: 564
64. Artzner A, Veber M, Clerc M, Levelut AM (1997) Liq Cryst 23: 27
65. Percec V, Pugh C (1989) Side Chain Liquid Crystal Polymers, Mc Ardle CB, Blackie
66. a) Feughelman M, Langridge R, Seeds WE, Stokes AR, Wilson HR, Hooper HCW, Wilkins MH, Barklay RK, Hamilton LD (1955) Nature 175: 834 b) Luzzati V (1963) Prog Nucleic Acid Res 1: 347
67. Godovski YK, Papkov (1989) Adv in Polymer Science 88: 130
68. Miller RD (1989) Angew Chem Int Ed Engl Adv Mater 8: 466
69. Miller RD (1989) Soc Chim Belg 98: 695
70. Miller RD, Hofer D, Fickes GN, Willson CG, Marinero E, Trefonas P, West R (1986) Polym Eng Sci 26: 1129
71. Kepler RG, Zeigler JM, Harrah LA, Kurz SR (1987) Phys Rev B25: 2818
72. Weber P, Guillon D, Skoulios A, Miller RD (1990) Liq Cryst 8: 825
73. Weber P, Guillon D, Skoulios A, Miller RD (1989) J Phys Paris 50: 793
74. Klemann BM, West R, Koutsky (1996) Macromolecules 29: 198
75. a) Karikari EK, Farmer BL, Miller RD, Rabolt JF (1993) Macromolecules 26: 3937 b) Klemann BM, West R, Koutsky JA (1993) Macromolecules 26: 1042 c) Asuke T, West R (1991) Macromolecules 24: 343
76. Koopmann F, Frey H (1996) Macromolecules 29: 3701
77. Papkov VS, Il'ina MN, Zhukov VP, Tsvankin D, Tur DR (1992) Macromolecules 25: 2033
78. Ungar G (1993) Polymer 34: 2050
79. Molenberg A, Siffrin S, Möller M, Boileau S, Teyssié D (1996) Macromol Symp 102: 199
80. Out G, Turetskii, Möller M, Oelfin D (1994) Macromolecules 27: 3310
81. Turetskii A, Out GJJ, Klok HA, Möller M (1995) Polymer 36: 1303
82. Out GJJ, Turetskii AA, Möller M, Oelfin D (1995) Macromolecules 28: 596
83. a) Fischer H, Ghosh SS, Heiney PA, Maliszewskyj NC, Plesnivy T, Ringsdorf H, Seitz M (1995) Angew Chem Int Ed Engl 34: 795 b) Seitz M, Plesnivy T, Schimossek K, Edelmann M, Ringsdorf H, Fischer H, Uyama H, Kobayashi S (1996) Macromolecules 29: 6560
84. Watanabe J, Takashina Y (1991) Macromolecules 24: 3423
85. Watanabe J, Sekine N, Nematsu T, Sone M, Kricheldorf HR (1996) Macromolecules 29: 4816
86. Percec V, Schlueter D, Ungar G (1997) Macromolecules 30: 645
87. Malher W, Guillon D, Skoulios A (1985) Mol Cryst Liq Cryst Lett 2: 111
88. Newkome GR, Moorefield CN, Vögtle F (1996) Dendritic Molecules: Concepts, Syntheses, Perspectives, VCH, Weinheim
89. Hawker CJ, Wooley KL, Fréchet JMJ (1993) J Chem Soc, Perkin Trans 1: 1287
90. Issbermer J, Moors R, Vögtle F (1994) Angew Chem, Int Ed Engl 33: 2413
91. a) Percec V, Kawasumi M (1992) Macromolecules 25: 3843 b) Percec V, Chu P, Ungar G, Zhou J (1995) J Am Chem Soc 117: 11441
92. Percec V, Johhansson G, Ungar G, Zhou J (1996) J Am Chem Soc 118: 9855
93. Pesak DJ, Moore JS (1997) Angew Chem Int Ed Engl 36: 1636
94. Lorenz K, Frey H, Stühn B, Mülhaupt (1997) Macromolecules 30: 6860
95. Cameron JH, Facher A, Lattermann G, Diele S (1997) Adv Mater 9: 398
96. (a) Giroud-Godquin AM, Maitlis PM (1991) Agew Chem Int Ed Engl 30: 375 (b) Espinet P, Esteruelas MA, Oro LA, Serrano JL, Sola E (1992) Coord Chem Rev 117: 215 (c) Bruce DW (1992) Inorganic Materials Wiley & sons, New York (d) Serrano JL (1996) Metallomesogens, VCH, Weinheim

97. (a) Abied H, Guillon D, Skoulios A, Weber P, Giroud-Godquin AM, Marchon JC (1987) Liq Cryst 2: 269 (b) Ibn-Elhaj M, Guillon D, Skoulios A, Giroud-Godquin AM, Maldivi P (1992) Liq Cryst 11: 731 (c) Attard GS, Templer RH (1993) J Mater Chem 3: 207 (d) Marchon JC, Maldivi P, Giroud-Godquin AM, Guillon D, Ibn-Elhaj M, Skoulios A (1992) In: Göpel W, Ziegler C (eds)Nanostructures Based on Molecular Materials. VCH, Weinheim, p 285 (e) Bonnet L, Cukiernik FD, Maldivi P, Giroud-Godquin AM, Marchon JC, Ibn-Elhaj M, Guilon D, Skoulios A (1994) Chem Mater 6: 31 (f) Baxter DV, Cayton RH, Chisholm MH, Huffman JC, Putilina EF, Tagg SL, Wesemann JL, Zwanziger JW, Darrington FD (1994) J Am Chem Soc 116: 4551

98. Mehrotra RC, Bohra R (1983) Metal Carboxylates, Academic Press

99. Ibn-Elhaj M, Guillon D, Skoulios A, Maldivi P, Giroud-Godquin AM, Marchon JC (1992) J Phys II France 2: 2237

100. Ibn-Elhaj, Guillon D, Skoulios A (1992) Phys Rev A 46: 7643

101. Cukiernik FD, Ibn-Elhaj M, Chaia ZD, Marchon JC, Giroud-Godquin AM, Guillon D, Skoulios A, Maldivi P (1998) Chem Mater 10: 83

102. Cotton FA, Matusz M, Zhong B (1988) Inorg Chem 27: 4368

103. Marcot L, Maldivi P, Marchon JC, Guillon D, Ibn-Elhaj M, Broer DJ, Mol GN (1997) Chem Mater 9: 2051

104. Paleos C, Tsiourvas D (1995) Angew Chem 107: 1839

105. Bunning JD, Lydon JE, Eaborn C, Jackson PM, Goodby JW, Gray GW (1982) J Chem Soc Faraday Trans I 78: 713

106. Fouquey C, Lehn JM, Levelut AM (1990) Adv Mater 2: 254

107. (a) Lattermann G, Staufer G (1990) Mol Cryst Liq Cryst 191: 199 (b) Staufer G, Schellhorn M, Lattermann G (1995) Liq Cryst 18: 519

108. Praefcke K, Marquardt P, Kohne B, Stephan W, Levelut AM, Wachtel E (1991) Mol Cryst Liq Cryst 203: 149

109. Beginn U, Lattermann G (1994) Mol Cryst Liq Cryst 241: 215

110. Beginn U, Lattermann G, Festag R, Wendorff JH (1996) Acta Polymer 47: 214

111. Pucci D, Veber M, Malthête J (1996) Liq Cryst 21: 153

112. Albouy PA, Guillon D, Heinrich B, Levelut AM, Malthête J (1995) J Phys II France 5: 1617

113. Bernhardt H, Kresse H, Weissflog W (1997) Mol Cryst Liq Cryst 301: 25

114. Ungar G, Abramic D, Percec V, Heck J (1996) Liq Cryst 21: 73

115. Goldmann D, Dietel R, Janietz D, Schmidt C, Wendorff JH (1998) Liq Cryst 24: 407

116. Bengs H, Ebert M, Karthaus O, Kohne B, Praefcke K, Ringsdorf H, Wendorff JH, Wüstefeld (1990) Adv Mater 2: 141

117. (a) Praefcke K, Singer D, Langner M, Kohne B, Ebert M, Liebmann A, Wendorff JH (1992) Mol Cryst Liq Cryst 215: 121 (b) Singer D, Liebmann A, Praefcke K, Wendorff JH (1993) Liq Cryst 14: 785

118. Zamir S, Singer D, Spielberg N, Wachtel EJ, Zimmermann H, Poupko R, Luz Z (1996) Liq Cryst 21: 39

119. Zheng H, Xu B, Swager TM (1996) Chem Mater 8: 907

120. Paulus W, Ringsdorf H, Diele S, Pelzl G (1991) Liq Cryst 9: 807

121. Percec V, Tomazos D, Heck J, Blackwell H, Ungar G (1994) J Chem Soc Perkin Trans 2: 31

122. Percec V, Heck J, Johansson G, Tomazos D, Kawasumi M, Chu P (1994) Mol Cryst Liq Cryst 254: 137

123. Zimmermann H, Poupko R, Luz Z, Billard J, Naturforsch Z (1985) 40a: 149

124. Malthête J, Collet A (1985) Nouv. J. de Chimie 9: 151; Lei L (1987) Mol Cryst Liq Cryst 146: 41; Poupko R, Luz R, Spielberg N, Zimmermann H (1989) J Am Chem Soc 111: 5094

125. Xu B, Swager TM (1993) J Am Chem Soc 115: 8879

126. Jakli A, Saupe A, Scherowsky G, Chen XH (1997) Liq Cryst 22: 309

127. Scherowsky G, Chen XH (1995) J Mat Chem 5: 417

128. Berardi R, Orlandi S, Zannoni C (1997) J Chem Soc Faraday Trans 93, 1493

129. Goodby JW (1991) In: Taylor GW (ed) Ferroelectric Liquid Crystals. Gordon and Breach Science Publishers, Philadelphia, p 99

130. Prost J, in *"Symmetries and Broken Symmetries"*, Ed. N. Boccara (1981), P. 159
131. Bock H, Helfrich W (1992) W Liq Cryst 12: 697
132. Scherowsky G, Chen XH (1994) Liq Cryst 17: 803
133. Bock H, Helfrich W (1995) Liq Cryst 18: 387
134. Heppke G, Lötzch D, Müller M, Sawade H, 6th International Conference on Ferroelectric Liquid Crystals, Brest (France), 20–24 July 1997
135. Scherowsky G, Chen XH, Levelut AM (1998) Liq Cryst 24: 157
136. Bock H, Helfrich W. *Liq. Cryst.*, 1995, 18, 707

Twist Grain Boundary (TGB) Phases

John W. Goodby

The Department of Chemistry, The University of Hull, Hull, HU6 7RX, UK
E-mail: J.W.Goodby@chem.hull.ac.uk

Chirality in self-organizing media can yield a variety of novel mesophases with unique properties and structures. At high levels of chirality, however, non-linear effects come into play. The effects of chirality in self-assembled structures can compete successfully with those of conventional self-assembling properties to yield frustrated structures. For instance, the desire for rod-like molecules to form layered mesophase structures can be frustrated by the molecules' needs to form twisted structures. This can result in a defect stabilized phase being formed, where screw dislocations punctuate the normal self-organized phase in a similar way to how lines of flux punctuate the superconducting phase of a type II superconductor. This novel frustrated structure is called the Twist Grain Boundary Phase, its discovery links the physical understanding of liquid crystals to that of superconductors.

Keywords: Liquid crystals, self-organizing systems, Chirality, Frustrated phases, Superconductivity, Abrikosov flux phases

Structure and Bonding, Vol. 95
© Springer Verlag Berlin Heidelberg 1999

1
Introduction

A liquid crystalline mesophase is a state of matter that is precariously balanced between the organized solid and the amorphous liquid [1, 2]. As a consequence liquid crystals, which are aptly described as Nature's delicate state of matter [1], are of fundamental scientific interest because they provide a wide variety of examples of the interplay between ordered and disordered systems, they demonstrate the intricate checks and balances between weakly attractive and repulsive intermolecular forces, and they give insights into the unusual structural characteristics of frustrated systems. In this article we explore all of these phenomena in relation to optically active, chiral systems.

Molecular shape anisotropy plays a very important role in determining whether or not liquid crystal phases will be formed, and indeed, when they are formed, which particular modification will be generated in preference to other mesophases [3, 4]. Molecules that form liquid crystals are usually carbon based, but they need not be. They often have preferred shapes, based on rods, bananas, boards or discs. In the following account, the properties of some unusual rod-like organic systems which have asymmetric structures thereby making them optically active and chiral will be described. In particular, the unusual self-organizing properties and the structures of the frustrated phases of the (R), (S), and racemic forms of 1-methylheptyl 4′-(4-n-tetradecyloxyphenylpropioloyloxy)biphenyl-4-carboxylate, **1** (see Fig. 1), will be related [5].

In the following, a description of the melting processes of liquid crystals will be given, and then the general structures of the nematic, smectic A and smectic C achiral and chiral phases that are involved in TGB phenomena will be discussed. When mesophases are formed by molecules (such as **1**) that have asymmetric or dissymmetric structures, a reduction in the environmental space symmetry occurs, which in some cases can induce the creation of helical

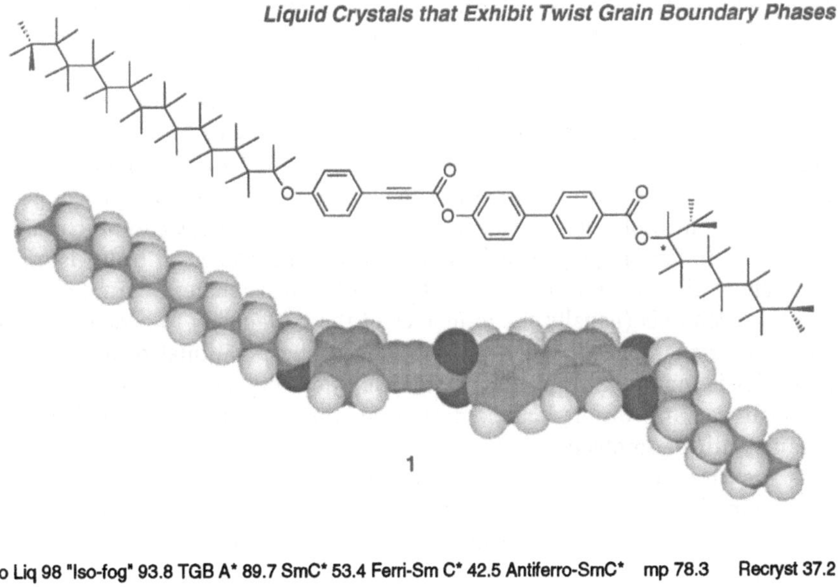

Liquid Crystals that Exhibit Twist Grain Boundary Phases

Iso Liq 98 "Iso-fog" 93.8 TGB A* 89.7 SmC* 53.4 Ferri-Sm C* 42.5 Antiferro-SmC*   mp 78.3   Recryst 37.2

Fig. 1. The Structure and transition temperatures of (R) and (S), 1-methylheptyl 4′-(4-n-tetradecyloxyphenylpropioloyloxy)biphenyl-4-carboxylate, **1**. The space filling structure is for the (S)-enantiomer

macrostructures. The desire to form a helical macrostructure can clash with the necessity of phases to form their normal structures where space is filled uniformly by the molecules. Such competitions result in structural frustrations occurring and hence novel defect stabilized phases being formed. In this chapter we will be concerned with the way in which twist distortions are localized in defects that periodically punctuate the normal smectic state. The physical properties of this defect stabilized phase are similar to those of the Abrikosov vortex state found in type II superconductors. We will investigate the analogies and draw the conclusion that the underlying physical concepts that govern superconductors and smectic liquid crystals are the same.

1.1
Melting Processes Involving Liquid Crystals

In any discussion concerning the structures of liquid crystal phases formed by rod-like molecules, a point is typically reached where the arguments center on what particular structural features define the "liquid-crystalline" state [4, 6]. For instance, when a conventional solid melts to a liquid, the strongly organized molecular array of the solid collapses to yield a disordered liquid where the molecules are free to translate, tumble, and rotate. Thus, at the melting point the molecules undergo large and rapid simultaneous changes in rotational, positional, and orientational order. However, when a melting process is mediated by the formation of liquid crystal phases there is usually a stepwise breakdown in the ordering of the molecules. The incremental steps of this decay occur with changing temperature, thus producing a variety of thermodynamically stable intermediary states between the solid and the liquid. Consequently, liquid crystals can be defined as those orientationally ordered phases that occur between the breakdown of positional/translational order on melting a solid, and the breakdown of orientational order on melting to a liquid. Thus, for melting involving calamitic liquid crystalline phases where the constituent molecules have rod-like shapes, the process can be characterized in the following way. The first step in the breakdown of order is for the relatively static lath-like molecules of the solid to oscillate, or rotate rapidly, about a given axis (usually the long axes of the molecules) to give a "smectic-like" soft-crystal phase. Secondly, the long-range positional ordering of the molecules is then lost to produce a smectic liquid crystal mesophase. Thirdly, the local packing order is then destroyed, but the orientational order still remains with the molecules reorganizing so that their long axes lie roughly in the same direction (known as the director of the phase) to give the nematic phase. Finally, this order breaks down to give the amorphous liquid. This description of the melting process for rod-like molecules is shown schematically in Fig. 2.

As with other states of matter, liquid crystal mesophases are indefinitely stable at defined temperatures and pressures. Mesophases observed above the melt point during the heating process are stable and termed enantiotropic,

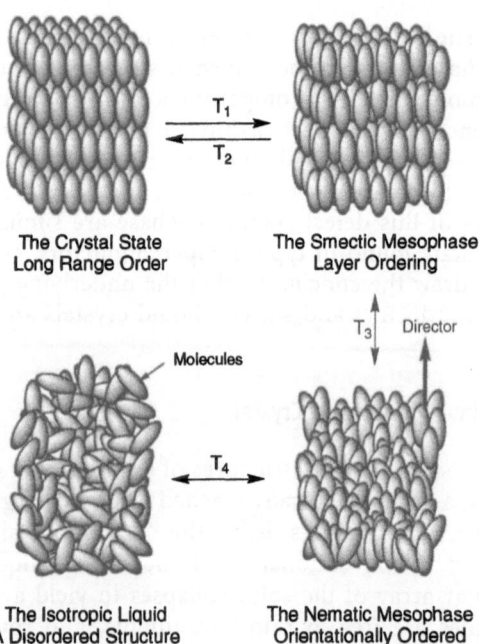

Fig. 2. The melting process for a calamitic liquid crystal, the rod-like molecules are shown as ellipses

whereas those phases that occur below the melt point on supercooling of the crystal are metastable and called monotropic. Transitions between the various liquid crystal mesophases invariably occur at well-defined temperatures, with little hysteresis observed between heating and cooling cycles.

1.2
Elastic Properties of Liquid Crystals

As liquid crystals do not have periodic ordering of their constituent molecules and because they have quasi-liquid natures they can be treated as media that have continuum/elastic properties [2, 7]. For example, the nematic phase, which is a one-dimensionally ordered fluid where the molecules are orientationally ordered but have no long-range positional order, can be treated as an elastic fluid. Given that there are as many molecules pointing in one direction as there are pointing in the opposite direction (i.e., a rotation of π), which implies for the director of the phase that $\mathbf{n} = -\mathbf{n}$, three possible elastic deformations of its structure are possible, these correspond to a splay, twist, and bend deformations as shown schematically in Fig. 3. These deformations have the following Frank elastic constants associated with them, respectively, k_{11}, k_{22}, and k_{33}, which are analogous to spring constants. The free energy per unit volume of a nematic liquid crystal can be written as follows [2]:

$$F_V = 1/2(k_{11}[\nabla \cdot \mathbf{n}]^2) + 1/2(k_{22}[\nabla \cdot \mathbf{n}]^2) + 1/2(k_{33}[\nabla \cdot \mathbf{n}]^2) \tag{1}$$

1.3
The Structures of the Achiral and Chiral Nematic (N/N*) Phases

The nematic phase is essentially a one-dimensionally ordered elastic fluid in which rod-like molecules are orientationally ordered, but where they have no long-range positional ordering [8, 9]. In this phase the rod-like molecules tend to align parallel to each other with their long axes all pointing roughly in the same direction. The average direction along which the molecules point is called the director of the phase, and is usually given the symbol $\mathbf{n}$. The rod-like molecules in the nematic phase are free to rotate about their long axes and to

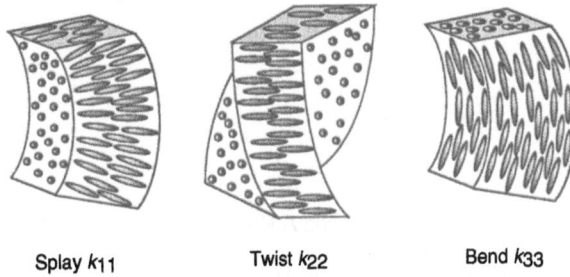

Splay k_{11} Twist k_{22} Bend k_{33}

Fig. 3. Bend and splay deformations in nematic liquid crystals

some degree about their short axes. The structure of the achiral nematic phase is depicted in Fig. 4.

In the bulk nematic phase, there are as many rod-like molecules pointing in one direction relative to the director as there are pointing in the opposite direction (a rotation of 180°), i.e. the molecules have a disordered head-to-tail arrangement in the phase. Consequently, the phase has rotational symmetry relative to the director, and therefore the phase is uniaxial. The degree to which the molecules are aligned along the director is termed the order parameter, which is defined by the equation

$$S = 1/2 \langle 3\cos^2 \theta - 1 \rangle \tag{2}$$

where θ is the angle made between the long axis of each individual rod-like molecule and the director. An order parameter of zero implies that the phase has no order at all (it is liquid-like), whereas a value of one indicates that the phase is perfectly ordered, i.e., all the long axes of the molecules are parallel to one another and to the director. For a typical nematic phase the order parameter has a value in the region of between 0.4 and 0.7 indicating that the molecules are considerably disordered.

When the nematic phase is composed of optically active materials (either a single component or a multicomponent mixture made up of chiral compounds or chiral compounds mixed with achiral materials), the phase itself becomes chiral and has reduced environmental space symmetry. The structure of the chiral nematic N* (or cholesteric) modification is one where the local molecular ordering is identical to that of the nematic phase, but in the direction normal to the director the molecules pack to form a helical macrostructure, see Fig. 5. As in the nematic phase the molecules have no long-range positional order, and no layering exists. The pitch of the helix can vary from about 0.1×10^{-6} m to almost infinity, and is dependent on optical purity and the "degree of molecular

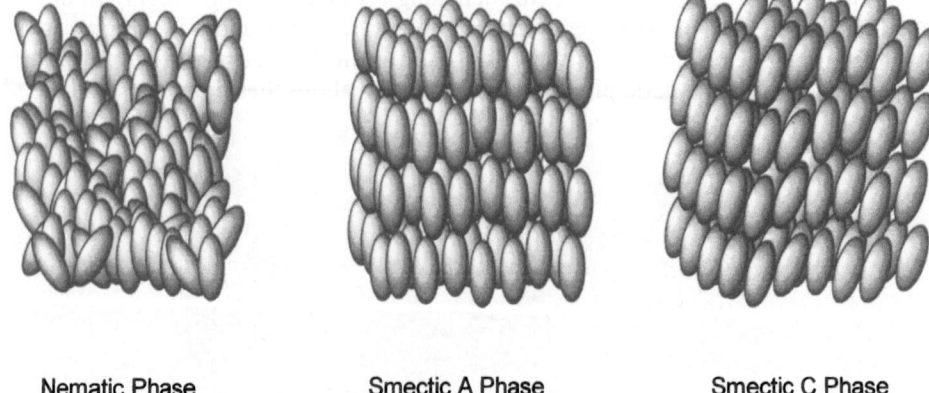

Nematic Phase   Smectic A Phase   Smectic C Phase

Fig. 4. The Structures of the achiral nematic, smectic A and smectic C phases. In each case there is no positional ordering of the rod-like molecules (shown as ellipses in the figure)

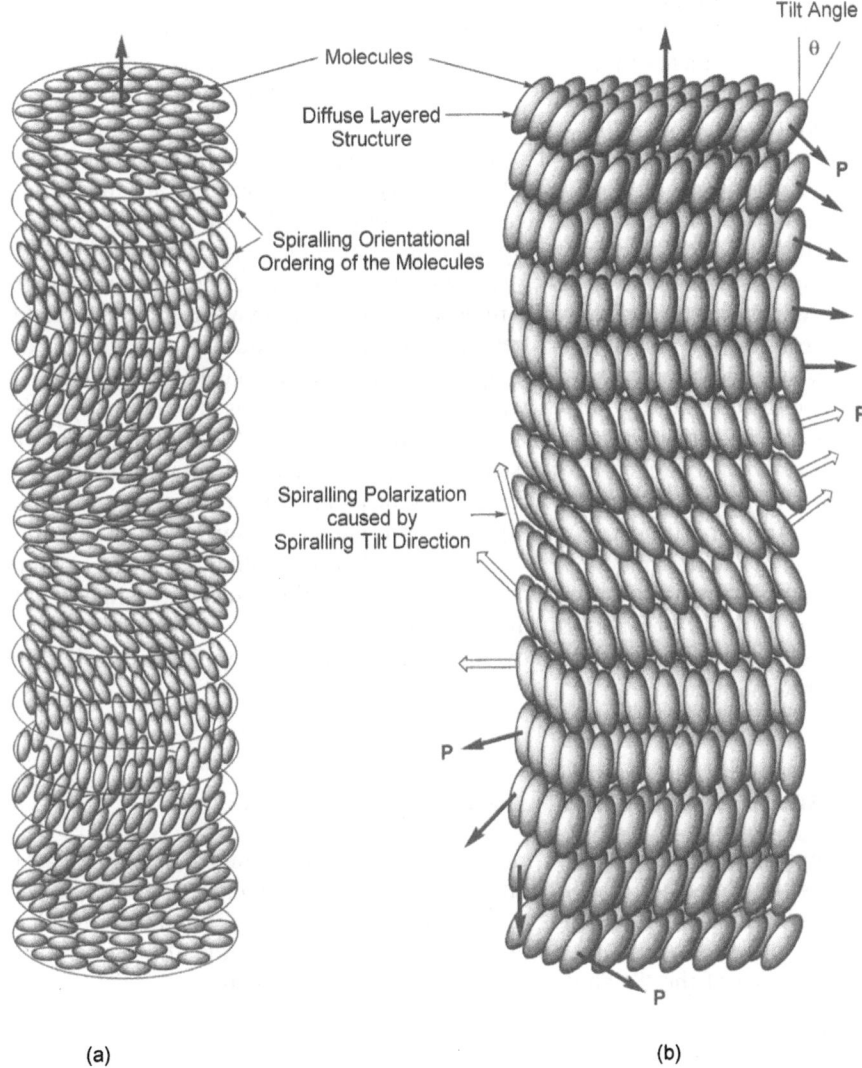

Fig. 5. Helical Structures of the chiral nematic and chiral smectic C* phases

chirality" of the material. The optic axis of the phase is parallel to the helical axis, and so the phase has negative birefringence and is optically uniaxial.

1.4
The Structures of the Achiral and Chiral Smectic A/A* Phases

In the smectic A phase the rod-like molecules are arranged in layers so that their long axes are on average perpendicular to the layer planes, see Fig. 4. The molecules are undergoing rapid reorientational motion about their long and short axes, and they are arranged so that there is no translational periodicity in the planes or between the layers [8]. Therefore, there is only short range ordering extending over a few molecular centers at most (~15–25 Å). Perpendicular to the layers the molecules are essentially arranged in a one dimensional density wave, therefore, the layers themselves must be considered as being diffuse. As a consequence, the concept of a lamellar mesophase is somewhat misleading because the layers are so diffuse that on a macroscopic scale they are almost non-existent. In actual fact, the molecules are arranged within the lamellae in such a way that they are often randomly tilted at slight angles with respect to the layer normal. This makes the layer spacing on average slightly shorter than the molecular length. Typically, the molecules are tilted anywhere up to about 14–15° from the layer normal. However, as the tilting is random across the bulk of the phase, the mesophase is also optically uniaxial, and has $D_{\infty h}$ symmetry.

The structure of the smectic A phase when it is composed of optically active material (i.e., smectic A*) remains the same as that for the achiral phase. The molecules are arranged in diffuse disordered layers, and there is no long-range periodic order. However, because of the molecular chirality, the environmental symmetry is reduced to D_∞ [10]. As a consequence, when an electric field is applied to a chiral smectic A* phase there will be a coupling of the electroclinic susceptibility to the field and the long axes of the molecules will tilt with respect to the layer planes. The tilt angle, for relatively low applied fields, varies linearly with the field. This linear electrooptic phenomenon is called the electroclinic effect.

1.5
The Structures of the Achiral and Chiral Smectic C/C* Phases

In the smectic C phase the rod-like molecules are arranged in diffuse layers where the molecules are tilted at a temperature-dependent angle with respect to the layer planes [8]. Thus the phase has C_{2h} symmetry When the smectic C phase is formed from the smectic A phase upon cooling, the temperature dependence of the tilt angle approximately takes the form

$$(\theta)_T = (\theta)_0 (T_C - T)^\alpha \tag{3}$$

where $(\theta)_T$ is the tilt angle at temperature T °C, $(\theta)_0$ is a constant, T_C is the smectic A to smectic C transition temperature, T is the temperature and α is an exponent which is usually set equal to 0.5 (typically the experimental value of the exponent is found to be less than 0.5). The molecules within the layers are locally hexagonally close-packed with respect to the director of the phase; however, this ordering is only short range, extending over distances of approximately 15–25 Å. Over large distances the molecules are randomly packed, and in any one domain the molecules are tilted roughly in the same direction in, and between, the layers (see Fig. 4). Thus, the tilt orientational ordering between successive layers is preserved over long distances.

When chiral materials are introduced into the smectic C phase it becomes optically active in a similar way to the chiral nematic phase, i.e., the smectic C* phase has similar optical activity properties and also possess a helical distribution of its molecules [6]. The helical macrostructure is generated by a precession of the tilt about an axis normal to the layers, as shown in Fig. 5. The tilt direction of the molecules in a layer above or below an object layer is rotated through an azimuthal angle relative to the object layer. This rotation always occurs in the same direction for a particular material, thus forming a helix. The helix can be either right-handed or left-handed depending on the chirality (absolute spatial configuration) of the constituent molecules. The pitch of the helix for most C* phases is commonly greater than one micrometer in length, indicating that a full twist of the helix is made up of many thousands of layers. Thus, the azimuthal angle is relatively small and is usually of the order of one-tenth to one-hundredth of a degree.

Let us now compare the environmental symmetries for the achiral and chiral smectic C phases [11]. For an achiral smectic C phase the space or environmental symmetry elements are (i) a mirror plane perpendicular to the planes of the layers and containing the long axes of the molecules, (ii) a twofold axis of rotation normal to the mirror plane and parallel to the layers, and (iii) a center of symmetry. Thus, the symmetry is classed as C_{2h}. However, when this phase contains optically active material, these symmetry elements are reduced to a single polar twofold axis parallel to the layer planes and normal to the vertical planes that contain the long axes of the molecules; consequently the phase has a reduced C_2 symmetry. The result of the packing of the dipolar regions of the molecules in these phases requires that a spontaneous polarization (P_s) acts along the C_2 twofold axis normal to the tilt direction, as predicted by Meyer [11, 12]. Since the smectic C* phase does not have a well-organized or rigid structure, the molecules can be reoriented by applying a field of known polarity [13]. This ability gives rise to the term *ferroelectric* to describe such tilted chiral phases. The presence of a spontaneous polarization in smectic liquid crystals is assumed to be due to a time-dependent coupling of the lateral components of the dipoles of the individual molecules with the chiral environment. Consequently, only the time-averaged projections of the dipole moments along the polar C_2 axis are effective in producing the macroscopic spontaneous polarization.

Another consequence of the formation of a helix in the smectic C* phases is that the spontaneous polarization direction, which is tied to the tilt orientational ordering, is itself spiraling in a direction normal to the layers. Thus, when the helix makes one full 360° turn, the polarization is averaged to zero. Therefore, in a bulk undisturbed phase which is not influenced by external forces the spontaneous polarization will average to zero, and the phase becomes helielectric [14].

2
Structure of the Twist Grain Boundary (TGBA*) Phase

Now that we have discussed all of the structures of the phases that play a part in the formation of twist grain boundary phases, let us now consider the events that can occur at a nematic to smectic phase transition. At a normal chiral nematic to smectic transition, the helical ordering of the chiral nematic phase collapses to give the layered structure of the smectic phase. However, for strongly chiral systems, there can be competition between the need for the molecules to form a helical structure due to their chiral packing requirements and the need for the phase to form a layered structure. The molecules relieve this frustration by forming a helical structure, where the axis of the helix is perpendicular to the long axes of the molecules (as in the chiral nematic phase), but at the same time they also form a lamellar structure as shown in

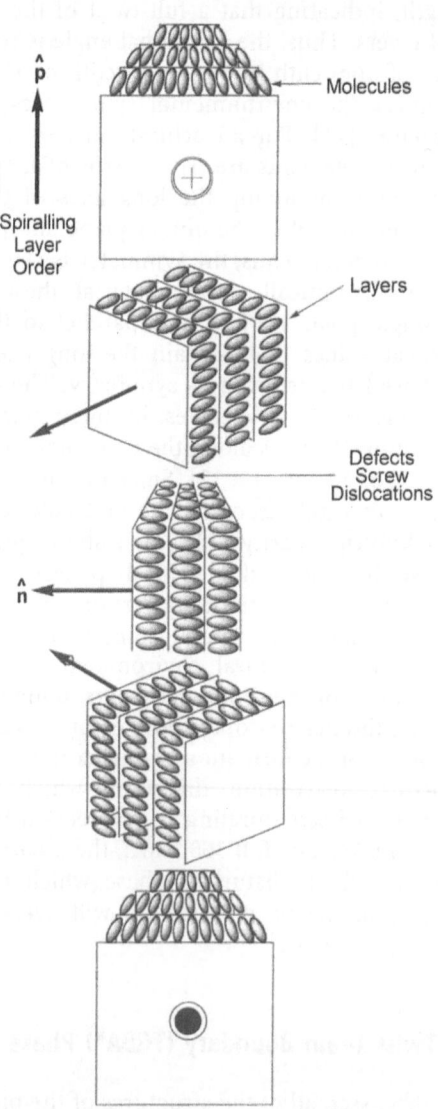

Fig. 6. Structure of the twist grain boundary smectic A* phase

Fig. 6. However, these two structures are incompatible with one another and cannot co-exist and the molecules still fill space uniformly without forming defects. The matter is resolved by the formation of a periodic ordering of screw dislocations which enables a quasi-helical structure to co-exist with a layered structure. This is achieved by having small blocks/sheets of molecules, which have a local smectic structure, being rotated with respect to one another by a set of screw dislocations, thereby forming a helical structure [15]. As the macroscopic helix is formed with the aid of screw dislocations, the dislocations themselves must be periodic. It is predicted that rows of screw dislocations in the lattice will form grain boundaries in the phase, see Fig. 7, and hence this structurally frustrated phase, which was theoretically predicted by Renn and Lubensky [15], was called the twist grain boundary (TGB).

In this analysis it must be emphasized that the TGB phase, is not simply a layered chiral nematic phase, and should not be confused with this concept. A layered chiral nematic phase simply cannot exist on a macroscopic scale, and it is a requirement that defects must be formed.

The analogues of the smectic A* and C* phases, the TGBA* and TGBC phases, have been detected at the chiral nematic to smectic A* and the chiral nematic or TGBA* to smectic C* transitions [16, 17]. Two helical forms of the TGB phases are possible, one where the helical structural arrangements of the blocks or sheets of the smectic phase are rational, i.e., the number of blocks required to form a 360° rotation of the helix is a whole number, in which case the phase is said to be "commensurate", or alternatively where the number of

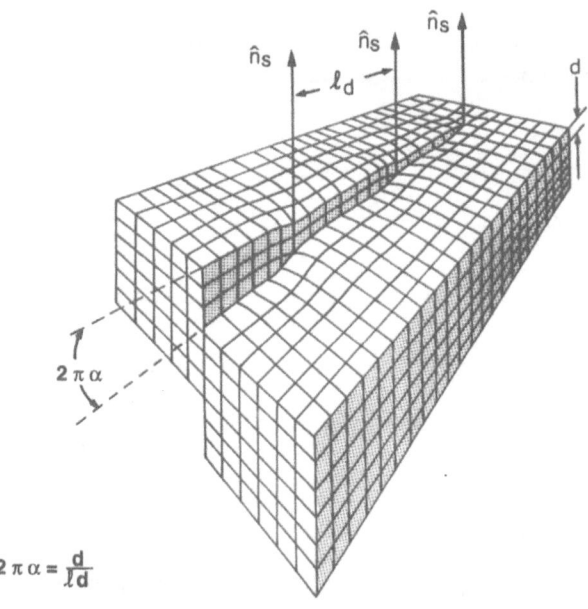

Fig. 7. A twist grain boundary made up of screw dislocations

blocks or sheets of the normal smectic phase forming one twist of the helix is not a whole number, in which case the phase is said to be "incommensurate", see Fig. 8. Furthermore, for the TGBC phase there are two further possible structural modifications, one where the molecules in the blocks or sheets are simply inclined to the layer planes with no interlayer twist, as in an achiral smectic C phase, and another where they are allowed to form helical structures normal to the layer planes as in a normal smectic C* phase. These two modifications have been given the preliminary designations TGBC and TGBC*, respectively.

3
The de Gennes Analogy with Superconductors

The presence, and to some degree the discovery, of TGB phases in liquid-crystalline systems stems from theoretical studies by de Gennes [18]. Through modeling of the N-A transition, de Gennes predicted that, for a second order nematic to smectic A phase transition, a defect stabilized phase could occur

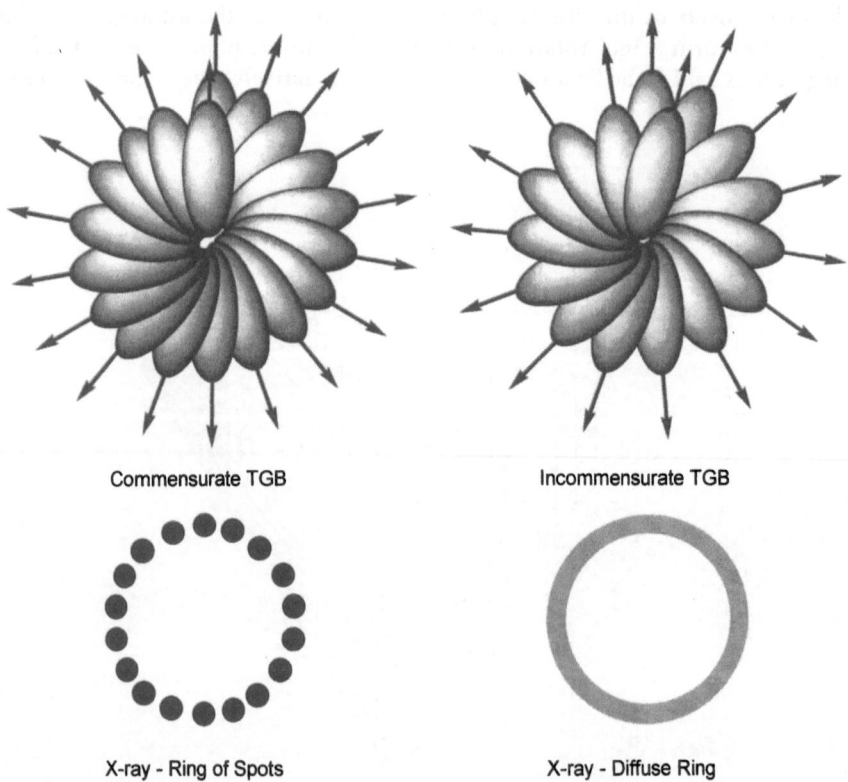

Commensurate TGB   Incommensurate TGB

X-ray - Ring of Spots   X-ray - Diffuse Ring

Fig. 8. Commensurate and incommensurate twist grain boundary phases

when the liquid crystal was subjected to twist or bend distortions. Furthermore, de Gennes also sought to unify his physical theory of the nature of this phase transition with similar theories for phase transitions in superconductors. In his analogy, de Gennes suggested that the twist and bend distortions could be incorporated into a layered smectic A structure via the presence of an array of screw or edge dislocations. The screw dislocations permeate the normal A phase in the form of a lattice, and this is similar to how the conducting phase permeates the superconducting phase to form a lattice of vortices in the Abrikosov flux phase of type II superconductors.

For superconductors Meissner and Ochsenfeld [19] showed that when a bulk superconducting material in its normal conducting phase is placed in a weak magnetic field, it will act as a perfect diamagnet with zero magnetic induction in the interior. When the material is cooled into its superconducting phase, the magnetic flux originally present will be ejected, a process that is called the Meissner effect, see Fig. 9. Thus, at the transition the lines of induction, B, are pushed out, and the bulk interior behaves in an applied external field as if inside the specimen B=0. The magnetization curve as a function of an applied magnetic field for a superconductor exhibiting a Meissner effect with a perfect diamagnetism is shown in Fig. 10(a). Materials that exhibit this form of behavior are called type I superconductors.

There are, however, other materials that do not exhibit perfect diamagnetism for which the superconducting properties tend to fall off more gently as the applied field is increased [20]. Up to a field, denoted as H_{c1} in Fig. 10(b), the material exhibits type I behavior, and above this point to a second critical field, labeled H_{c2}, the superconducting properties slowly decrease in strength. Between the lower critical field H_{c1} and the upper critical field H_{c2} the flux density $B \neq 0$ and the Meissner effect is said to be incomplete, see Fig. 10(b). In the region between the two critical fields the superconductor is threaded by flux lines and is said to be in a vortex state (or an Abrikosov phase). The term vortex state describes the circulation of superconducting currents in vortices

TYPE-I BEHAVIOUR (MEISSNER EFFECT):

NORMAL   SUPERCONDUCTOR

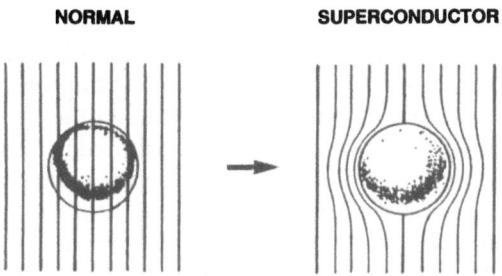

Fig. 9. Meissner effect in a sphere of superconducting material cooled in a constant applied magnetic field. At the transition from the normal to the superconducting phase the lines of induction are ejected from the sphere

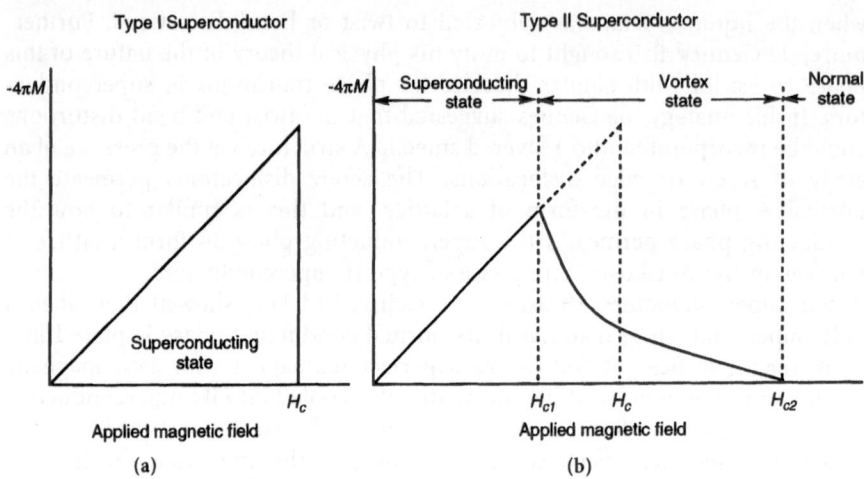

Fig. 10. a Magnetization versus applied magnetic field for a bulk type I superconductor exhibiting a complete Meissner effect; **b** Superconducting magnetization curve of a type II superconductor

throughout the bulk specimen, however, there are no chemical or crystallographic differences between the normal and superconducting regions in this state. The vortex state is stable because the penetration of the applied field into the superconducting material causes the surface energy to become negative.

In a superconducting state the external field is damped as we enter into the superconductor, and consequently the extent to which the field penetrates is called the *London penetration depth* (λ_T) [21]. For a pure superconductor the field is exponentially damped, and therefore the penetration distance is defined as the distance over which the field decreases by the factor of e^{-1}. Thus, the London penetration distance is a fundamental length that characterizes a superconducting material. Another independent length of equal importance is the coherence length, ω. The coherence length is the distance within which the gap parameter, which is related to the Fermi electron gas, cannot change drastically in a spatially varying magnetic field. The coherence length first appeared in a pair of phenomenological equations called the Landau-Ginsburg equations (which also follow from the Bardeen, Cooper and Schrieffer, BCS, theory of superconductors) [22]. The equations describe the structure of the transition layer between normal and superconducting phases which are in contact with one another. The coherence length and the penetration distance depend on the mean free path of the electrons measured in the normal state, and the ratio $\kappa = \lambda_T/\nu$ is called the Landau Ginsburg parameter. For a type I superconductor $\xi \gg \lambda_T$ and for type II superconductor $\lambda_T \ll \xi$ see Fig. 11, and the resultant situation about a vortex in the Abrikosov phase is shown in Fig. 12.

de Gennes argued that in a similar way to how magnetic flux penetrates type II superconductors via a lattice of vortices, it was possible to incorporate twist and bend distortions into a layered smectic A structure via the presence of an array of screw or edge dislocations to give a liquid crystalline equivalent

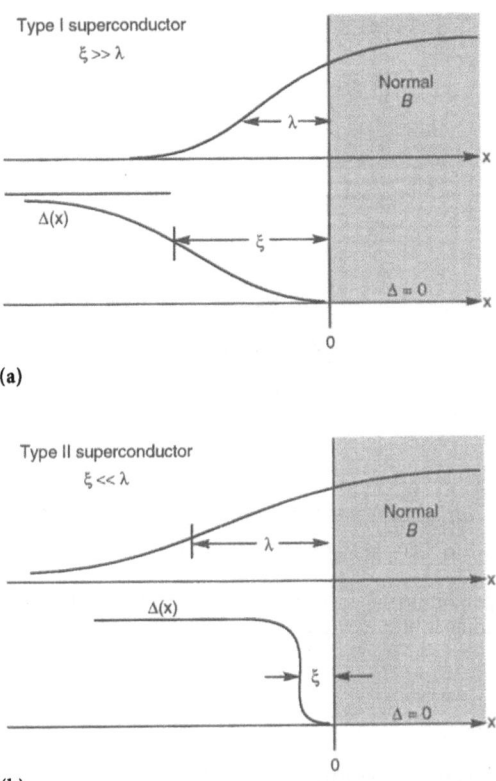

(a)

(b)

Fig. 11. The variation of the magnetic field and the energy gap parameter at the interface of superconducting and normal regions of (a) type I and (b) type II superconductors

of the Abrikosov phase [18]. He predicted for a second order nematic to smectic A transition that, (i) the transition temperature would lowered if twist and bend distortions were imposed (and as a consequence we might expect lower transition temperatures for chiral over achiral systems – which was experimentally verified), and (ii) the Frank coefficients k_{22} and k_{33} of the nematic phase should show strong anomalies in the pretransitional region. In his analogy with superconductors de Gennes defined two correlation lengths, (a) the smectic correlation length (ξ) which is analogous to the correlation length in superconductors, and (b) the penetration length of the twist or bend deformations (λ_T) which is analogous to the London penetration depth. As with superconductors the ratio ($\kappa = \lambda_T/\xi$) is equivalent to the Landau-Ginsburg parameter. Two solutions were found:

$$\kappa < 1/\sqrt{2} \text{ and } \kappa > 1/\sqrt{2}. \tag{4}$$

For $\kappa < 1/\sqrt{2}$ a normal phase is found, but for $\kappa > 1/\sqrt{2}$ a dislocation-stabilized "Abrikosov" phase becomes possible. Thus, the smectic A free energy density

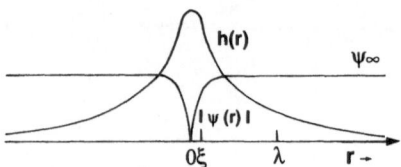

GINSBURG PARAMETER $K = \dfrac{\lambda_T}{\xi}$

TYPE II BEHAVIOUR IF $K > \dfrac{1}{\sqrt{2}}$.

SUPERCONDUCTOR VORTEX

ψ (r) : **LOCAL SUPERCONDUCTOR ORDER PARAMETER**

h (r) : **LOCAL MAGNETIC INDUCTION**

Fig. 12. The local superconductor order parameter and the local magnetic induction in the vicinity of a vortex in an Abrikosov phase.

can be given by:

$$F_A = F_F + a|\psi|;^2 + c|(\nabla - iQ_0\eta)\psi|^2 + 1/2(b|\psi|^4) \qquad (5)$$

and making the following associations:

$$\psi \Rightarrow (\text{Cooper pair wave function}) \qquad (6)$$

$$\eta \Rightarrow \mathbf{A}(\text{vector potential}) \qquad (7)$$

$$h = \eta(\nabla \times \eta)(\text{for twist}) \text{ and } h = \eta \times (\nabla \times \eta)$$
$$(\text{ for bend}) \Rightarrow (\text{magnetic induction}) \qquad (8)$$

The smectic A free energy reduces to the Landau-Ginsburg superconductor free energy provided that $k_s = 0$ for splay and $k_t = k_b$ for twist and bend. The full extent of the relationships between liquid crystals and superconductors in the de Gennes analogy are summarized in Table 1 [23].

This analogy was taken further by Renn and Lubensky [15] to incorporate chiral systems. At the smectic A* to chiral nematic phase λ_T diverges and κ becomes large, hence the formation of a dislocation-stabilized structure in chiral systems becomes possible, see Fig. 13. In the case of chiral phases, Renn and Lubensky showed that the intermediary phase would be stabilized by the incorporation of screw dislocations, see Fig. 14(a).

A screw dislocation is similar in structure to that of a double twist cylinder which is the basis for the formation of Blue Phases [24]. The equivalent of the

Table 1. The analogy between superconductors and liquid crystals (after Lubensky [23])

Superconductor	Liquid Crystal
ψ = Cooper pair amplitude	ψ = density wave amplitude
A = vector potential	n = nematic director
B = $\nabla \times$ A=magnetic induction	κ_0 = n· $\nabla \times$ n = twist
normal metal	nematic phase
normal metal in a magnetic field	chiral nematic (N*) phase
Meissner phase	smectic A phase
Meissner effect	twist expulsion
London penetration depth λ_T	twist penetration depth λ_T
superconducting coherence length ω	smectic correlation length ω
vortex (magnetic flux tube)	screw dislocation
Abrikosov flux lattice	twist grain boundary phase

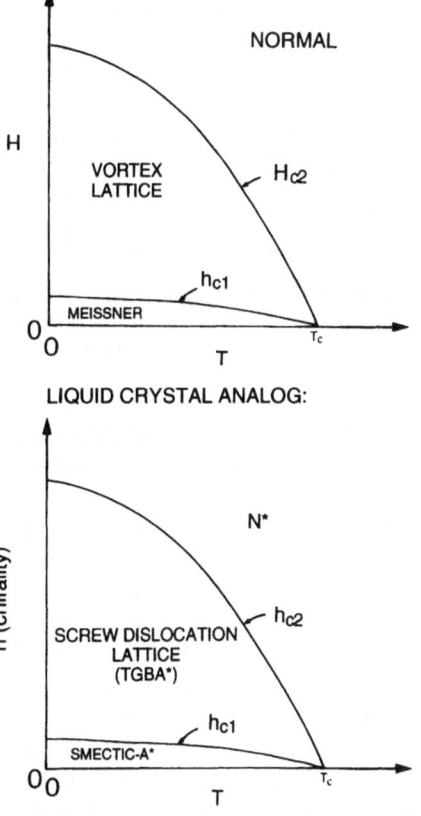

PHASE DIAGRAM OF A TYPE-II SUPERCONDUTOR:

LIQUID CRYSTAL ANALOG:

Fig. 13. Analogous phase diagrams for the Vortex state in superconductors and the screw dislocation lattice state in liquid crystals

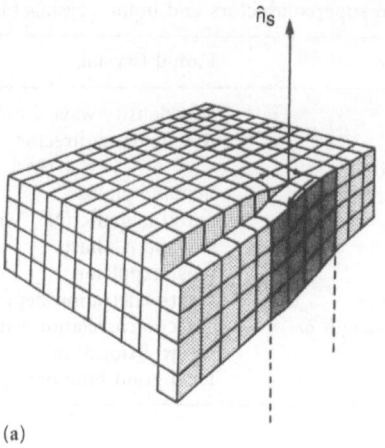

(a)

SMECTIC-A SCREW DISLOCATION

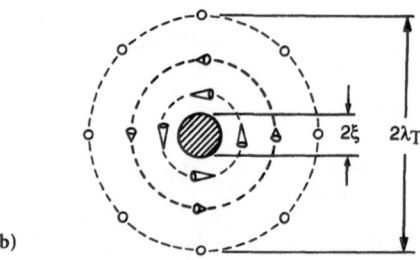

(b)

Fig. 14. a The Structure of a screw dislocation in the smectic A* phase; **b** Looking down the screw dislocation, the twist is comparable to the penetration depth and the coherence length is comparable to the correlation length of the smectic A* phase

London penetration depth, λ_T, is the penetration length of the twist distortion, and the equivalent of the coherence length, ω, is the correlation length of the smectic A* phase. In a screw dislocation, the core of the defect approximates to the smectic A* correlation length and a twist of 90° to the penetration depth, see Fig. 14(b).

As we are dealing with a chiral system the incorporation of screw dislocations (Fig. 15) would be expected to be periodic, with the distance between defects being related to the twist between the blocks of the normal smectic A* phase, $2\pi\alpha$, and inversely related to the layer spacing, d. As a consequence, the screw dislocations are strung together in chains thereby forming grain boundaries and because the phase is helical the grain boundaries are also periodic, see Fig. 16. Hence the defect stabilized phase that is be formed has a lattice of defects (screw dislocations), and its structure is akin to that of the vortex state found in superconductors. Subsequently, Renn and Lubensky developed their models further to incorporate the chiral

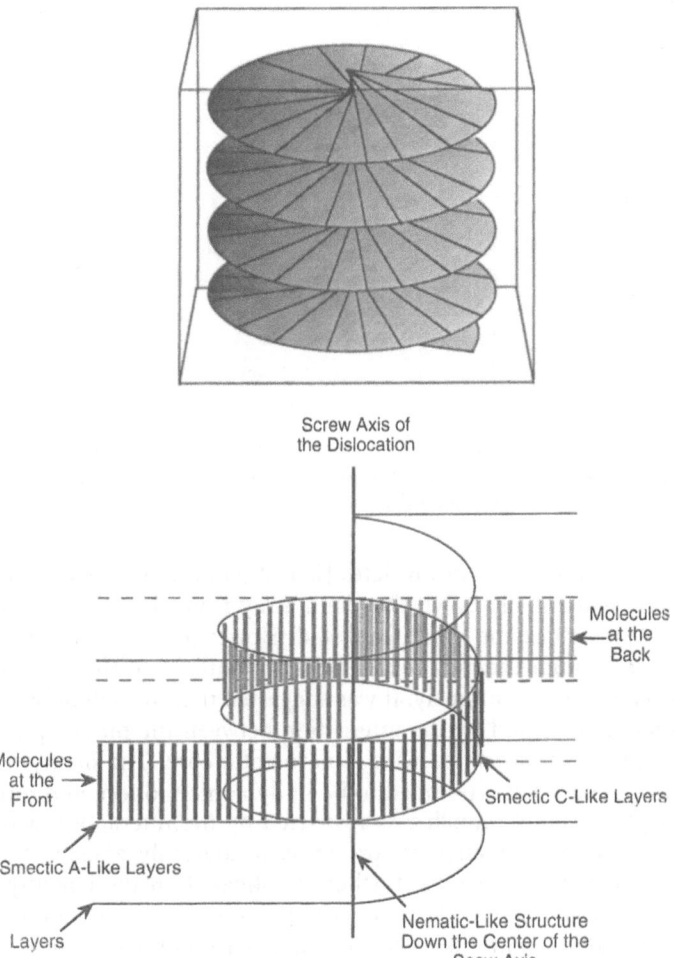

Fig. 15. A twist grain boundary in the Abrikosov phase of liquid crystals

nematic to smectic C* transition; the phase predicted to be formed in this case was called the TGBC phase [25].

4
Anatomy of the Discovery of the TGB Phase

The twist grain boundary smectic A* phase was discovered serendipitously at Bell Laboratories in 1987. Its discovery followed the 'back-tracking' of a number of decisions made concerning the development of ferroelectric liquid crystals for display device applications.

During the mid-1970s we examined the effect of methyl branching in the terminal alkyl chains on the formation of smectic C phases in the methylalkyl

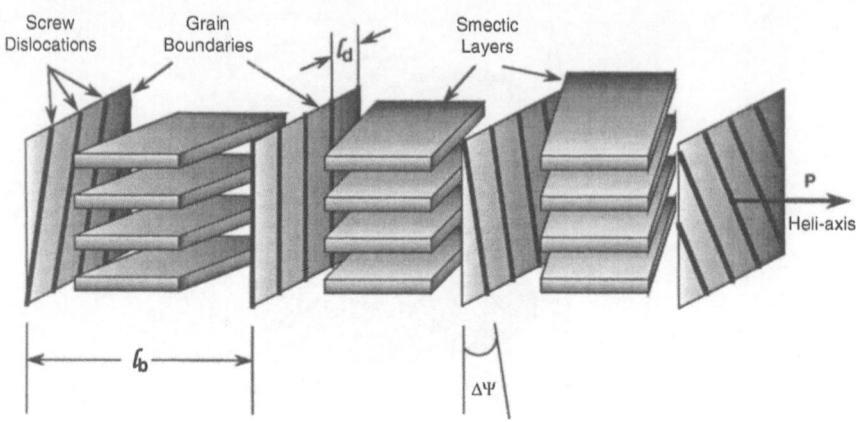

Fig. 16. The helical arrangement of the twist grain boundaries showing the relationship between the block/sheet thickness of the smectic layers and the distance between the screw dislocations (after Renn and Lubensky)

4'-n-octyloxybiphenyl-4-carboxylates [26]. We found that as the methyl group was moved sequentially down the chain away from the aromatic core, so the smectic C phase became more stabilized. However, conversely as the methyl branch was brought nearer to the rigid aromatic core the smectic C phase stability fell away precipitously. It was suggested that this fall off was due to the steric hindrance caused by the interaction between the methyl group and the ester functionality associated with the aromatic core. The methyl branch and the aliphatic chain were assumed to be in dynamic motion, and thus once the motion of the methyl branch was restricted by the interaction with the core, then due to its greater effective presence, it adversely affected the intermolecular packing. This effect was further investigated for the 1-methylalkyl 4'-n-octyloxybiphenyl-4-carboxylates, 2, by positioning the methyl branch adjacent to the core and extending the aliphatic chain (m) on the peripheral side of the branch. When the length of the peripheral chain is extended the rotational motion of the methyl group is damped and so the disruptive effects of steric hindrance grows. Thus, smectic A and smectic C phases were found to be depressed as the chain was extended.

$$C_8H_{17}O-\text{biphenyl}-C(=O)-O-\overset{C_mH_{2m+1}}{\underset{CH_3}{\mid}}$$

2

During the 1980s the development of ferroelectric liquid crystals continued at Bell Laboratories, and the above property-structure correlations suggested to us that, for the development of smectic C and smectic C* materials which would be suitable for use in applications of ferroelectric displays, it would not be wise to investigate 1-methylalkyl-substituted systems because of the detrimental effects that this moiety had on mesophase stability. In addition, as some of the materials for ferroelectric mixtures had to be chiral, it also suggested that the use of chiral end groups such as 1-methylheptyl (which is readily available via (R)- or (S)-2-octanol as an intermediate/starting material) should be avoided. However, in 1986 at the 11th International Liquid Crystal Meeting at Berkeley USA, scientists at Chisso Co reported work on the synthesis of ferroelectric smectic C materials that possessed chiral 1-methylheptyl moieties in their structures [27]. This prompted us to reinvestigate the use of the chiral and achiral 1-methylalkyl moiety in the development of ferroelectric materials.

Initially we chose to examine the properties of materials that had longer core systems, relative to the biphenyls, which we had previously examined, because we expected them to have higher transition temperatures. Thus we prepared the racemic and chiral 1-methylalkyl 4'-(4-n-alkoxybenzoyloxy)biphenyl-4-carboxylates, 3 [28, 29].

These compounds were prepared at AT&T Bell Laboratories in 1986–7 and they were the first materials that we encountered which, we were later to discover from the work of Fukuda and his colleagues, exhibited antiferroelectric and ferrielectric properties [30]. However, at the time, we also found that the chiral materials exhibited unusual physical properties, for example we discovered that the mesophase behaviors of the racemic and chiral systems did not tally, and we also found that there was a precipitous drop in the value of the spontaneous polarization as a function of the reduced temperature from the Curie point (i.e., this was the point at which the antiferroelectric properties were manifested). In addition, we also found some oddities in the phase behavior at the isotropic liquid to smectic A* transition. Shortly after this study we attempted to raise the transition temperatures further by extending the aromatic core. It was judged that the inclusion of a short conjugative spacer in the core would help improve on the smectic A* to smectic C* transition temperatures and that at the same time we might also reduce melting points by disrupting the packing of the materials in the solid state. Subsequently, the (R)- and (S)-1-methylheptyl 4'-n-(4-alkoxyphenylpropioloyloxy)biphenyl-4-carboxylates, 4, were prepared and examined by Maurice Waugh et al. [16, 31].

   Initially the octyl to dodecyl compounds were prepared and these were
found to exhibit relatively normal behavior, i.e. smectic A* phases were found
for the lower homologues with smectic A* and ferroelectric smectic C* phases
occurring for the higher members. However, when the tetradecyl homologue
was examined in the polarizing transmitted light microscope, an iridescent
helical mesophase was observed which upon cooling underwent a further
phase transition to a ferroelectric smectic C* phase. In addition, this
compound was also found to exhibit antiferroelectric and ferrielectric phases.
   As this compound was one of the higher homologues in the series, and
because we knew that the earlier homologues did not exhibit a chiral nematic
phase, it was clear that the new phase also could not be a chiral nematic phase.
In addition, it was clear from the formation of the defect structures seen in the
microscope that the phase first formed from the isotropic liquid possessed a
helix, see Plate 1, which had its heli-axis at right angles to the heli-axis in the
lower temperature chiral ferroelectric smectic C* phase. This simple observation meant that if the phase was a lamellar smectic phase then the helix
would have to be formed, inconceivably, in a direction parallel to the layers.
Synthesis of the achiral variant confirmed that the phase formed first on
cooling from the isotropic liquid was indeed a smectic A phase, and thus we
immediately knew that we had found a smectic A* phase where the helical
macrostructure formed in the planes of the layers, and thus the helix must

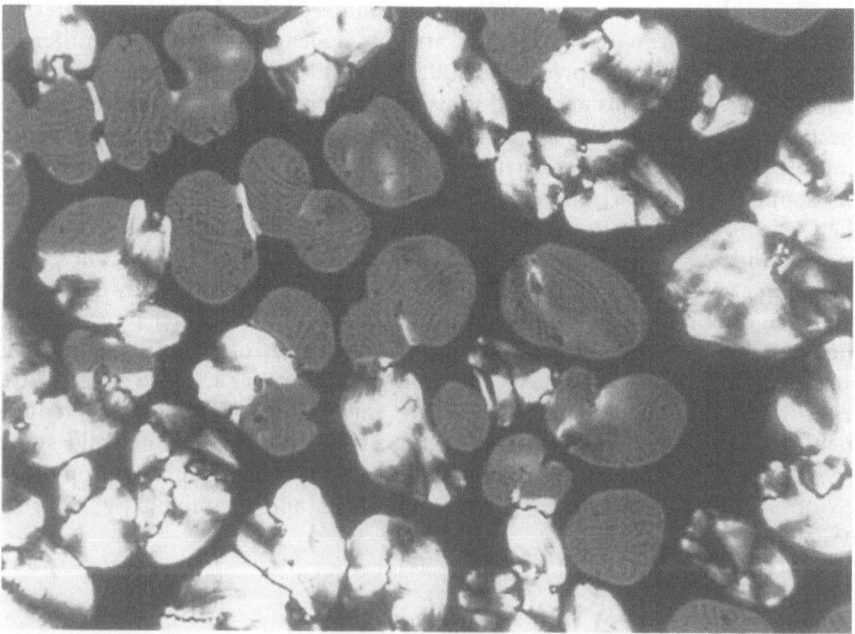

Plate 1. The defect texture of the TGBA* phase, under crossed polars, formed on cooling
from the isotropic liquid (×100). The textures show both platelets and Grandjean planar
regions (associated with a helical macrostructure)

somehow disrupt the layers of the mesophase. Mixture work between the (R)- and the (S)-isomers also showed that when they were mixed together in a 1:1 ratio, a normal smectic A phase was obtained. Plate 2 shows the defect textures found when the two enantiomers are allowed to flow together under a glass cover-slip on a microscope slide. Down the center of the contact region where the two materials meet and mix there is the formation of a normal smectic A phase which exhibits a focal-conic texture. Away from the contact boundary the texture reverts to being homeotropic, platelet or Grandjean planar. This study confirmed the identity of the phase first formed from the isotropic liquid as smectic A in nature.

Initially attempts were made to build a model of the structure of the mesophase using large sheets of velum; however, twisting of the layered structure of the phase was not easy achieved. Subsequent discussions with Jay Patel and Ron Pindak, however, led to the creation of various models of the phase where the helical structure was allowed to develop through the incorporation of defects. As this model was being developed a remarkable coincidence occurred. Ron Pindak produced a preprint from Tom Lubensky of a manuscript by Scott Renn and Tom Lubensky which presented a theory

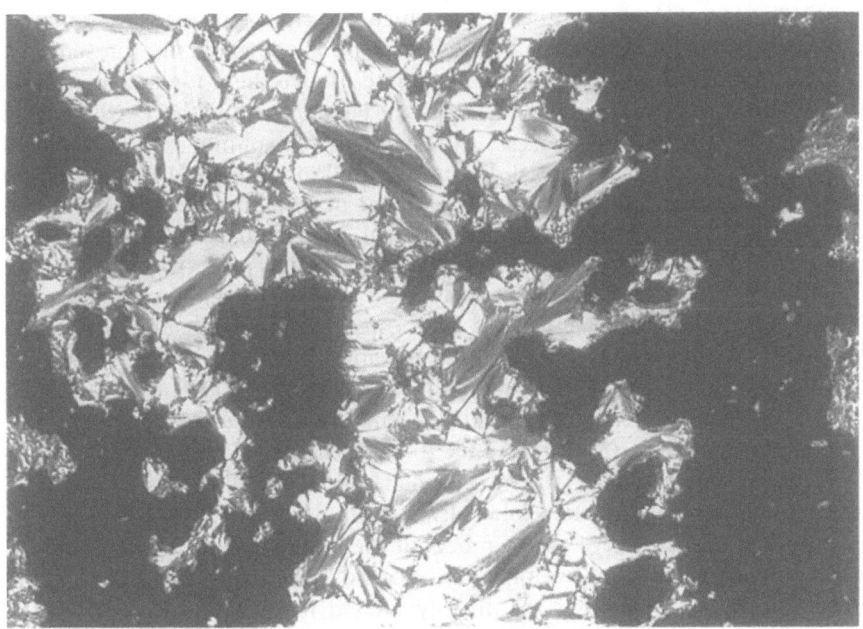

Plate 2. The contact regions between the (R)- and the (S)-1-methylheptyl 4'-(4-n-alkoxyphenylpropioloyloxy)biphenyl-4-carboxylates under a cover-slip on a microscope slide and between crossed polars (×100). The central region is where the two materials met and mixed. This region exhibits the normal focal-conic texture of the achiral smectic A phase. The left- and right-hand sides of the plate show the homeotropic and filamentary textures of the chiral TGBA* phase

relating the physics of twisted smectic phases to the physics of the formation of Abrikosov (type II) superconductor phases.

Thus, without planning or any prior knowledge of the seminal contribution to the subject made by Pierre de Gennes in 1972, we had arrived through serendipitous experimentation at a point where our work complemented and supported the recently reported beautiful theoretical model of Renn and Lubensky.

4.1
Experimental Evidence for the Presence of TGB Phases

In this section we will discuss how the nature of the TGBA* phase was elucidated, characterized and its structure investigated.

4.1.1
Synthesis, Transition Temperatures, and Phase Behavior

As noted above, the twist grain boundary phases discovered in low molar mass materials at Bell Laboratories were found in the optically active variants of the 1-methylheptyl   4'-(4-n-alkoxyphenylpropioloyloxy)biphenyl-4-carboxylates (nP1M7's) [16, 31], the syntheses for which are shown in Fig. 17. Commercially available 4-hydroxybenzaldehyde can be alkylated in the presence of anhydrous potassium carbonate in butanone to give the 4-n-alkoxybenzaldehyde. This product is treated in a Corey-Fuchs reaction [32], which is a modified Wittig reaction process, with triphenylphosphine, carbon tetrabromide, and zinc to give the β,β-dibromo-4-n-alkoxystilbene. Upon lithiation using butyllithium and subsequent treatment of the lithium salt with carbon dioxide, the 4-n-alkoxyphenylpropiolic acid intermediate is obtained. The biphenyl substrate is prepared as the carbonate by first protecting the hydroxy function of 4'-hydroxybiphenyl-4-carboxylic acid with methyl chloroformate [33]. The free carboxylic moiety is then esterified via a Mitsunobu reaction [34] with the appropriate ((R)-, (S)- or racemic) 2-octanol in the presence of diethyl azodicarboxylate, triphenylphosphine, and dry THF. The resulting carbonate ester is deprotected using a mixture of aqueous ammonia and ethanol to give (R)-, (S)- or racemic 1-methylheptyl 4'-hydroxybiphenyl-4-carboxylate. This product is then esterified with a 4-n-alkoxyphenylpropiolic acid in the presence of dicyclohexylcarbodiimide and 4-(N,N-dimethylamino)pyridine to give one of the members of the family of 1-methylheptyl 4'-(4-n-alkoxyphenylpropioloyloxy)biphenyl-4-carboxylates (nP1M7's). The optical purities of the enantiomers are directly related to the optical purity of the starting 2-octanol. For the optically active materials prepared the enantiomeric excess was in the region of 0.98.

The phase transitions for this series were examined and determined by thermal optical microscopy and differential scanning calorimetry, the results for which are given in Table 2. The transition temperatures (°C) for the series are plotted as a function of increasing n-alkoxy chain length in Fig. 18. It can

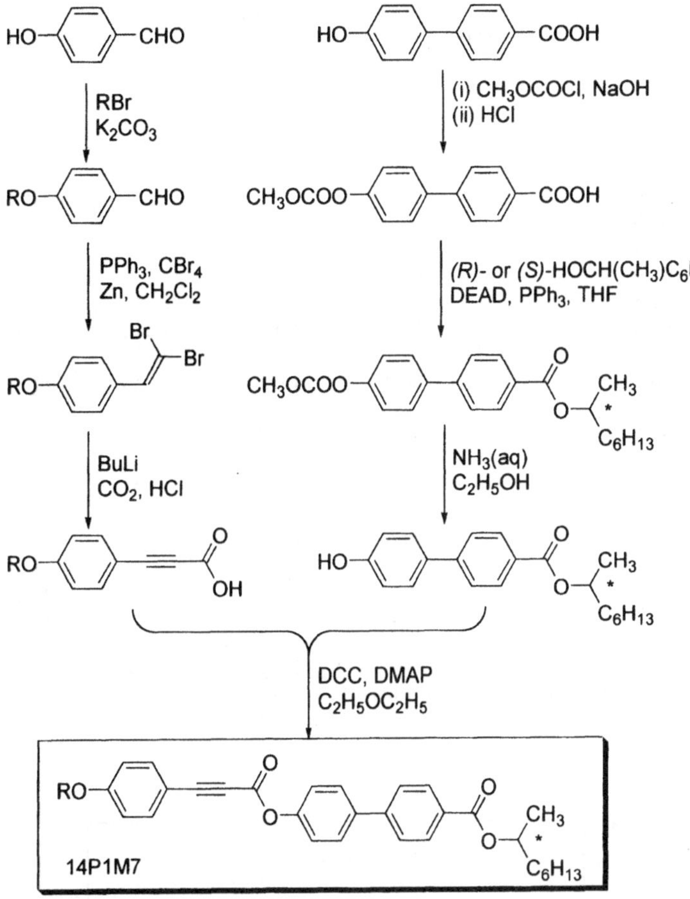

Fig. 17. The general synthetic pathway to the 1-methylheptyl 4'-(4-n-alkoxyphenylpropioloy-loxy)biphenyl-4-carboxylates (nP1M7's)

be seen that, as the series is ascended, the smectic A* to smectic C* transition temperatures rise sharply. Over the same interval of chain length the smectic A* to isotropic liquid transition temperatures fall, thereby decreasing the temperature range of the smectic A* phase with increasing chain length. For the n-hexadecyloxy homologue this effect results in a direct smectic C* to isotropic liquid transition. Thus, the smectic A* phase senses an increasing degree of fluctuations from both the isotropic liquid and the smectic C* phase as the series is ascended.

4.1.2
Calorimetric Studies of TGB Phases

If we examine the variation in the clearing point enthalpy with respect to chain length we find that the values fall sharply as the n-alkoxy chain length is

Table 2. Transition temperatures (°C) and clearing point enthalpies (cal g⁻¹) for the (R)- and (S)-1-methylheptyl 4′-(4-n-alkoxyphenylpropioloyloxy)biphenyl-4-carboxylates (nP1M7's)

n	Iso to TGBA*	SmA*Iso to SmC* TGBA*	SmA* TGBA* SmC*†	orSmC* toSmC*$_{ferri†}$	toSmC*$_{ferri†}$ SmC*$_{anti}$†	tomp
8	97.5					81.3
ΔH	1.6					11.3
9	97.5					81.3
ΔH	1.4					12.1
10	98.0		(76.3)			85.4
ΔH	1.5					12.3
11	95.0		(78.1)	(56.0)	(44.6)	85.1
ΔH	0.8					12.8
12	96.3		86.0	(57.3)	(48.0)	82.4
ΔH	0.9					11.4
13	94.1		88.3	(57.1)	(43.7)	81.6
ΔH	0.5					10.8
14	93.8		89.7	(53.4)	(42.5)	78.3
ΔH	0.3					11.0
15	91.6		90.6	[50.0]		76.5
ΔH	0.2					11.4
16		99.0				73.4
ΔH		0.4				10.5

Where ( ) denotes a monotropic transition; [ ] denotes a virtual transition (observed on fast supercooling); † denotes a phase transition where the enthalpy was too small to be measured.

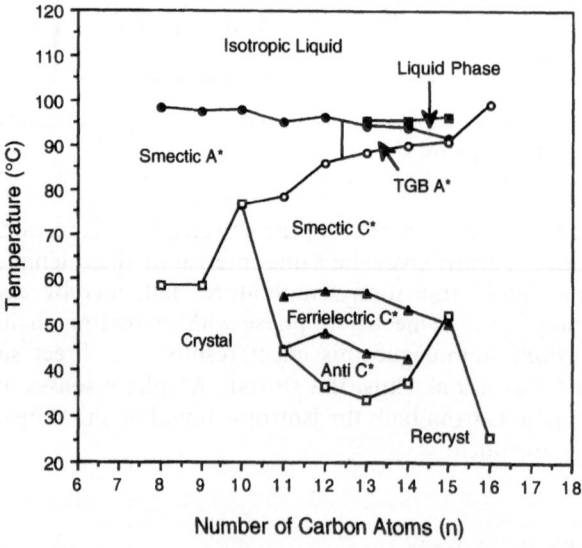

Fig. 18. Transition temperatures (°C) shown as a function of n-alkoxy chain length for the (R)- and (S)-1-methylheptyl 4′-(4-n-alkoxyphenylpropioloyloxy)biphenyl-4-carboxylates (nP1M7's)

increased. Thus, the clearing point transitions appear to become more weakly first order in nature with increasing n-alkoxy chain length. The sharp decrease in enthalpy values for the clearing points (see Fig. 19) coupled with the decreasing smectic A* range on ascending the series satisfy the criteria set out by de Gennes for the emergence of dislocation-stabilized phases, i.e., the transitions should become nearer to being second order in nature and strong pretransitional effects must be felt for defect-stabilized phases to occur. Thus, the small enthalpy values combined with the relatively small projected temperature ranges of the smectic A* phases supported the formation of TGBA* phases for the n-tridecyl to n-pentadecyl homologues. Fig. 19 shows how the clearing point enthalpies fall dramatically with increasing chain length, and it is interesting to note the sudden increase in value for the n-hexadecyloxy homologue which exhibits a direct isotropic to smectic C* transition, i.e., the clearing point transitions become more strongly first order again as the smectic C* modification becomes the dominant phase.

These studies also showed two other interesting phenomena, firstly that certain members in the homologous series exhibit ferrielectric and antiferro-electric phases, and secondly for the materials that exhibit TGBA* phases, a novel transition was found to occur in the isotropic liquid. The ferri- and anti-ferroelectric phases appear first for the n-undecyl homologue on ascending the series, and disappear once the chain length reaches sixteen carbon atoms in length.

Above the clearing point transitions the (R)- and (S)-1-methylheptyl 4'-(4-n-alkoxyphenylpropioloyloxy)biphenyl-4-carboxylates also exhibit enthalpies in the temperature range of the isotropic liquid. These transitions are detected as broad diffuse peaks by differential scanning calorimetry in both heating and cooling cycles. No effects are observed in the polarized microscope for this transition, however, suggesting that the two liquid states have very similar

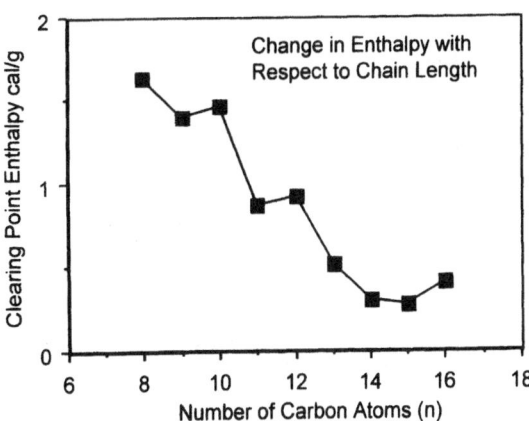

Fig. 19. The enthalpies of the clearing points (cal g^{-1}) (°C) shown as a function of n-alkoxy chain length for the (R)- and (S)-1-methylheptyl 4'-(4-n-alkoxyphenylpropioloyloxy)biphe-nyl-4-carboxylates (nP1M7's)

properties and any ordering of the molecules in the lower temperature liquid phase must be weak. The presence of this new "quasi-liquid phase" was verified by X-ray diffraction and optical rotary studies, see Fig. 20.

When the clearing point enthalpies of the homologous series were examined, it was found that the enthalpy peak that occurs in the isotropic liquid was only observed near to transitions to TGB phases. For example, Fig. 21 shows the clearing point enthalpies taken on cooling for the related (S)-1-methyloctyl 4'-(4-n-alkoxyphenylpropioloyloxy)biphenyl-4-carboxylates [35]. It can be seen that, as the alkoxy chain is extended, the broad enthalpy peak in the isotropic liquid regime becomes more pronounced, and when TGB phases are present the peak is well-separated from the liquid to liquid crystal enthalpy peak.

4.1.3
Effects of Optical Purity

It is interesting to note that when these materials are prepared in their racemic forms the liquid to liquid phase transition, discussed above, disappears. Moreover, the clearing temperature moves to a higher value and the magnitude of the associated enthalpy also increases, see Fig. 22 for the racemic and (S)-1-methyloctyl 4'-(4-n-tetradecyloxyphenylpropioloyloxy)bip-henyl-4-carboxylates [5], and Fig. 23 for the analogous racemic, (R)- and (S)-1-methylheptyl 4'-(3-fluoro-4-n-tetradecyloxyphenyl-propioloyloxy)biphenyl-4-carboxylates [5, 36].

For each of the homologues that exhibit TGB phases, if the clearing point enthalpy for the liquid to TGB phase transition is added to the value for the liquid to liquid transition, then the total value is approximately equal to that for the smectic A to isotropic transitions in the equivalent racemate, see Table 3 for the (R), (S), and racemic forms of the 1-methylheptyl 4'-(4-n-

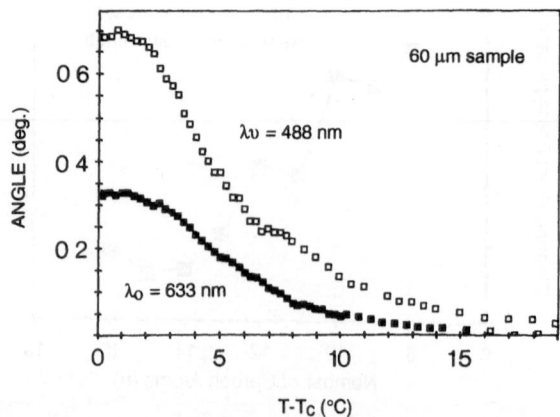

Fig. 20. The optical rotation taken in the isotropic liquid near to the TGB phase of (R)- and (S)-1-methylheptyl 4'-(4-n-tetradecyloxyphenylpropioloyloxy)biphenyl-4-carboxylate

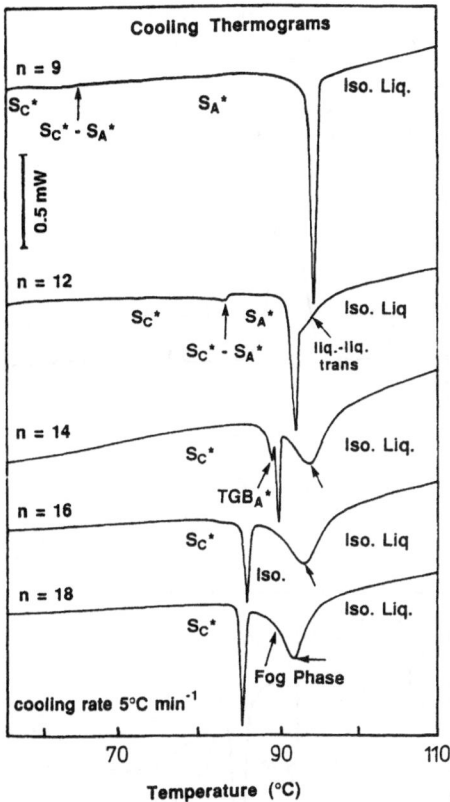

Fig. 21. Cooling thermograms taken by differential scanning calorimetry for the (S)-1-methyloctyl 4'-(4-n-alkoxyphenylpropioloyloxy)biphenyl-4-carboxylates

Table 3. Transition temperatures (°C) and enthalpies (cal g^{-1}) for the (R), (S), and racemic forms of the 1-methylheptyl 4'-(4-n-alkoxyphenylpropioloyloxy)biphenyl-4-carboxylates (nP1M7's)

n	liq to liq	Iso to TGBA*	TGBA* to SmC*
13	95.4	94.1	88.3
ΔH	0.6	0.5	0.06
15	96.3	91.6	90.6
ΔH	0.9	0.2	0.08
14	95.7	93.8	89.7
ΔH	0.9	0.3	0.04
Racemate		*Iso to SmA*	*SmA to SmC*
14		97.7	90.3
ΔH		1.24	0.05

(R) and (S) Enantiomers e.e. ≈ 0.98 for both isomers.

alkoxyphenylpropioloyloxy)biphenyl-4-carboxylates (nP1M7's). This is further evidence to support the view that the liquid to liquid transition is real and not an artifact of the experiment.

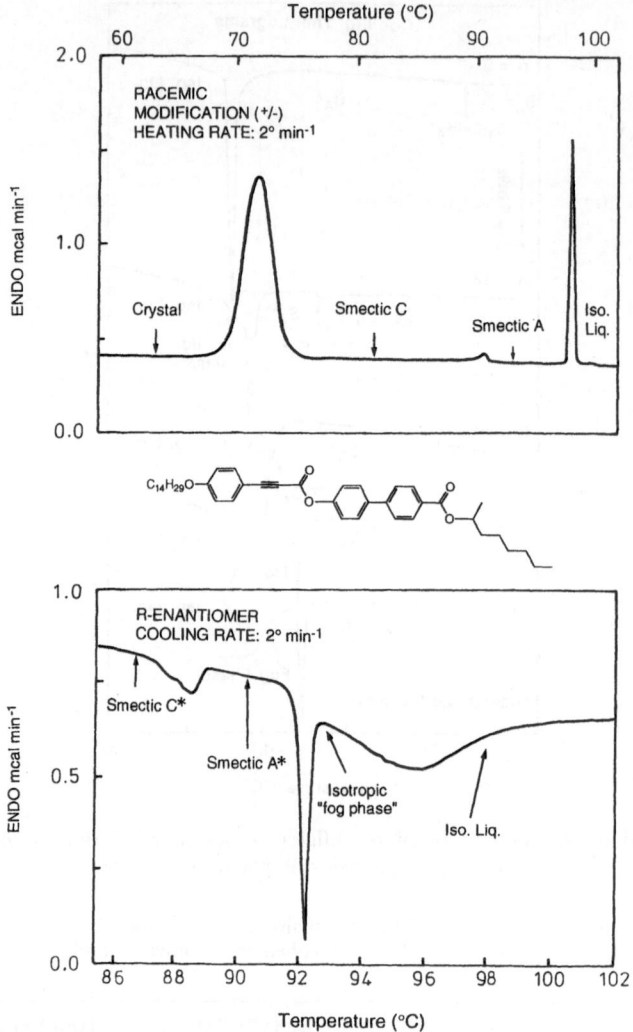

Fig. 22. Cooling thermograms taken by differential scanning calorimetry the racemic and (S)-1-methyloctyl 4′-(4-n-tetradecyloxyphenylpropioloyloxy)biphenyl-4-carboxylates

Table 3 shows that the clearing point temperatures are lower for the optically active 1-methylheptyl 4′-(4-n-alkoxyphenylpropioloyloxy)biphenyl-4-carboxylates in comparison to their racemates (almost 4 °C for the tetradecyloxy homologue) which is in agreement with de Gennes' theoretical model. These results clearly show that chirality and optical purity can markedly affect transition temperatures in liquid crystal systems.

The presence of a liquid to liquid transition in these systems is one of the more intriguing aspects of the affects that chirality has in ordered fluid systems. The question of why this transition should occur, and what the

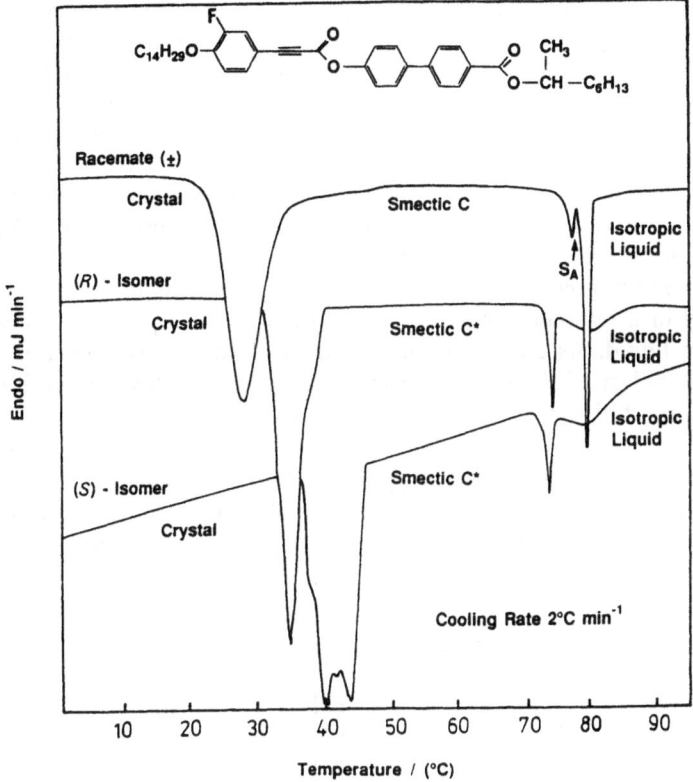

Fig. 23. Cooling thermograms taken by differential scanning calorimetry for the racemic, (R)- and (S)-1-methylheptyl 4'-(3-fluoro-4-n-tetradecyloxyphenylpropioloyloxy)biphenyl-4-carboxylates

nature/structure of the liquid phase which is present between the TGB phase and the isotropic liquid are questions that are still to be resolved. However, it is possible that the intermediary liquid phase is similar in nature to the Blue Fog phase (or Blue Phase III) which is found to occur between Blue Phase II and the isotropic liquid. Intuitively, it appears that the intermediary liquid phase has short-range ordering of the molecules, and that when the TGB phase melts either the lattice of defects melts first leaving clusters of smectic A* cybotactic-like groups surrounded by the normal liquid, or the smectic A* regions of the TGB phase melt first leaving a network of screw dislocations floating in an amorphous liquid. This type of phase could be described as an entangled flux phase [37].

4.1.4
Optical Activity

The optical activity inherent in TGB phases can be demonstrated by measuring the wavelength dependency of the transmitted intensity of circularly polarized light. Fig. 24 shows the dependency for the first TGB material (R)-1-methylheptyl 4'-(4-n-tetradecyloxyphenylpropioloyloxy)biphenyl-4-carboxylate (14P1M7) [38]. This result was obtained using a sample cell that was placed in a microfurnace, which was regulated to ±10 mk, and light from a 75 Watt xenon arc lamp was passed through a monochromator and appropriate filters and then focused onto the sample. The range of incident wavelengths was varied between 500 to 1600 nm, and the transmitted intensity detected by either a Si or Ge photodiode. The results shown in the inset of Fig. 24 are for a 20 μm thick sample held at 92.8 °C; the spectra were normalized to the source intensity and plotted as a ratio of the measured intensity to the corresponding intensity in a reference spectrum taken when the sample was heated to the isotropic liquid. As shown in the inset, the transmission spectra for right circularly polarized light is characterized by a Bragg reflection band centered at λ_m with a full width half maximum of 0.12 λ_m. Since the sample reflected right circularly polarized light it confirmed that the TGBA* phase of (R)-1-methylheptyl 4'-(4-n-tetradecyloxyphenyl-propioloyloxy)biphenyl-4-carboxylate (14P1M7) exhibits a right hand helical structure (which is in agreement with the Gray and McDonnell rule relating helical twist sense to molecular

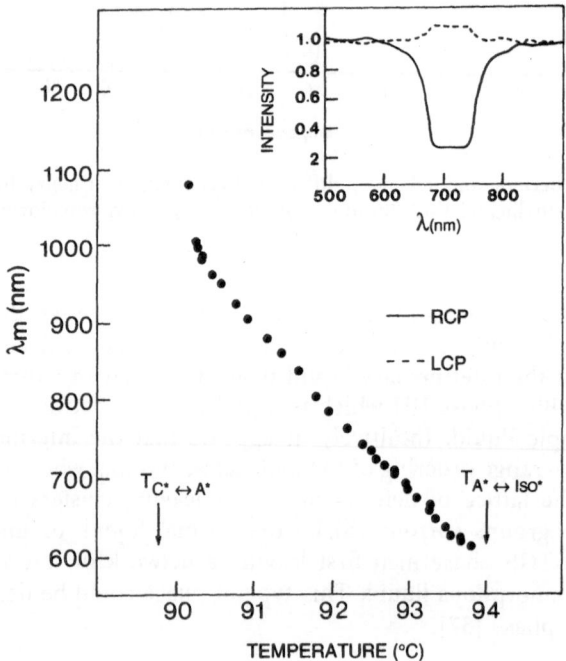

Fig. 24. The wavelength of the Bragg reflection, λ_m, of circularly polarized light shown as a function of temperature for (R)-1-methylheptyl 4'-(4-n-tetradecyloxyphenylpropioloy-loxy)biphenyl-4-carboxylate (14P1M7). The inset shows a typical transmission spectra for right circularly polarized light (solid line), and left circularly polarized light (dashed line) at 92.8 °C

structure). The absolute value of the pitch, λ_0, is related to λ_m by $\lambda_m = \lambda_0 \cdot n$, where $n \approx 1.6$, which is the average refractive index of the material in the TGBA* phase. In Fig. 24 λ_m is plotted as a function of temperature, and except for pretransitional effects, λ_m increases linearly with decreasing temperature, which corresponds to a linear increase in the pitch from 0.38 µm to 0.63 µm over approximately 4 °C. Thus, probing the TGBA* phase with circularly polarized light demonstrates that the phase has a helical macrostructure, for the (R)-isomer the helix is right handed, the pitch of the helix is sub-micron, and the helix unwinds with falling temperature indicating that the defects (screw dislocations) are not permanently pinned and that their density must change as a function of temperature.

4.1.5
X-ray Diffraction Studies

Synthesis, the investigation of racemates, mixture studies, calorimetry, and optical rotary studies show that (a) the smectic A* phase (i.e., the TGBA* phase) can be helical, (b) the handedness of the helicity is dependent on the absolute spatial configuration of the molecules, (c) the smectic A* to isotropic liquid phase transition approaches being second order, and (d) there are strong pretransitional effects in the isotropic liquid before the phase transition to the TGBA* phase. X-ray diffraction of unaligned and aligned specimens, in addition, allows for the probing of the layered structure of the twisted smectic A* phase (the presence of layer order clearly distinguishes the twisted A* phase from the possibility of having a chiral nematic phase).

The first X-ray diffraction studies performed on the TGBA* materials were carried out by Pindak et al. [16, 38] on the dodecyl (C_{12}), tetradecyl (C_{14}), and the hexadecyl (C_{16}) homologues of the (R)-1-methylheptyl 4'-(4-n-alkoxy-phenylpropioloyloxy)biphenyl-4-carboxylates (nP1M7's). X-ray diffraction studies were performed using an 18-kW rotating anode based spectrometer with the specimens prepared in 1 mm glass capillaries. A vertically bent pyrolytic graphite (002) crystal was used to focus CuKα X-rays to a 0.5 × 2.0 mm spot size, and the scattered radiation was then analyzed using slits and a flat pyrolytic graphite crystal. Thus the instrumental resolution was 0.01 Å^{-1} full width at half maximum in the scan direction, and 0.02 Å^{-1} in the transverse scattering plane.

The temperature dependence of the layer spacing was determined, via 2θ scans, for the three enantiomers as shown in Fig. 25. The layer spacing for each of the three enantiomers increases continuously on heating, and for the dodecyloxy and tetradecyloxy members saturates to a relatively constant value in the twisted (TGB) smectic A* phase. The layer spacing was found from these studies to be commensurate with molecular length.

The insert in Fig. 25 shows the layer peak scans within the smectic C*, the twisted (TGBA*) smectic A*, and the isotropic liquid of the tetradecyloxy homologue. Within experimental error the layer peak widths for the smectic

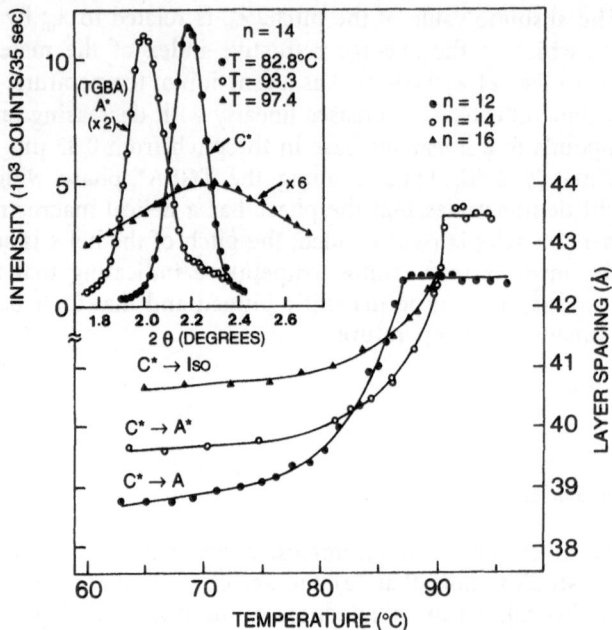

Fig. 25. The temperature dependence of the layer spacing for the dodecyl (C_{12}) (*filled circles*), tetradecyl (C_{14}) (*open circles*), and the hexadecyl (C_{16}) (triangles) homologues of the (R)-1-methylheptyl 4'-(4-n-alkoxyphenylpropioloyloxy)biphenyl-4-carboxylates (nP1M7's). The inset shows scans through the layer peak for the tetradecyloxy homologue at temperatures within the smectic C* phase (*filled circles*; T = 82.8 °C), the TGBA* phase (*open circles*; T = 93.3 °C), and just into the isotropic liquid (*triangles*; T = 97.4 °C)

phases are similar and the integrated intensities differ only by a factor of two. This work implied that the correlations of the layers in the twisted (TGB) smectic A* phase extend a minimum distance of ten layers, thereby establishing the fact that the helical phase is smectic and not chiral nematic in type.

The scan carried out just in the liquid phase of the tetradecyloxy homologue showed a diffuse peak with a peak position that was comparable to the layer peak position measured well within the temperature range of the smectic C* phase. Thus, this study showed that there were strong smectic-like fluctuations occurring in the liquid state before the TGBA* is formed.

Diffraction studies of aligned samples were used by Srajer et al. [38] to investigate the orientation of the layers with respect to the heli-axis and to investigate whether the rotation of the blocks of the smectic layers with respect to the heli-axis was commensurate or incommensurate. Aligned specimens were obtained by cooling into the smectic C* phase and heating into the TGBA* phase. In the smectic C* phase the layer scattering peaks were used to determine the orientation of the normal to the glass plates of the sample cell, n_G, and then in the TGBA* phase the moment transfer vector, Q_s, was set to a fixed value $|Q_s| = 2\pi/d$ where d was the TGBA* layer spacing. A β scan was defined as a rotation of the sample cell about an axis perpendicular to both n_G and Q_s with $\beta = 0$ when $n_G \perp Q_s$. Since n_G coincides with the helical pitch axis, **P**, the β scan reveals the desired layer orientation, see Fig. 26. Although the scattering peak is broad, it clearly has a peak at $\beta = 0$, thus confirming the fact that the layers are oriented parallel to the pitch axis as shown schematically in the illustration at the top of Fig. 26.

Subsequently, high resolution X-ray scans were performed by Srajer et al. at the National Synchrotron Light Source at Brookhaven National Laboratories. In this case Srajer et al took $Q_{||}$ and $Q_\perp$ as the momentum transfer parallel and perpendicular to the molecular director, and Q_t as orthogonal to $Q_{||}$ and $Q_\perp$. By doing this, the smectic A* layer peaks occur along the $Q_{||}$ axis at $\pm Q_o = 2\pi/d$, where d is the layer spacing. The helical rotation of the blocks of the smectic layers turns the smectic A* scattering peaks into a ring of scattering in the ($Q_{||}$ and Q_t) plane as shown in the inset in Fig. 27. Since the grain boundaries do not disrupt the smectic layering in the direction normal to the layers and the layer fluctuations are not included, the structure factor in the ($Q_{||}$ and Q_t) plane is a δ-function along the radius Q_o. The grain boundaries were, however,

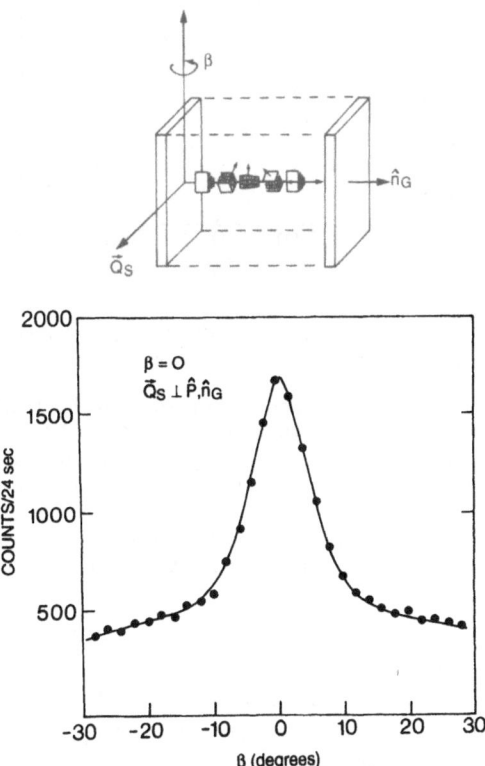

Fig. 26. Scattering intensity as a function of β for a fixed moment transfer $|Q_s|$. **P** is the pitch axis of the helix and **n** is the molecular director. The "blocks" of smectic layers have an infinite extent in the direction transverse to the pitch axis

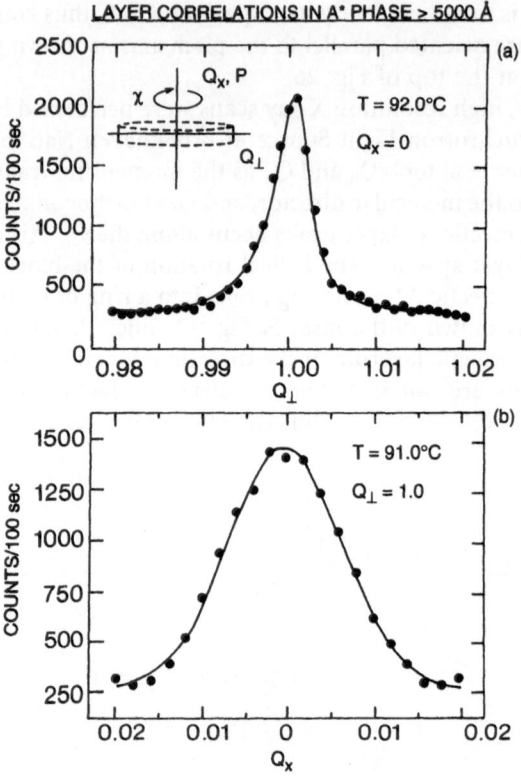

$Q_\perp$ RESOLUTION = 5.4 x 10^{-4} Å^{-1} (FWHM)

$Q_\perp$ LINESHAPE RESOLUTION-LIMITED $\Rightarrow$

LAYER CORRELATIONS IN A* PHASE > 5000 Å

GAUSSIAN FIT Q_X LINESHAPE GIVES L^{-1} = 0,033 Å^{-1}
NO χSTRUCTURE

Fig. 27. *Top* a $Q_{||}$ scan taken at a temperature of 92 °C in the TGBA* phase of (R)-1-methylheptyl 4′-(4-n-tetradecyloxyphenylpropioloyloxy)biphenyl-4-carboxylate (14P1M7). The solid line through the data points is a two dimensional convolution of the ($Q_{||}$, Q_t) instrumental resolution with the predicted TGB cylindrical structure factor which is shown schematically in the inset. $Q_{||}$ is in units of $2\pi/d$=0.146 Å^{-1}. *Bottom* a $Q_\perp$ scan taken at a temperature of 91 °C in the TGBA* phase. The solid line through the data points is a fit to a Gaussian. The resolution width for $Q_\perp$ is=1.1×10^{-5} Å^{-1}, and $Q_\perp$ is in units of 1.676 Å^{-1}

found to disrupt the layering in the direction parallel to the smectic layers so the ring of scattering is broadened in the $Q_\perp$ direction into a cylinder of scattering. It was also expected that if the rotation of the smectic blocks was a rational fraction of 2π, then the cylinder of scattering would develop further structure around its circumference.

A typical $Q_{||}$ scan ($Q_\perp$ = 0) is shown in the upper part of Fig. 27. The solid line is a fit of the data to a two-dimensional convolution in the ($Q_{||}$ and Q_t) plane. The width of the peak is essentially determined by the instrumental resolution with a slight ($\sim$10%) asymmetric broadening arising from off-axis scattering. The radial δ-function structure factor which describes the scattering was found to be, within instrumental resolution, in agreement with the predictions made for the model TGBA* phase provided by Renn and Lubensky. Furthermore, this experiment enabled Srajer et al. to set a lower limit for the layer correlations in the TGBA* phase to be 5000 Å.

The second scan, depicted in the lower half of Fig. 27, shows a $Q_\perp$ scan with $Q_\parallel = 1.0$. The solid line is a fit of the data to a Gaussian line shape of characteristic width equal to 0.033 Å^{-1}. Both the Gaussian line shape and its width were found to be in agreement with the Renn and Lubensky model of the TGBA* phase.

The structure around the cylinder of scattering was examined using a χ scan. However, no structure around the circumference was found, indicating that for (R)-1-methylheptyl 4'-(4-n-tetradecyloxyphenylpropioloyloxy)biphenyl-4-carboxylate (14P1M7) the rotation of the smectic "blocks" is not a rational function of 2π, and hence the phase in this case is said to be incommensurate. However, the length scales in relation to the locations of the screw dislocations of the structure of the TGBA* phase were not suitable for probing by X-ray diffraction, and so an alternative technique had to be sought.

4.1.6
Freeze-Fracture Studies of TGB Phases

The final evidence for the formation of an "Abrikosov flux lattice" of screw dislocations in liquid crystals was achieved by Zasadzinski et al. [39] via the visualization of the screw dislocations of (R)- and (S-)1-methylheptyl 4'-(4-n-tetradecyloxyphenylpropioloyloxy)-biphenyl-4-carboxylates using freeze-fracture transmission electron microscopy. Freeze-fracture transmission microscopy (TEM) is an essential tool for visualizing the TGBA* phase at sufficient resolution in order to resolve the molecular organization.

The preparation of a specimen of the TGBA* phase suitable for replication was achieved by rapidly quenching the sample of the phase at a rate such that the structure was preserved in the resulting frozen state. Thus a 0.1 to 0.5 µl drop of 14P1M7 was sandwiched between two copper planchettes coated with poly(1,4-butylene terephthalate) that had been unidirectionally buffed in order to give an aligned sample. The thickness of the sample was not controlled and was allowed to vary from 10 to 50 µm. The sample was then heated to 92.5 °C, before being plunged into liquid propane cooled to –190 °C by liquid nitrogen. The cooling rate was estimated to be approximately 5000 °C per second. The quenched samples were fractured at –170 °C and 10^{-8} torr, and were replicated with a 1.5 nm thick shadowing film of platinum deposited at an angle of 45° with respect to the fracture surface, followed by a 15 nm film of carbon deposited normal to the fracture surface. The replicas were removed from the copper planchettes and cleaned and then examined using a transmission electron microscope operating at 80 kV. The freeze fracture process is shown schematically in Fig. 28.

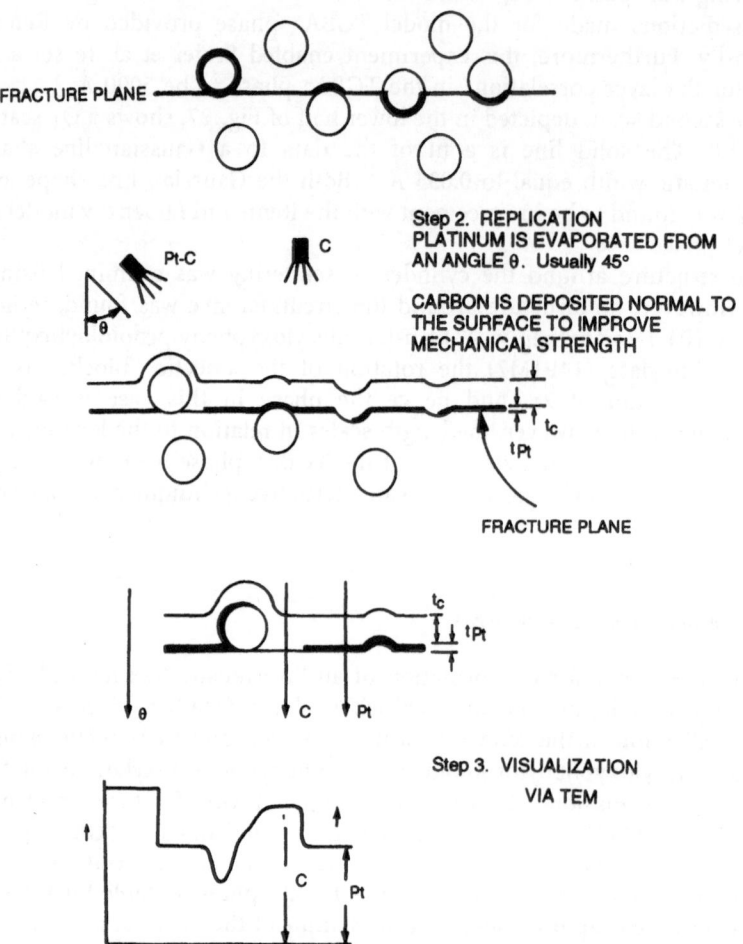

Fig. 28. Schematic representation of the steps involved in the freeze fracture procedure

A typical fracture pattern for the TGBA* phase with a surface cut nearly parallel to the pitch direction is shown in Plate 3. The surface has a regular undulating texture similar to that observed for chiral nematic phase, which also agrees with the surface contours predicted by a simple model for the surface fracture of a chiral nematic. Measurement of the minimum undulating repeat distance gives a value of 0.25 to 0.30 μm, which corresponds to one half of the pitch. Thus, the pitch is about 0.5 to 0.6 μm, which is consistent with the value determined from the optical measurements described earlier.

An oblique fracture through the TGBA* phase is shown in Plate 4. Here in addition to the gentle undulating fracture surface, which is similar to that of a

Plate 3. A typical fracture pattern for the TGBA* phase with a surface cut nearly parallel to the pitch direction, the surface has a regular undulating texture similar to that observed for the chiral nematic phase

chiral nematic phase, discrete layering is visible (shown by the arrows). These are the smectic layers and they are parallel to the heli-axis (which is in agreement with the X-ray diffraction studies described earlier). At the bottom of each trough of the undulating fracture surface (open arrow), the layers are roughly normal to the fracture direction, and the 4.1 ± 0.3 nm (41 Å) spacing measured from optical diffraction from the TEM negative is consistent with the layer spacing determined from X-ray diffraction.

Plate 5 shows a high magnification view of the fracture. The characteristic "river" fracture pattern, which terminates abruptly at the defect core, is characteristic of screw dislocations (see arrows). From measurements of the shadow width, it can be concluded that the screw dislocations have a Burgers vector of a single layer step, which is consistent with the Renn and Lubensky model of the TGBA* phase. The screw dislocations form the twist grain boundaries that cause the orientation of the layers to change abruptly with respect to the fracture surface. The dislocations are spaced at about 15 nm apart, and the apparent layer spacing changes in discrete vertical bands from right to left as in the model of the TGBA* phase. Thus, the beautiful freeze

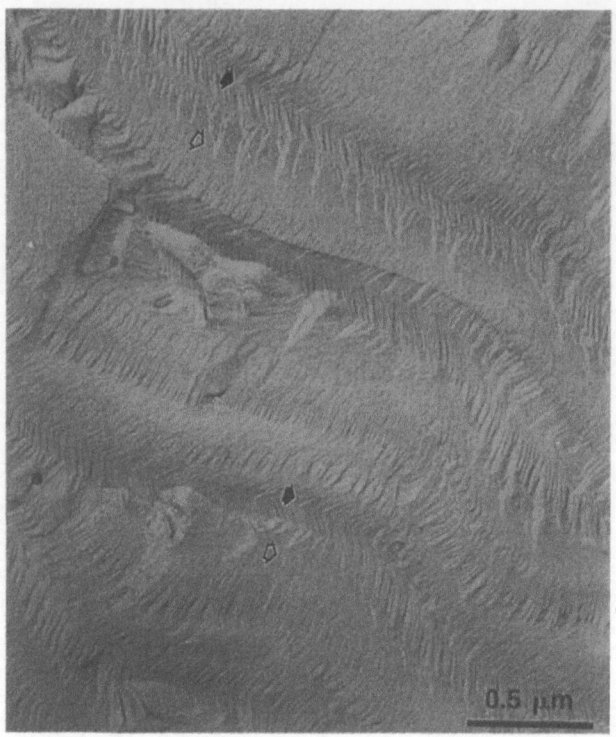

Plate 4. An oblique fracture through the TGBA* phase. Here in addition to the gentle undulating fracture surface, which is similar to that of a chiral nematic phase, discrete layering is visible (shown by the *arrows*)

fracture studies of the (R)- and (S-)1-methylheptyl 4'-(4-n-tetradecyloxy-phenylpropioloyloxy)biphenyl-4-carboxylates performed by Zasadzinski complement the also beautiful X-ray diffraction studies of Srajer and Pindak and confirm that the Renn and Lubensky theoretical model of the TGBA* phase is indeed the correct one.

4.2
Early Inferences of the Presence of TGB Phases

Looking back, the earliest written report concerning the possibility of having a smectic phase with twisted layers comes from the work of Wolfgang Ullrich Müller at the Technischen Universität Berlin in 1974 [40]. In his thesis entitled "Verhalten cholesterischer Mesophasen unter dem Einfluss von Phasenumwandlungen", Müller examined the mesophase behavior at the chiral nematic to smectic A* transition for various mixtures of cholesteryl oleoyl carbonate (COC), 5, and cholesteryl chloride (CC), 6. Müller came to the conclusion from his work that the smectic phase must have a helical structure and that the heli-axis must lie in the plane of the layers, and the structure must also have

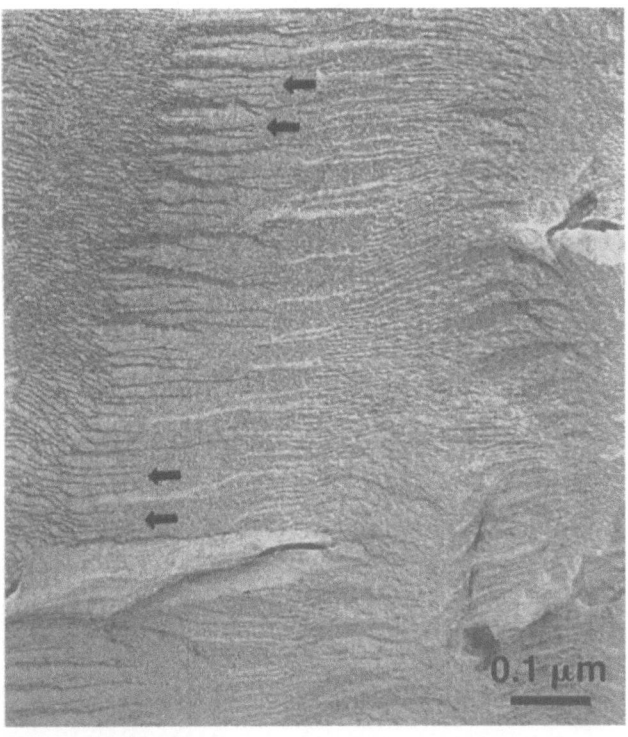

Plate 5. A high magnification view of the fracture surface, the freeze fracture image in this case shows screw dislocations. *Arrows* highlight the screw dislocations and the smectic grains rotate discontinuously from left to right. The dislocations are spaced every 15 nm, and the shadowing shows that the dislocations have a unit Burgers vector

5

discontinuities associated with it. However, Müller did not enter into a model based on a lattice of defects. Fig. 29 shows a sketch of one of the figures in Müller's thesis which shows a twisted smectic A* structure.

"Abrikosov-like" phases are also believed to be exhibited by some optically active side-chain liquid crystal polymethacrylates where the mesogenic side-group is derived from cholesterol [41]. The polymer, of structure 7, was reported by Freidzon et al. to exhibit a twisted layered structure in its smectic

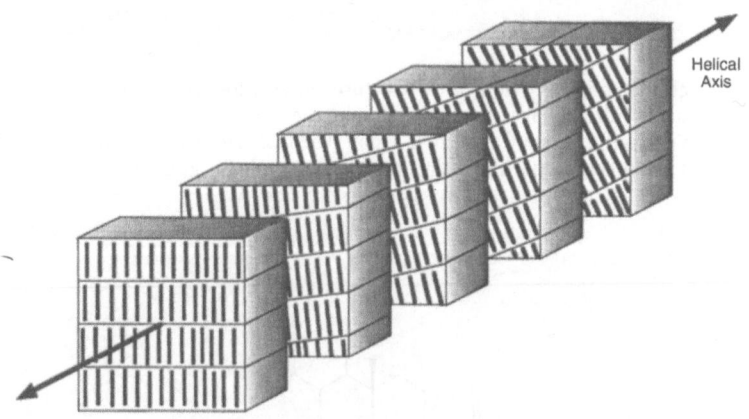

6

7

Fig. 29. The twisted smectic layering proposed by Wolfgang Ullrich Müller

A* phase, which the authors likened to a layered chiral nematic phase, see Fig. 30. Clearly such a structure cannot exist without the creation of defects and hence the ensuing formation of a lattice defect structure. However, as with Müller's work the possibility of the formation of an Abrikosov flux-type lattice was not entered into, but presumably this polymer must also exhibit a twist grain boundary phase like low molar mass materials. Possibly, as the liquid crystal phase develops from either the liquid or the crystal states, the main chain of the polymer prevents a cholesteric phase from forming and instead

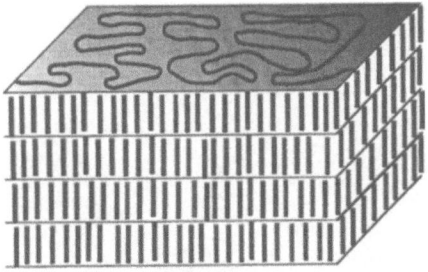

Smectic A Polymer

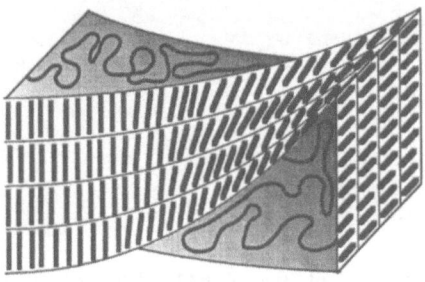

Twisted A* Polymer

Fig. 30. The normal layered structure of the smectic A phase of a side chain liquid crystal polymer, and the proposed twisted structure of the layered smectic A* phase of a chiral side chain polymer. The backbone of the polymer in each case is shown meandering between the layers

induces the formation of a smectic A* phase. However, the strong chirality of the side group in turn forces the layers to twist to give an "Abrikosov phase".

4.3
Defect Textures of TGB Phases

The TGBA* phase can be readily identified via thermal optical polarized transmitted light microscopy at magnifications of approximately ×100 [5, 16, 31, 42]. For example, in a simple experiment when a TGBA* material, sandwiched between clean glass slides, is heated into the isotropic liquid and cooled into its liquid crystal state, two distinct textures can be obtained. One texture is a platelet texture that takes the form of mosaic domains separated from each other by grain boundaries. The platelet texture resembles that of Blue Phase I with disclination lines occurring across the platelets, see Plate 6. The other texture is a Grandjean plane texture which appears iridescent and similar to that of the chiral nematic phase. As the Grandjean plane texture grows, striations within the texture are seen, which possibly correspond to integral multiples of the pitch of the helix of the phase (see Plate 7). Rotation

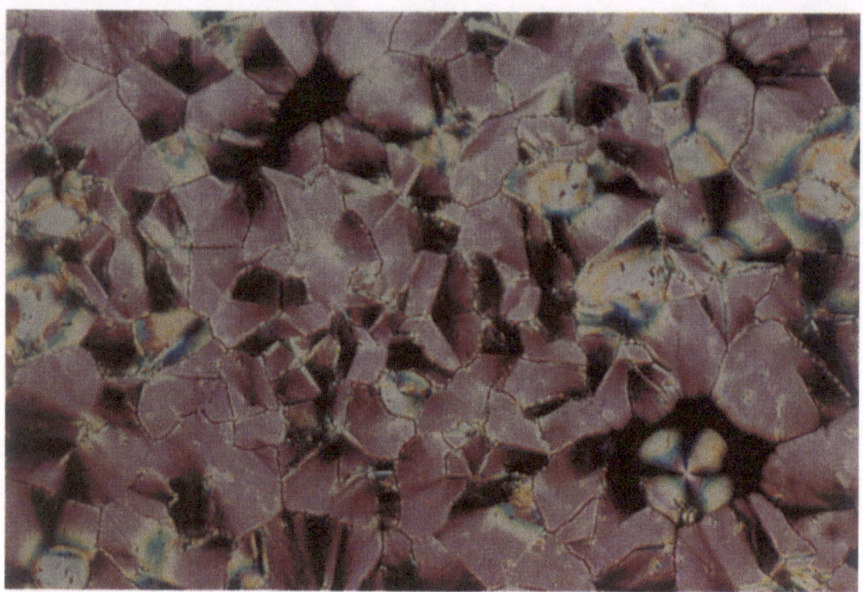

Plate 6. The platelet texture of the TGBA* phase of (S)-1-methylheptyl 4'-(4-n-tetradecyloxyphenylpropioloyloxy)biphenyl-4-carboxylate (×100)

of the upper polarizer of the microscope demonstrates that the phase is helical and gives the handedness, right or left, of the twist direction.

The formation of the TGBA* phase from the isotropic liquid can also be observed in free standing films. The defects formed in this case tend to be filamentary in nature, see Plate 8. Such a texture is not normally seen in free standing films of other liquid crystals, and therefore the presence of this defect texture is diagnostic for the characterization of TGB phases. In addition to the observation of such filamentary textures in free standing films, filaments were also observed embedded in homeotropically aligned smectic A* phases, see Plate 9.

Structures for the filaments in doped polymer systems have been proposed by Gilli and Kamayé [43]. They conceived that the filaments could be composed of double twist cylinders in a similar way to how double twist cylinders form the basis for the defect structures of Blue Phases. Fig. 31 shows the concept of a double twist cylinder where the long axes of the molecules in the middle of the cylinder are parallel to the central axis of the cylinder, but on moving away from the center of the cylinder the long axes of the molecules rotate in a preferred direction. If such double twist cylinders are then embedded into a normal layered smectic A* phase, then filaments will be observed in the microscope. However, unlike Blue Phases, this model requires that the cylinders are relatively large, i.e., they can be observed under low magnification in the microscope whereas they occur on a much smaller scale in Blue Phases. Figure 32 shows a number of possible arrangements for the

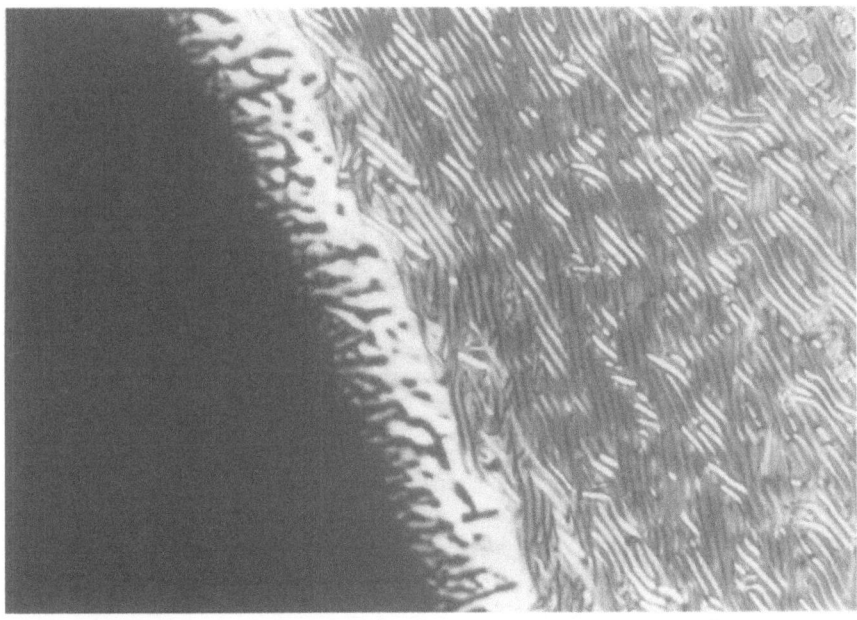

Plate 7. Pitch bands associated with the TGBA* phase at a temperature near to the clearing point of (S)-1-methylheptyl 4′-(4-n-tetradecyloxyphenylpropioloyloxy)biphenyl-4-carboxylate (×100)

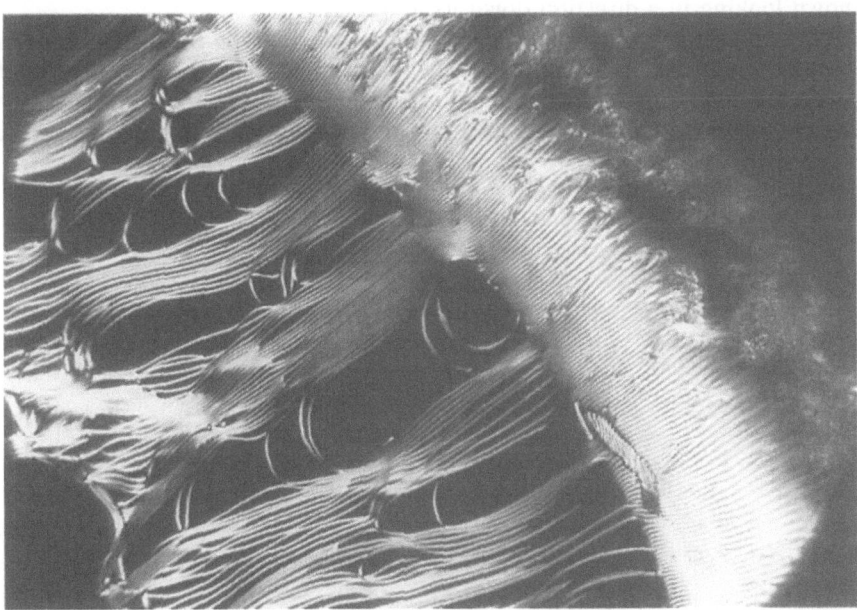

Plate 8. The filamentary texture formed in a free standing film of the TGBA* phase of (S)-1-methylheptyl 4′-(4-n-tetradecyloxyphenylpropioloyloxy)biphenyl-4-carboxylate (×100)

Plate 9. The filamentary texture of the TGBA* phase of (S)-1-methylheptyl 4'-(4-n-dodecyloxybenzoyloxy)biphenyl-4-carboxylate embedded in a homeotropic preparation (×100)

formation of filaments in TGBA* phases. In each of the sketches the filament is shown looking in a direction down its long axis.

Gilli and Kamayé went on to build the analogy with type II superconductors into their models of filaments. Firstly they noted that the coexistence of vortices in the bulk of the superconducting volume results from the existence of small regions that make possible a partial penetration of the magnetic flux

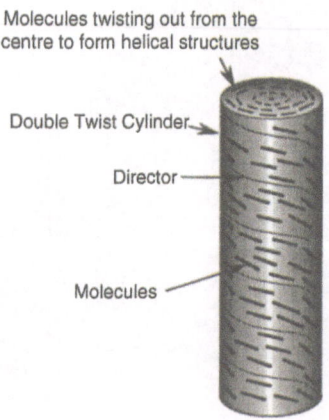

Molecules twisting out from the centre to form helical structures

Double Twist Cylinder

Director

Molecules

Fig. 31. Structure of a double twist cylinder (the molecules are shown as rods)

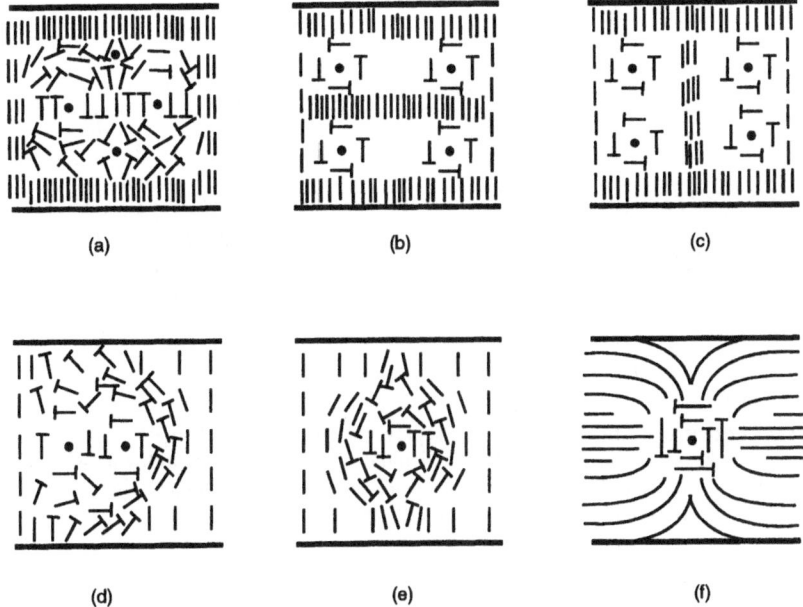

Fig. 32a–f. Various models for the filamentary texture of TGBA* phases. The molecules are shown as short lines, or nails when they are tilted out of the page. In each case the filaments are shown looking down the axis of the double twist cylinder (after Gilli and Kamayé [43])

lines into superconducting domains, thus negative energy surfaces would be expected. They then transposed this case to that of TGB phases by using screw dislocations as the analogues of the vortices, and it was found that the periphery of the double twist tube was a perfect candidate to support smectic correlations and twist coexistence.

For the conventional defect texture of the TGB phase, the +1 screw dislocations are straight and therefore can be considered as constituting a planar twist wall. Gilli and Kamayé built a comparable structure but this time with the planar twist wall of the TGB phase taking the form of a cylindrical surface. By taking a number of concentric cylinders together the classical double twist tube of the nematic phase was turned into a smectic tube or filament, see Fig. 33(a) and (b). When such double twist cylinders are embedded in a homeotropically aligned smectic A* matrix, a finite density lattice of edge dislocations in the strongly distorted regions joining the unperturbed homeotropic areas to the external surface of the double twist domain is required.

Gilli and Kamayé realized that there still remained the open question about the possibility of building a hypothetical double cylinder, but without an associated array of edge dislocations. They suggested that the unrealized strong thickness variation of the layers seen in Fig. 33(b) invalidated this last hypothesis. The complete model would be expected to involve a large number of such screw dislocations, and that the energy density of the edge dislocations would be strongly lowered in the proximity of the chiral nematic phase.

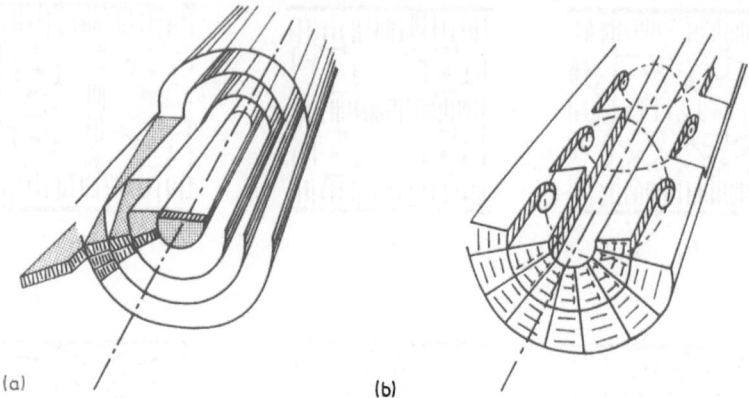

(a)   (b)

Fig. 33. a Concentric cylindrical twist walls separating different chiral smectic regions. Inside each region the layer surface adopts a helicoidal shape. **b** Each cylindrical twist wall is comparable to that found in a conventional TGB phase, a helical shape being given to the screw dislocations constituting them (after Gilli and Kamayé [43])

5
Other TGB Materials – Property/Structure Correlations

5.1
Low Molar Mass Systems

Since the discovery of the first examples of liquid crystalline materials which exhibit "Abrikosov flux phases" a wider range of compounds and mixtures of materials that exhibit this phase have been discovered. Typically, most materials tend to have somewhat similar structures to those of the original TGB compounds. In the following sections a brief overview of some of these materials is given.

A listing of the homologous series of phenyl propiolates for which some members exhibit TGB phases is given in Fig. 34. The study of propiolate systems has involved changing the length of the terminal alkoxy chain, the lengths of the alkyl chains attached to the chiral center, positioning a variety of halogen substituents at the chiral center [42, 44], and locating fluorine substituents in the aromatic core [5, 36]. Typically, the effect of chain lengthening at the chiral center is detrimental to the formation of liquid crystal phases and TGB phases in particular [35]. When the alkoxy chain is varied in length TGB phases persistently appear at chain lengths around 12 to 16 carbon atoms in length. When halogens are located at the chiral center it appears that as the size of the halogen is increased the stability of the TGB phase is decreased, see Table 4 [5]. In addition, halogen substituents at the chiral center tend to introduce chiral nematic to TGBA* to smectic A* phases sequences as opposed to the isotropic to TGBA* to smectic C* sequences found in the alkyl-substituted systems [44]. Lateral fluorine substitution in the aromatic core on the other hand has the effect of introducing TGBC phases into the phase sequence.

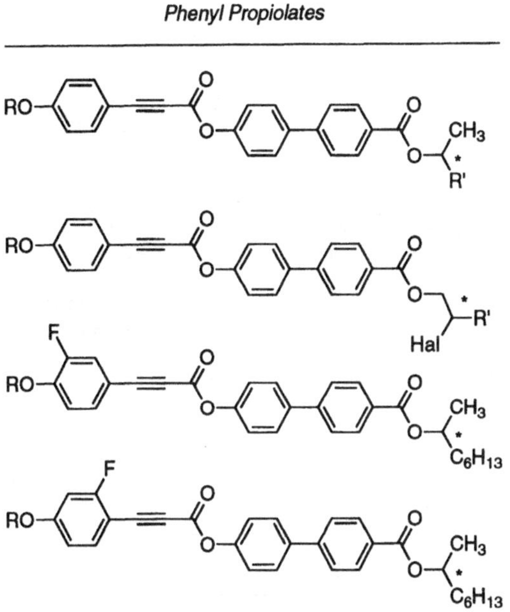

Phenyl Propiolates

Fig. 34. Families of phenyl propiolates for which some members exhibit "Abrikosov flux phases"

Table 4. Transition temperatures (°C) for the (S)-2-halo-4-methylpentyl 4'-(4-n-nonyloxyphenylpropioloyloxy)biphenyl-4-carboxylates

X	Iso to BP	Iso to N*	BP to N*	N* to SmA*	N* to TGBA*	TGBA* to SmA*	SmA* to SmC*	mp
F	142.0		141.0		113.4	111.3	94.4	68.4
Cl	129.4		128.3		100.4	96.8	79.9	67.8
Br		193.0		123.6			94.4	72.1

When the terminal ester functionality in the phenyl propiolates is replaced with a carbonyl group, i.e., a keto function, TGB phases disappear altogether indicating that the increased size of the lateral dipole at terminus to the core has detrimental effects on the stability of TGB phases [35]. These small variations in structure show how delicate the stability of the TGB phase is, nevertheless, most chiral nematic to smectic phase transitions are accompanied by a TGB phase no matter how transient it is.

It is interesting to note that the closely related aliphatic substituted benzoate esters which are analogous in structure to the phenylpropiolates (the original TGB materials), see Fig. 35, do not readily exhibit TGB phases, with fewer compounds in these series showing such properties. Conversely, analogs with halogens located at the chiral center appear to sustain the formation of TGB phases, see Tables 5 and 6. This may be because the halogen compounds exhibit chiral nematic phases, whereas their aliphatic counterparts do not.

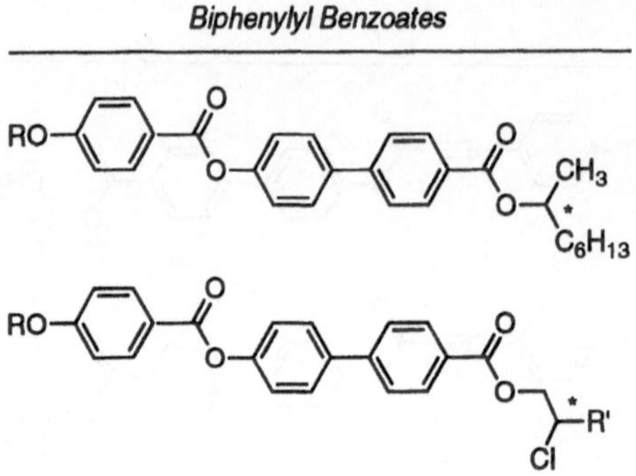

Biphenylyl Benzoates

Fig. 35. Families of biphenylyl benzoates for which some members exhibit "Abrikosov flux phases"

Table 5. Transition temperatures (°C) for the (S)-2-chloro-4-methylpentyl 4′-(4-n-alkoxybenzoyloxy)biphenyl-4-carboxylates

n	Iso to BP	BP to N*	N* to TGBA*	TGBA* to SmA*	Iso to SmA*	SmA* to SmC*	mp
7	156.1	153.2	144.9	144.2		119.2	73.4
8	155.4	148.3	146.8	146.0		115.1	75.0
9	150.4	145.5	145.4	144.6		121.0	74.9
10	149.7	146.8	145.6	144.8		121.9	64.4
11	148.1	146.8	145.8	145.0		124.2	69.4
12	145.6		145.0	144.8	144.1	122.0	68.8
13					142.5	121.7	73.5
14					142.0	121.4	76.7

Table 6. Transition temperatures (°C) for the (S)-2-chloro-4-methylpentyl 4′-(4-n-alkoxyphenylpropioloyloxy)biphenyl-4-carboxylates

n	Iso to BP	BP to N*	N* to TGBA*	TGBA* to SmA*	SmA* to SmC*	mp
9	129.4	128.3	100.4	96.8	79.9	67.8
12	122.7	122.5	108.1	106.8	93.2	73.3
14	119.9	116.4	110.9	109.7	98.1	68.5
16	117.5	116.3	114.6	113.6	101.2	59.9
18	115.4	114.9	114.2	113.4	99.9	56.1

Probably the most interesting materials to exhibit TGB phenomena are the tolanes prepared by Nguyen and coworkers [45–47]. In their exhaustive and beautiful studies of materials that exhibit TGB phases, Nguyen et al. have prepared many series of tolanes that are substituted at one end with a 4-n-alkoxybenzoyloxy group and at the other terminus with an ester or ether moiety that carries a chiral functionality, see Figs. 36 and 37. Their early work

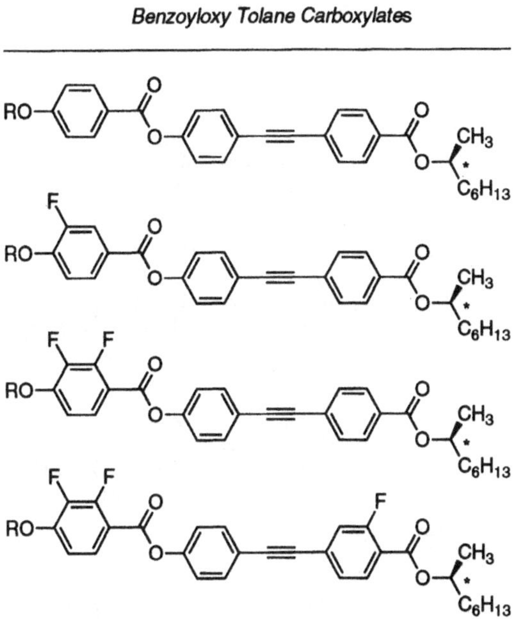

Benzoyloxy Tolane Carboxylates

Fig. 36. Families of unsubstituted and fluoro-substituted 1-methylheptyl 4-(n-alkoxyben-zoyloxy)tolane-4-carboxylates for which some members exhibit "Abrikosov flux phases"

was on systems with the chiral 1-methylheptyl group as the chiral moiety, lateral fluorine substituents were then introduced into the lateral positions of the aromatic core to produce some exciting effects. The introduction of lateral fluorine substitution induced the formation of TGBC phases, commensurate and incommensurate TGB phase behavior with respect to the rotation of the "smectic blocks" about the heli-axis, Blue Phase III – II – I – TGBA* – TGBC phase sequences, and multicritical [48] and tricritical points in phase diagrams. These phenomena have been extensively studied by Nguyen, Barois, Navailles, Pindak and others [49–51].

Zugenmaier et al. [52] also reported materials, of general structure **8**, that were similar to those of Nguyen et al. with one principle exception; the alkynic linkage of the tolane was hydrogenated. Not surprisingly the materials exhibited TGBA* phases over temperature ranges of about 0.3 °C. The heptyl and octyl homologues exhibited smectic C*, smectic I* and smectic F* phases in addition to the TGBA* phase.

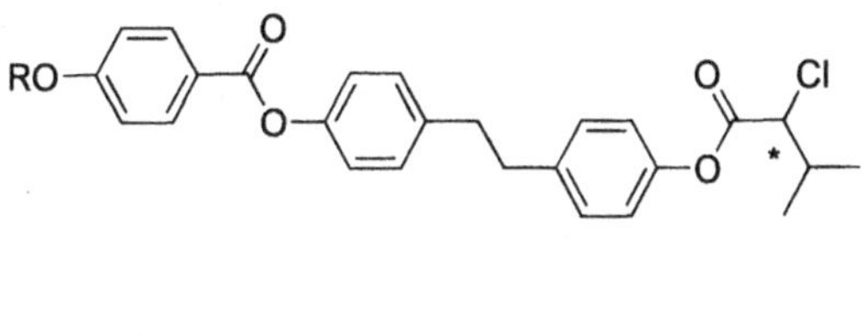

8

R=heptyl SmF* 86.1 SmI* 92.4 SmC* 120.8 TGBA* 121.1 N* 133.2 Iso Liq
R=octyl SmF* 83.7 SmI* 92.9 SmC* 125.6 SmA* 129.2 TGBA* 129.4 N* 134.5 Iso Liq

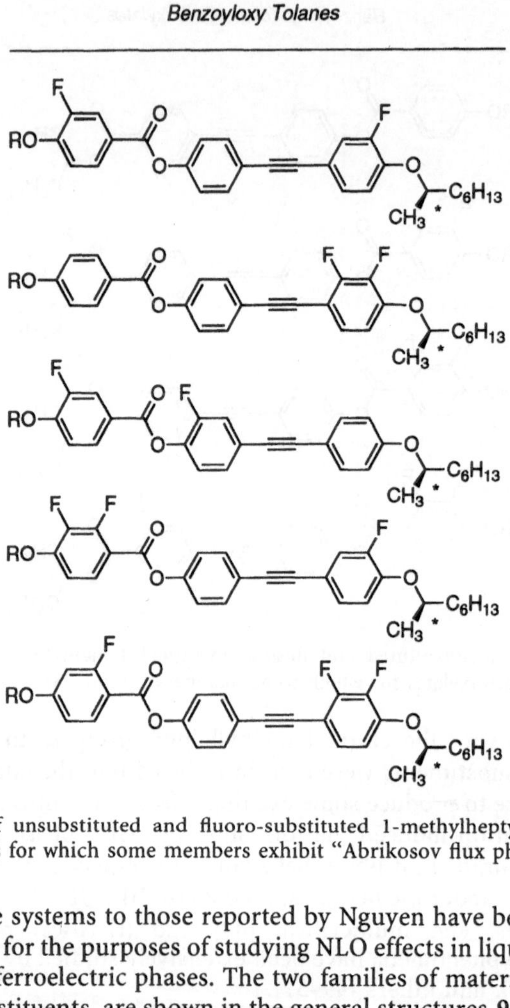

Benzoyloxy Tolanes

Fig. 37. Families of unsubstituted and fluoro-substituted 1-methylheptyloxy 4-(n-alkoxy-benzoyloxy)tolanes for which some members exhibit "Abrikosov flux phases"

Similar tolane systems to those reported by Nguyen have been prepared by Walba et al. [53] for the purposes of studying NLO effects in liquid crystals, and in particular in ferroelectric phases. The two families of materials, which have lateral nitro-substituents, are shown in the general structures **9** and **10**. Most of the ethers, **9**, are reported to exhibit TGBA** phases, which have relatively large temperature ranges, often far in excess of the compounds discussed above. In addition for the R = pentyl, R' = decyl, and R = hexyl, R' = decyl homologues, TGBC phases with wide temperature ranges were obtained. In this work the electrical field properties of the TGBC phase were investigated.

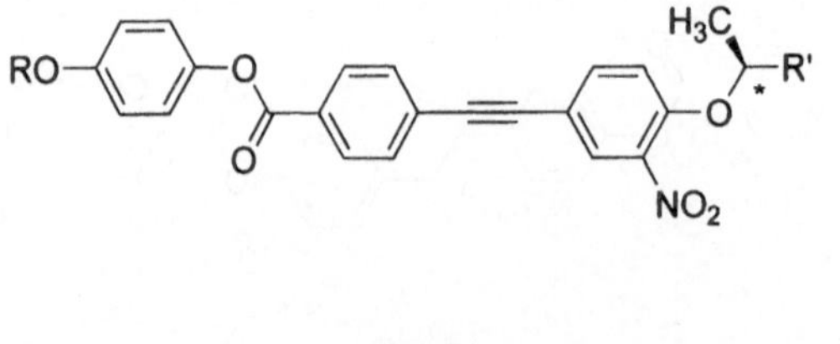

9

R=pentyl, R'=decyl SmX 48 TGBC 66 TGBA* 84 N* 86 Iso Liq
R=hexyl, R'= decyl SmX 46 TGBC 66 TGBA* 81 N* 82 Iso Liq

The second family of materials, **10**, were shown to exhibit chiral nematic to TGBC phases sequences in addition to the more conventional chiral nematic – TGBA* – TGBC sequence. For the R = hexyl, R′ = decyl and R = hexyl, R′ = dodecyl homologues TGBC phases with ranges of up to 35 °C were obtained.

10

R=hexyl, R′=decyl SmX 50 TGBC* 85 N* 120 Iso Liq
R=hexyl, R′=dodecyl SmX 48 TGBC* 62 N* 98 Iso Liq

TGBC phases with large temperature ranges were also obtained in a remarkable series of materials first prepared by Takatoh [54] and later added to by Lamb [55]. The materials synthesized were derivatives of terphenyl. Using Sharpless chiral epoxidation of hexen-2-ol they made liquid crystals with terminal oxirane moieties which introduced two sequential chiral centers into the terminal aliphatic chain, see general structure **11**.

11

R=ethyl SmC*59 TGBC 91.5 N*$_{(r)}$ 106 N*$_\infty$ 110 N*$_{(l)}$ 156.6 BP 156.7 Iso Liq

These terphenyls exhibited Blue Phases (I, II, and III), chiral nematic, TGBC, and smectic C* phases and, in addition, the materials also exhibited helix inversions in their chiral nematic phases. In some remarkable cases the TGBC phases extended over temperature ranges of more than 65 °C. Thus, unlike the TGBA* phase, the TGBC phase appeared to be more thermally stable. Often TGBA* phases are found to extend only over a few degrees at most, and about ten degrees at best; conversely, the TGBC phase appears to have much greater stability. The fact that the introduction of the TGBA* phase into liquid crystal systems depends on fluctuations of neighboring phases and approaching second order phase transitions, as predicted by de Gennes, Renn, and Lubensky, it might be expected that the temperature range of the TGBC phase would also be short; however, the wide temperature ranges of these phases cannot be so simply explained.

Other systems with multiple chiral centers have been prepared, for example Wu and Hsieh [56] reported the family of materials given by general structure **12**. The earlier members of this series of compounds (i.e., up to the

undecyl homologue) exhibit chiral nematic, TGBA*, smectic A*, and smectic
C* phases, whereas the higher homologues exhibit chiral nematic, TGBA*, and
TGBC phases.

                                                                                          12

R=octyl   Crystal 109.1 (SmC* 106.1)SmA* 143.7 TGBA* 146.4 N* 147.9 Iso Liq
R=tetradecyl Crystal 80.5 TGBC 119.4 TGBA* 132.8 N* 136.5 Iso Liq

5.2
TGB Phases in Mixtures

As noted earlier TGB phases were probably observed first in mixtures of
derivatives of cholesterol. They have also been observed in mixtures of a liquid
crystalline achiral host that had been doped with a chiral non-liquid crystalline
material. The resulting chiral mixtures were found to exhibit TGBA* and
TGBC phases over temperature ranges of a few degrees. However, the most
unusual study carried out on mixtures was performed by Vill et al. [57] on the
materials 13 and 14. When mixed these compounds exhibit re-entrant
phenomena for the TGBA* phase. For binary mixtures between 0.1 and
0.3 mole fraction of compound 13, a remarkable Blue Phase, chiral nematic,
TGBA*, smectic A*, TGBA*, chiral nematic phase sequence was observed on
cooling (i.e., a BP – N* – TGBA* – SmA* TGBA* – N* phase sequence).
Fig. 38 shows the full phase diagram for binary mixtures as a function of the
increasing content of compound 13. TGB phases appear in the phase diagram
over roughly the same concentration range as Blue Phases appear, indicating
that the presence/formation of these two structurally frustrated phases may be
connected. A second inference can be made from the phase diagram
concerning the permanence of the screw dislocations in the "Abrikosov"
phase. It appears that the screw dislocations are not stable and as the pitch of
the helical structure changes with temperature or concentration, so too does
the density of the defects. Thus the defects must be mobile and readily anneal
or are subsumed or created easily by the bulk phase.

                                                                                          13

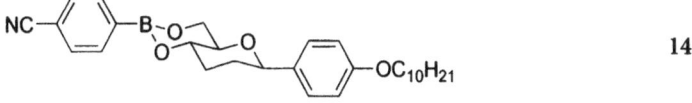

14

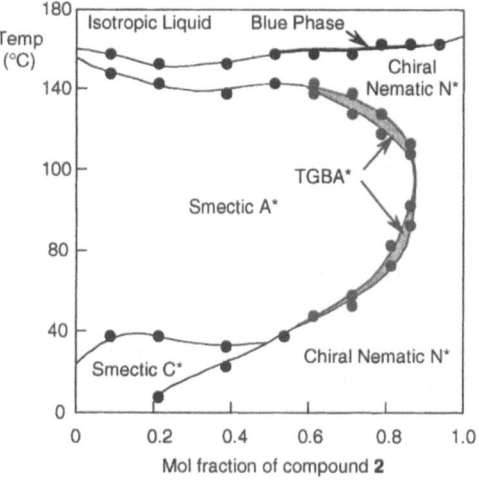

Fig. 38. Phase diagram for binary mixtures of compounds **13** and **14** as function of the increasing content of compound **13** (after Vill, Tunger and Peters [57])

5.3 Polymer TGB Systems

Most of the materials described so far are low molar mass materials; however, larger molecules, dimers, oligomers and polymers can also exhibit TGB phases. Compound **15** reported by Nguyen [58] is not a dimer or an oligomer, but is a supermolecule that combines two mesogenic groups in its structure, i.e., a tolane unit derived from previous studies on TGB phases and a cyanobiphenyl unit that is found in conventional mixtures used in twisted nematic display devices. This strange dichotomous compound exhibits a Blue Phase, a chiral nematic, a TGBA*, and a smectic A* phase.

15

Crystal 117 SmA* 138 TGBA* 140 N* 166 BP 167 Isotropic Liquid

As noted earlier form the work of Freidzon [41], side chain liquid crystal polymers derived from cholesterol can also apparently exhibit TGB phases. However, work on non-steroid systems also reveals that TGB phases can also be formed in typical chiral polymers. For example the polymethacrylate, structure **16**, exhibits chiral nematic, TGBA*, smectic A*, and ferroelectric smectic C* phases.

16

Iso 160 N* 149 TGBA* 140 SmA* 73 SmC* DP=15 γ=2.65

Conversely, it is interesting to note that the directly analogous polyacrylate, **17**, does not possess such disordered phases and therefore does not exhibit a TGB phase [59].

17

Iso 194 SmA* 152 SmC* 109 Sm2 75 Sm1 DP=12 γ=1.32

Similarly, side chain liquid crystalline polymethacrylates and polysiloxanes carrying chiral mesogenic side groups, where the mesogenic core has two aromatic rings, can apparently exhibit TGB phases. Kozlovsky and Demikov recently reported [60] that polymers **18** and **19**, probably exhibit TGB phases, where the "so-called TGB" phase is claimed to have a poorly birefringent non-characteristic texture which upon shearing gives a transparent pattern. They summarized the properties of their probable TGB phase as follows:

- The phase exists in a board temperature range and can be frozen into a glass.
- The heat of the clearing point transition ($\approx$7–9 kJ mol^{-1}) is typical for a smectic to isotropic transition, and two orders of magnitude higher than for Blue phase isotropic transitions.

- The phase shows poor birefringence beyond a narrow wavelength range close to the adsorption edge.
- The phase has well-developed circular dichroism with a maximum in the near UV range and has a selective reflection in the same range.
- The phase forms highly transparent textures.
- The phase possess a short range ordering (8–23 nm).
- Pyroelectric measurements show the absence of ferroelectric properties, thereby indicting that the phase is not tilted.

Thus, they conclude from these results that the phase probably has a TGB-like structure.

18

Glass 30 SmX 64 Iso

and

19

Glass 24 SmX 47 Iso

6
Other TGB Phenomena

TGB phases appear in a wide range of natural phenomena, particularly in biological systems. There are suggestions that TGB-like behavior might be found in self-assembled structures of natural and synthetic polymers such as DNA, PBLG and xanthan, in viruses, in bone and cuticles etc. [61, 62]. There are also indications that TGB-like phenomena might be used to investigate crystal growth [63]. In addition, a number of TGB variants have been discovered which have considerably extended the TGB family of phases, and have demonstrated that TGB phenomena is universal.

6.1
TGBC Phases

There are a number of possible structures for TGBC phases. The following structural features need to be considered with respect to each possibility. Firstly, the orientation of the layers can be either tilted with respect to the heli-axis or parallel, secondly, the local spontaneous polarization of the smectic C* layers can be either parallel or perpendicular to the heli-axis, thirdly, the smectic "blocks" can have a local helical structure caused by a precession in the tilt of the molecules perpendicular to the layers or alternatively the twist can be expelled to the grain boundaries, and fourthly, the rotation of the blocks about the heli-axis can be either rational (commensurate) or irrational (incommensurate).

As noted earlier, the first tilted version of the TGBA* phase, was discovered in some chiral derivatives of tolane, **20**, by Nguyen et al. Very careful optical and X-ray diffraction studies on this phase by Navailles, Barois and Nguyen [17, 48–51] elucidated the structure to be one where the molecules remain perpendicular to the heli-axis of the phase and the planes of the layers are tilted. In the smectic "blocks" the smectic C* twist is expelled to the grain boundaries, and thus locally the phase is ferroelectric (i.e., the symmetry is reduced to C_2). The local spontaneous polarization, associated with the smectic "blocks" is rotated as the blocks rotate about the heli-axis. Consequently, the TGBC phase does not possess a spontaneous polarization as it is averaged to zero and the phase becomes helielectric. In addition the X-ray scattering results showed that the phase is rational (commensurate) rather than incommensurate, with approximately 17 smectic "blocks" making up one turn along the heli-axis (i.e., one pitch length). Fig. 39 shows the local Structure in the TGBC phase where the long axes of the molecules are perpendicular to the heli-axis.

20

R=$C_{11}H_{23}$Cryst 51.7 SmC* 100.6TGBC   101.3   TGBA*105.1 N* 111.9BP 112.7 IsoLiq
R=$C_{12}H_{25}$Cryst 51   SmC* 102.6TGBC   103   N* 110.5 BP 111.7 IsoLiq

The commensurability of the TGBC phase was investigated by two dimensional intensity maps recorded in the plane perpendicular to the pitch for oriented samples (i.e., samples with rubbed aligning surfaces) mounted with their flat walls perpendicular to the beam. For the dodecyl homologue, at 130 °C in the chiral nematic phase the diffraction pattern displays a continuous ring corresponding to a spacing of 37.5 Å. The intensity was found to be uniform over the ring, see Fig. 40. Below 102.8 °C, in the TGBC

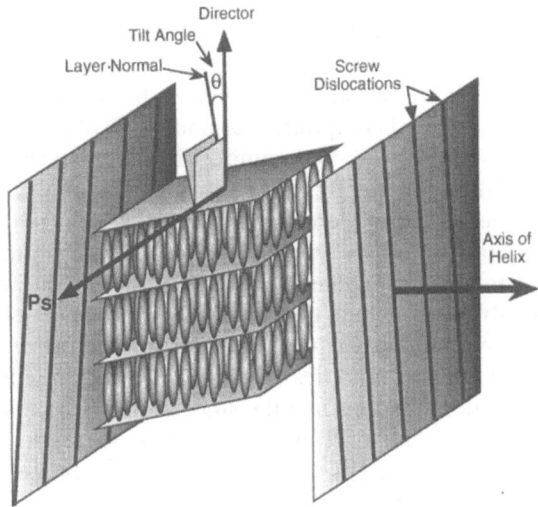

Fig. 39. Structure of the TGBC phase showing one smectic sheet and two sets of grain boundaries

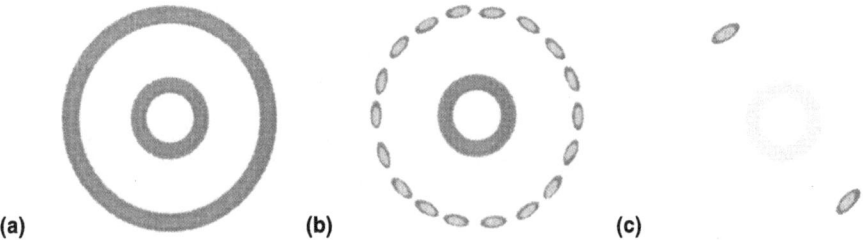

(a)   (b)   (c)

Fig. 40a–c. A schematic representation of two dimensional intensity maps (**a**) in the nematic phase, (**b**) in the TGB phase and (**c**) in the smectic C phase of 3-fluoro-4-[(R or S)-1-methylheptyoxy]-4′-(4-dodecyloxy-2,3-difluorobenzoyloxy)tolane

phase a continuous but now modulated ring is observed, and at a temperature of 102.15 °C the ring appears to be split into 18 equidistant spots revealing a commensurate TGB structure. Upon cooling from the chiral nematic phase, the number of spots changes successively from 16 to 18 to 20. Upon heating from the smectic C* phase there is a hysteresis of about a quarter of a degree for the 18 to 20 transition, however, no state with 16 diffraction spots is observed. At 101.7 °C a smectic C* phase develops, and the ring of diffraction spots is replaced by two intense Bragg spots in a direction that makes an angle $\theta \approx 20°$ with respect to the director at the surface, i.e., with respect to the rubbing direction of the glass plates. Opposite tilt angles ($-\theta$), and coexistence of both sets (+ and $-\theta$) were also observed. The value of θ was found to agree well with optical measurements of the tilt angle.

6.2
3-D Modulated Structures of TGBC Phases

Modulated Structures were found in TGB phases by Madhusudana et al. [64] who investigated mixtures of compounds 21 and 22. TGB phases were found in compositions with ≈5 to 45% of compound 22 in the mixture. The TGB phase was shown to be TGBA* in type. By using optical diffraction, X-ray diffraction, electrical field studies etc., they were able to show that near to the transition to the smectic C* phase another type of TGB phase was introduced. They proposed that the intermediate phase was TGBC in nature, however from their studies they were able to conclude that the TGB blocks had a two dimensional undulating structure such that a square grid was formed. They called this new phase an undulated UTGBC modification. The UTGBC phase is characterized by helical axes both along and normal the smectic C* layers, and it is expected to occur naturally between the TGBA* phase, in which the helical axis is parallel to the smectic A* layers, and the smectic C* phase, in which it is normal to the smectic C* planes.

21

22

6.3
Columnar and Smectic Order in TGB Phases

Recently Ribeiro et al. [65] reported on the synthesis and characterization of a variety of tolanes that had optically active sulfinate groups. Some of these compounds, see 23, were found to possess a phase that exhibited oily-streak textures typical of chiral nematic phases and also defect pattern associated with columnar phases (the earlier photomicrograph Plate 6 for 14P1M7 is similar).

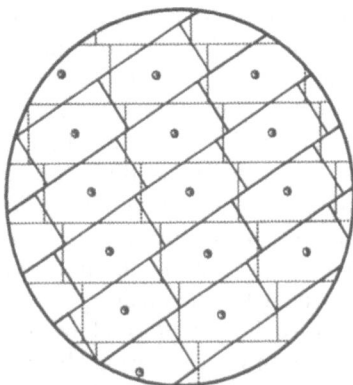

Crystal 72 M 98 Iso Liq

For compound **23** the uncharacterized phase, denoted as M, was found to be exhibited over a fairly wide temperature range. Ribeiro et al. found that the X-ray diffraction patterns associated with the intermediary phase were typical of a disordered smectic modification. The d spacing of 45.3 Å was found to be a little bit shorter than the calculated molecular length which is consistent with the presence of a smectic A phase. As the material also exhibits a Grandjean plane texture it was concluded that the phase must be a TGB modification. However, the platelet/fan texture exhibited by the mesophase conversely suggests that a columnar ordering must be present as well.

Ribeiro et al. suggested the following model (see Fig. 41) in order to resolve these conflicting observations. First, they considered the situation for two adjacent smectic slabs separated by a grain boundary with an angle of α between the two slabs. Then the important feature they noted for this representation was the presence of common areas between the two adjacent slabs, which mark the places where the smectic layers go continuously from one slab to the next. These interfaces were expected to form twisted ribbons of molecules. As two-dimensional columnar ordering exists only in the grain boundaries, it was assumed that the columns are interrupted and merge back into the smectic layers. However, as the lateral dimensions are fixed, the

Fig. 41. Grain boundary between two adjacent smectic slabs. The array of equidistant screw dislocations arranges the smectic layers into ribbons or columns in a direction perpendicular to the figure

columns have a constant density and it is therefore possible to track them from one boundary to the next, which ensures that the formation of developable domains occurs in the classical way for columnar structures made of continuous columns.

6.4
Lyotropic TGB Phases

Lyotropic liquid crystals are principally systems that are made up of amphiphiles and suitable solvents or liquids. In essence an amphiphilic molecule has a dichotomous structure which has two halves that have vastly different physical properties, in particular their ability to mix with various liquids. For example, a dichotomous material may be made up of a fluorinated part and a hydrocarbon part. In a fluorinated solvent environment the fluorinated part of the material will mix with the solvent whereas the hydrocarbon part will be rejected. This leads to microphase separation of the two systems, i.e., the hydrocarbon parts of the amphiphile stick together and the fluorinated parts and the fluorinated liquid stick together. The reverse is the case when mixing with a hydrocarbon solvent. When such systems have no bend or splay curvature, i.e., they have zero curvature, lamellar sheets can be formed. In the case of hydrocarbon/fluorocarbon systems, a mesophase is formed where there are sheets of fluorocarbon species separated from other such sheets by sheets of hydrocarbon. This phase is called the Lα phase. In the Lα phase the molecules are orientationally ordered but positionally disordered, and as a consequence the amphiphiles are arranged perpendicular to the lamellae. The Lα phase of lyotropics is therefore equivalent to the smectic A phase of thermotropic liquid crystals.

Many naturally occurring lyotropic systems possess chiral molecules, for example cell membranes are the equivalent of 2-D Lα phases. Thus, there is the potential to have twist built into the structure in a similar way to how it can be built into smectic A* phases to give TGBA* modifications. Kamien and Lubensky [66] have developed a theoretical model for such an event. In their modeling, which is directly analogous to that for TGBA* phases, they describe the formation of a defect-stabilized phase where the defects are screw dislocations. The screw dislocations in this case, however, have slightly different structures to those of thermotropic phases. The dislocation contains the solvent or liquid along the line of the screw axis, i.e., down the middle of the defect, see Fig. 42. In the defects of the thermotropic TGB phase, the cores of the defect are essentially nematic and made up of molecules of the mesogen.

So far no examples of lyotropic TGB phases have been discovered; however, there are many different types of vesicles formed by lyotropic systems, and the possibility that some have structures based on TGB phases should not be ruled out.

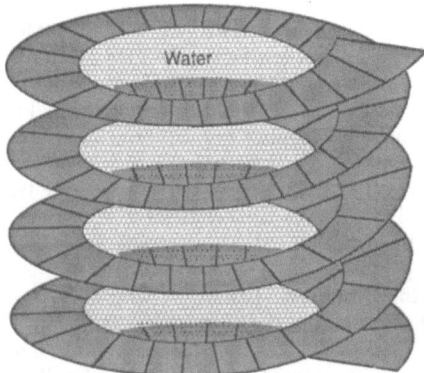

Fig. 42. A screw dislocation in a lyotropic Lα phase. The core of the defect is made up of solvent or liquid

7
Summary

The discovery of TGB phases links the physical understanding of phase transitions in liquid crystals to those of superconductors, and demonstrates, yet again, the abilities of liquid crystals to provide an experimental arena for the exploration of new theoretical models concerning the structures of condensed phases and their accompanying phase transitions. In the case of TGB phases, the experimental verification and characterization of the structure of TGB phases utilizes a wide variety of physical and chemical techniques. In addition, although not discussed in detail, TGB sructures are not only found in synthetic organic chemical systems, but also in naturally occurring biological structures. Thus, TGB phases can be considered as a natural phenomenon embracing physics, chemistry and biology. Their existence is a fundamental result of physical competition between different structural forces.

Acknowledgements. I would like to thank, in particular, Drs. R Pindak and J Zasadzinski for supplying me with very many drawings, photographs and references, and for many elucidating discussions. I would also like to thank my research colleagues; Drs. Patel, Slaney, Takatoh, Booth, Toyne, Waugh, Nishiyama, Cladis and Dunmur for their help and collaboration in my research on TGB phases. I would also like to thank Drs. Navailles, Nguyen, Barois, and Madhusudana for sending me preprints and/or details of their research work. I would also like to thank Lucert Bell Laboratories and the Engineering and Physical Science Research Council for support.

8
References

1. Collings PJ (1990) Liquid Crystals: Natures Delicate Phase of Matter, Princeton University Press, Princeton
2. Collings PJ, Hird M (1997) Introduction to Liquid Crystals: Chemistry and Physics, Taylor and Francis, London

3. Demus D (1989) Liq Cryst 5: 75
4. Goodby JW (1994) In: Bloor D, Brook RJ, Flemings MC, Mahajan S (eds) The Encyclopedia of Advanced Materials, Pergamon Press, Oxford, p 1325-1334
5. Goodby JW, Nishiyama I, Slaney AJ, Booth CJ, Toyne KJ (1993) 14: 37
6. Goodby JW, Blinc R, Clark NA, Lagerwall ST, Osipov MA, Pikin SA, Sakurai T, Yoshino K, Zeks B (1991) Ferroelectric Liquid Crystals: Principles, Properties and Applications, Gordon and Breach, Philadelphia, p 99-123
7. Leslie F (1998) In: Gray GW, Demus D, Spiess HW, Goodby JW, Vill V (eds) Handbook of Liquid Crystals, Vol I, Wiley-VCH, Weinhiem
8. Leadbetter AJ (1987) In: Gray GW (ed) Thermotropic Liquid Crystals, Wiley, Chichester, p 1-27
9. Goodby JW (1998) In: Gray GW, Demus D, Spiess HW, Goodby JW, Vill V (eds) Handbook of Liquid Crystals, Vol II, Wiley-VCH, Weinhiem
10. Goodby JW, Blinc R, Clark NA, Lagerwall ST, Osipov MA, Pikin SA, Sakurai T, Yoshino K, Zeks B (1991) Ferroelectric Liquid Crystals: Principles, Properties and Applications, Gordon and Breach, Philadelphia, p 249-316
11. Meyer RB, Liebert L, Strzelecki L, Keller P (1975) J Phys Lett France 36: 69
12. Meyer RB (1977) Mol Cryst Liq Cryst 40: 33
13. Clark NA, Lagerwall ST (1980) Appl Phys Lett 36: 899
14. Brand HR, Cladis PE, Finn PL (1985) Phys Rev 31 A: 361
15. Renn SR, Lubensky TC (1988) Phys Rev A 38: 2132
16. Goodby JW, Waugh MA, Stein SM, Chin E, Pindak R, Patel JS (1989) Nature 337: 449
17. Renn SC (1992) Phys Rev A 45: 953; Navailles L, Barois P, Nguyen HT (1993) Phys Rev Lett 71: 545
18. de Gennes PG (1972) Sol State Commun 10: 753
19. Meissner W, Ochsenfeld (1933) Naturwiss 21: 787
20. de Gennes PG (1966) Superconductivity of Metals and Alloys, Benjamin; Abrikosov AA (1957) JETP 5: 1174
21. London F, London H (1935) Proc Roy Soc (London) A149: 72
22. Bardeen J, Cooper LN, Schrieffer JR (1957) Phys Rev 106: 162: ibid 108: 1175; Ginzburg VL, Landau LD (1950) Zh Eksperim i Teor Fiz 20: 1064
23. Lubensky TC (1995) Physica A 220: 99
24. Crooker PP (1989) Liq Cryst 5: 751
25. Renn SR, Lubensky TC (1991) Mol Cryst Liq Cryst 209: 349
26. Goodby JW, Gray GW (1976) Mol Cryst Liq Cryst 37: 157
27. Terashima K, Ichihashi M, Kikuchi M, Furukawa K, Inukai T (1986) Presented at the IIth International Liquid Crystal Conference, San Francisco
28. Goodby JW, Chin E (1988) Liq Cryst 3: 1245
29. Goodby JW, Patel JS, Chin E (1992) J Mater Chem 2: 197
30. Chandani ADL, Hagiwara T, Suzuki Y, Ouchi Y, Takezoe H, Fukuda A (1988) Jpn J Appl Phys 27: 729
31. Goodby JW, Waugh MA, Stein SM, Chin E, Pindak R, Patel JS (1989) J Am Chem Soc 111: 8119
32. Corey EJ, Fuchs PL (1972) Tetrahedron Lett 36: 3769
33. Chin E, Goodby JW (1986) Mol Cryst Liq Cryst 141: 311
34. Mitsunobu O (1981) Synthesis 1
35. Nishiyama I (1992) PhD Thesis, University of Hull
36. Goodby JW, Nishiyama I, Slaney AJ, Vuijk J, Booth CJ, Styring P, Toyne KJ (1994) Mol Cryst Liq Cryst 243: 231
37. See for example Nelson DR (1988) Phys Rev Lett 60: 1973; Gammel PL, Bishop DJ, Dolan GJ, Kwo JR, Murray CA, Schneemeyer LF, Waszczak (1987) Phys Rev Lett 59: 2592
38. Srajer G, Pindak R, Waugh MA, Goodby JW, Patel JS (1990) Phys Rev Lett 64: 1545
39. Ihn KJ, Zasadzinski JAN, Pindak R, Slaney AJ, Goodby JW (1992) Science 258: 275
40. Müller WU (1974) PhD Thesis, Technischen Universität Berlin

41. Freidzon YaS, Tropsha YeG, Tsukruk VV, Shilov VV, Lipatov YuS (1987) J Polym Chem (USSR) 29: 1371; Freidzon YaS, Kharitonov AV, Shibaev VP, Plate NA (1985) Eur Polym J 21: 211
42. Slaney AJ, Goodby JW (1991) J Mater Chem 1: 5
43. Gilli JM, Kamayé (1992) Liq Cryst 4: 545
44. Slaney AJ (1992) PhD Thesis, University of Hull
45. Nguyen HT, Bouchta A, Navailles L, Barois P, Isaert N, Tweig RJ, Maaroufi A, Destrade C (1992) J Phys II France 2: 1889
46. Bouchta A, Nguyen HT, Achard MF, Hardouin F, Destrade C, Teig RJ, Maaroufi A, Isaert N (1992) Liq Cryst 12: 575
47. Li MH, Laux V, Nguyen HT, Gigaud G, Barois P, Isaert N (1997) 23: 389
48. Anakkar A, Daoudi A, Buisine JM, Isaert N, Delattre T, Nguyen HT, Destrade C (1994) J Therm Anal 41: 1501
49. Petit M, Barois P, Nyugen HT (1996) Europhys Lett 36: 185
50. Isaert N, Navailles L, Barois P, Nguyen HT (1994) J Phys II France 4: 1501
51. Navailles L, Barois P, Nguyen HT (1993) Phys Rev Lett 71: 545; Navailles L, Pindak R, Barois P, Nguyen HT (1995) Phys Rev Lett 74: 5224; Navailles L, Barois P, Nguyen HT (1994) 72: 1300
52. Dierking I, Giebelmann F, Zugenmaier P (1994) 17: 17
53. Shao R, Pang J, Clark NA, Rego JA, Walba DM (1993) Ferroelectrics 147: 255
54. Takatoh K, Goodby JW, Pindak R, Patel JS (1994) Proc 20th Jpn Liq Cryst Conf, Nagoya, 1G-304
55. Lamb A, Goodby JW unpublished results
56. Wu SL, Hsieh WJ (1996) 21: 783
57. Vill V, Tunger HW, Peters D (1996) Liq Cryst 20: 547
58. Faye V, Barois P, Nguyen HT, Laux V, Isaert N (1996) New J Chem 20: 283
59. Bolton EC, Smith PJ, Lacey D, Goodby JW (1992) Liq Cryst 12: 305
60. Kozlovsky M, Demikhov E (1996) Mol Cryst Liq Cryst 282: 11
61. see for example Bouligand Y (1971) J Microscopie 11: 441; Livolant F, Bouligand Y (1986) J Phys France 47: 1813
62. See for example Bouligand Y (1978) In: Blumstein A (ed) Liquid Crystalline Order in Polymers, Academic Press, New York, Ch 8
63. Cladis PE, Slaney AJ, Goodby JW, Brand HR (1994) Phys Rev Lett 72: 226
64. Pramod PA, Pratibha R, Madhusudana NV (1997) Current Science 73: 761
65. Ribeiro AC, Dreyer A, Oswald L, Nicoud JF, Soldera A, Gillon D, Galerne (1994) J Phys II France 4: 407
66. Kamien RD, Lubensky TC (1997) J Phys II France 7: 157

Liquid Crystal Dimers

Corrie T. Imrie

Department of Chemistry, University of Aberdeen, Meston Walk, Old Aberdeen,
AB24 3UE Scotland
E-mail: c.t.imrie@abdn.ac.uk

A liquid crystal dimer is composed of molecules containing two conventional mesogenic groups linked via a flexible spacer. These materials show quite different behaviour to conventional low molar mass liquid crystals and in particular their transitional behaviour exhibits a dramatic dependence on the length and parity of the flexible spacer. In this review a comprehensive overview of the relationships between molecular structure and liquid crystallinity in dimers is provided. This includes a description of the novel modulated and intercalated smectic phases exhibited by dimers.

Keywords: Liquid crystal dimers, Structure-property relationships, Modulated smectic phase, Intercalated smectic phase

1
Introduction

The vast majority of low molar mass liquid crystals are composed of molecules consisting of a single semi-rigid anisometric or mesogenic core attached to which are one or two terminal alkyl chains. In essence, it is the anisotropic interactions between the cores, normally consisting of phenyl rings linked through short unsaturated linkages, that give rise to the observation of liquid crystalline behaviour while the alkyl chains tend to reduce the melting point. Indeed for many years it was widely assumed that such a molecular structure was a prerequisite to the observation of liquid crystallinity [1]. It came as a great surprise, therefore, that during the 1980s a wide range of non-conventional molecular structures were shown to support liquid crystallinity [2].

Of all these new low molar mass liquid crystals discovered during the 1980s, one class that attracted particular attention and which still remains the focus of much research, are the so-called liquid crystal dimers [3]. A liquid crystal dimer is composed of molecules containing two conventional mesogenic groups linked by a flexible spacer, most commonly an alkyl chain. Thus, liquid crystal dimers contravened the accepted structure-property relationships for low molar mass mesogens by consisting of molecules having a highly flexible core rather than a semi-rigid central unit. In this respect liquid crystal dimers represent an inversion of the conventional molecular design for low molar mass mesogens. Several names have been used to refer to these materials including dimesogens or Siamese twins but these have all now been superseded by the preferred term liquid crystal dimer.

Liquid crystal dimers are, however, by no means a new class of materials, having been discovered some 70 years ago by Vorländer [4]. This first series of liquid crystal dimers were the α,ω-bis(4-alkyloxyphenyl-4'-azophenyl)alkane dioates,

$$C_mH_{2m+1}O\text{—}\langle\text{O}\rangle\text{—}N\text{=}N\text{—}\langle\text{O}\rangle\text{—}O.OC(CH_2)_nCO.O\text{—}\langle\text{O}\rangle\text{—}N\text{=}N\text{—}\langle\text{O}\rangle\text{—}OC_mH_{2m+1}$$

m = 1, 2; n = 1-8

1

The importance of this discovery was apparently overlooked and liquid crystal dimers did not re-emerge until the 1970s and their rediscovery by Rault et al. [5]. Again this report made little impact. Indeed, it was not until Griffin and Britt [6] described the transitional properties of a series of diesters,

$$C_mH_{2m+1}O\text{—}\langle\text{O}\rangle\text{—}\overset{O}{\overset{\|}{C}}\text{-}O\text{—}\langle\text{O}\rangle\text{—}O(CH_2)_{10}O\text{—}\langle\text{O}\rangle\text{—}O\text{-}\overset{O}{\overset{\|}{C}}\text{—}\langle\text{O}\rangle\text{—}OC_mH_{2m+1}$$

m = 2-10

2

that liquid crystal dimers became the focus of much research activity. At the root of this interest, at least initially, was the proposal made by Griffin and Britt that liquid crystal dimers could serve as model compounds for the technologically important semi-flexible main chain liquid crystal polymers.

A semi-flexible main chain liquid crystal polymer is composed of mesogenic units separated by flexible spacers, normally alkyl chains [7]. These polymers are not only of interest for their application potential [8]; they are also of major fundamental interest because of their unusual liquid crystalline properties. It is well known, for example, that the transitional behaviour of a semi-flexible main chain liquid crystal polymer shows a dramatic dependence on the length and parity of the flexible spacer linking the mesogenic units [9]. Other fascinating behaviour includes the observation of a nematic-nematic transition [10] and the occurrence of alternating smectic phases [11–15].

These polymers, therefore, provide a demanding challenge to our understanding of self-assembly in condensed phases and to begin to interpret their behaviour at a molecular level requires both experimental investigations and the development of molecular theories. The inherent structural heterogeneity of a polymeric system, however, greatly complicates these tasks. An alternative approach in developing a molecular understanding of polymers involves the use of monodisperse low molar mass compounds whose behaviour encapsulates the essential physics of the polymeric system. Griffin and Britt [6] argued that for semi-flexible main chain polymers the fundamental repeat unit contains two mesogenic units linked via a spacer. This proved to be correct and the dimers do indeed serve as useful model compounds for the polymers [16]. In particular, and as we shall see later, the transitional behaviour of dimers exhibits a dramatic dependence on the length and parity of the flexible spacer in a manner strongly reminiscent to that observed for the polymeric systems. More recently it has been suggested that dimers may also be used to model the behaviour of side group liquid crystal polymers [17].

The initial interest in liquid crystal dimers was triggered, therefore, by the similarity of their behaviour to that of the semi-flexible main chain liquid crystal polymers. It soon became apparent, however, that the dimers are of significant fundamental interest in their own right and exhibit quite different behaviour to conventional low molar mass liquid crystals. These studies, for example, have resulted in the discovery of a new family of intercalated smectic phases. This review focuses upon the novel behaviour of dimers and how it may be understood at a molecular level.

2
Classification of Liquid Crystal Dimers

The various structural possibilities for dimers are shown schematically in Fig. 1 and these can be sub-divided into two broad groups: symmetric dimers in which the two mesogenic units are identical and non-symmetric dimers which contain two different mesogenic moieties. These two groups can be further sub-divided according to the molecular geometry of the mesogenic groups. For example, in a symmetric calamitic liquid crystal dimer the two

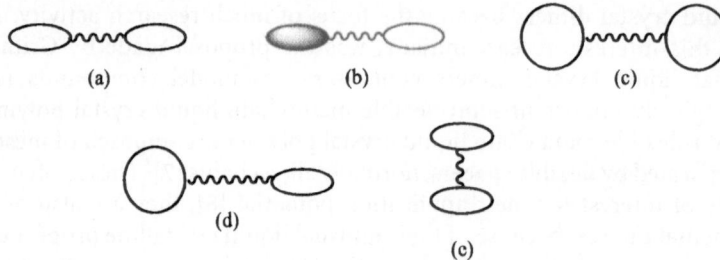

Fig. 1a–e. Sketches of the possible molecular architectures for liquid crystal dimers: **a** symmetric calamitic dimer; **b** non-symmetric calamitic dimer; **c** symmetric discotic dimer; **d** non-symmetric discotic-calamitic dimer; **e** laterally linked symmetric dimer. *Ellipses* represent rod-like mesogenic groups, *circles* disc-like mesogenic units and *wavy lines* denote the flexible spacers

mesogenic units are identical and rod-like in shape. In the overwhelming majority of dimers reported in the literature the mesogenic units are linked in terminal positions although a small number of laterally linked dimers have been reported in which the mesogenic units are connected via lateral positions. A number of chiral liquid crystal dimers have also been reported in which the chiral centre is located either in a terminal chain or in the spacer.

3
Structure Property Relationships in Liquid Crystal Dimers

3.1
Symmetric Calamitic Liquid Crystal Dimers

3.1.1
The Dependence of the Transitional Properties on the Length and Parity of the Flexible Spacer

The most extensively studied series of liquid crystal dimers are the α,ω-bis(4′-cyanobiphenyl-4-yloxy)alkanes[18–28],

$$NC-\!\langle O \rangle\!-\!\langle O \rangle\!-O(CH_2)_nO-\!\langle O \rangle\!-\!\langle O \rangle\!-CN \qquad 3$$

$$n = 1\text{-}12$$

and these shall be used to illustrate the characteristic behaviour of nematogenic dimers. The acronym used to refer to this series is BCBOn in which n refers to the number of methylene groups in the flexible spacer. The particular interest in the BCBOn series arises, in part, because they may be considered to be the dimeric analogues of the 4-*n*-alkyloxy-4′-cyanobiphenyls,

$$NC-\!\langle O \rangle\!-\!\langle O \rangle\!-OC_nH_{2n+1} \qquad 4$$

which are amongst the most widely studied series of conventional low molar mass liquid crystals. The acronym used to refer to this series is nOCB in which n refers to the number of carbon atoms in the terminal chain.

The first twelve members of the BCBOn series have been characterised [18] and only nematic behaviour is observed. Figure 2 shows the dependence of the melting temperatures, T_m, and the nematic-isotropic transition temperatures, T_{NI}, on the number of methylene groups, n, in the flexible spacer for the BCBOn series. It can be seen that T_{NI} depends critically on the length and parity of the spacer. Specifically, a very large odd-even effect is apparent in which the even members of the series exhibit the higher values of T_{NI} although this alternation is attenuated on increasing n [18]. An odd-even effect is also often observed for conventional nematogens containing a single semi-rigid core and terminal chains, but to highlight the difference in the magnitude of these effects, Fig. 2 also shows the clearing temperatures for the analogous conventional or monomeric low molar mass series, the 4-*n*-alkyloxy-4'-cyanobiphenyls. The clearing temperatures of the monomeric compounds are considerably lower than those of the dimers and their dependence on the number of carbon atoms in the terminal chain is very much weaker than that seen on varying the length and parity of the spacer in the dimers.

The melting points of the BCBOn series also exhibit a pronounced odd-even effect as n is varied (see Fig. 2). This behaviour is not observed for conventional nematogenic materials for which the dependence of T_m on the number of methylene groups in the terminal chains is quite haphazard. It was suggested that for nematic dimers this may indicate that the change in the

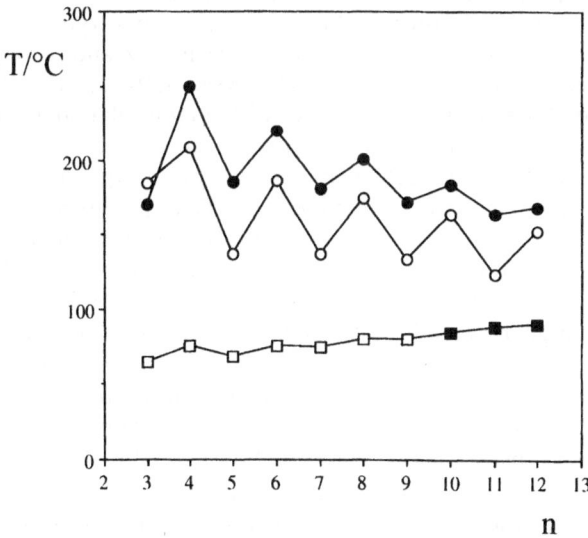

Fig. 2. The dependence of the melting points (○) and the nematic-isotropic transition temperatures (●) on the number of methylene units in the flexible spacer, n, for the BCBOn series [18]. Also shown are the nematic-isotropic transition temperatures (□) and the smectic A-isotropic transition temperatures (■) for the nOCB series

conformation statistical weights of the spacer on melting is small for even membered spacers but large for odd membered compounds [18]. An alternative explanation, however, considers enthalpic effects. Thus, at the root of this odd-even effect is possibly the difficulty that the odd membered compounds, with their bent conformations, experience in packing efficiently into a crystalline structure as compared with the more elongated even membered dimers; the molecular shapes of dimers is discussed in detail later. As we will see, however, this regular dependence of the melting point on the length of the spacer does not extend to smectogenic series for which the melting point behaves in an irregular manner.

The entropy changes associated with the nematic-isotropic transition, $\Delta S_{NI}/R$, for the BCBOn series also show a pronounced alternation as the length and parity of the spacer is varied (see Fig. 3) [18]; again, it is the even members which exhibit the higher values. Indeed for the early members of the series $\Delta S_{NI}/R$ is almost four times larger for even members as compared with those for the odd members. Unlike the alternation seen for T_{NI} in Fig. 2, the odd-even effect in $\Delta S_{NI}/R$ does not appear to attenuate on increasing the length of the spacer. Although it may be argued that it is attenuated in a relative sense since the values of $\Delta S_{NI}/R$ increase with increasing n but the difference between the n and (n+1) homologues appears approximately constant. For comparative purposes, the clearing entropies for the monomeric nOCB series are also shown in Fig. 3 [17]. The propyl to nonyl members of this series exhibit a nematic-isotropic transition and the associated entropy change is considerably smaller than that observed for either the odd or even membered dimers.

The pronounced alternations in both the nematic-isotropic temperature, T_{NI}, and the associated entropy change, $\Delta S_{NI}/R$, on varying the length and parity of the spacer in the BCBOn series (see Figs. 2 and 3, respectively) is archetypal behaviour for liquid crystal dimers [29–36]. This behaviour is most often attributed to the dependence of the molecular shape on the parity of the spacer considered in the all-*trans* conformation. Specifically, in an even membered dimer the mesogenic groups are antiparallel whereas in an odd membered dimer they are inclined (see Fig. 4). This structure for even membered dimers is then considered to be more compatible with the molecular organisation found in the nematic phase than for the odd membered dimers and it is this greater compatibility which results in, for example, the higher nematic-isotropic entropies found for the even membered dimers. Such an argument neglects the flexibility of the spacer and a more realistic interpretation of the dependence of the transitional properties on the parity of the spacer certainly includes a wide range of conformations and not solely the all-*trans* conformation [3]. In the isotropic phase approximately half the conformers of an even membered dimer are essentially linear whereas for an odd membered dimer just 10% are linear. There exists a synergy between conformational and orientational order and hence at the transition to the nematic phase for even membered dimers many of the bent conformers are converted to a linear form. This enhances the orientational order of the nematic phase, resulting in a larger nematic-

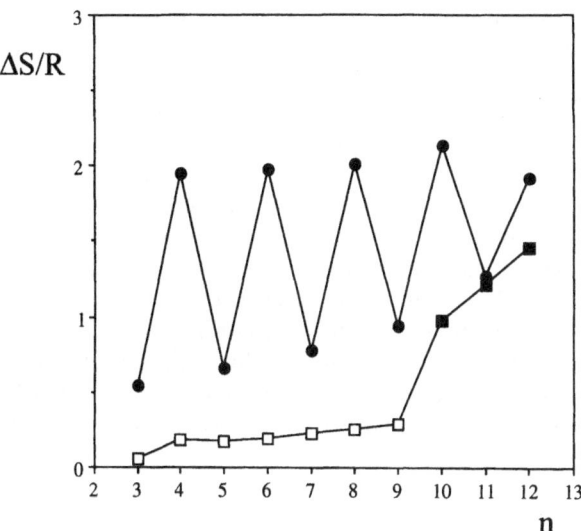

Fig. 3. The dependence of the entropy change associated with the nematic-isotropic transition (●) on the number of methylene units in the flexible spacer, n, for the BCBOn series [18]. Also shown are the entropy changes associated with the nematic-isotropic (□) and the smectic A-isotropic transitions (■) for the nOCB series [17]

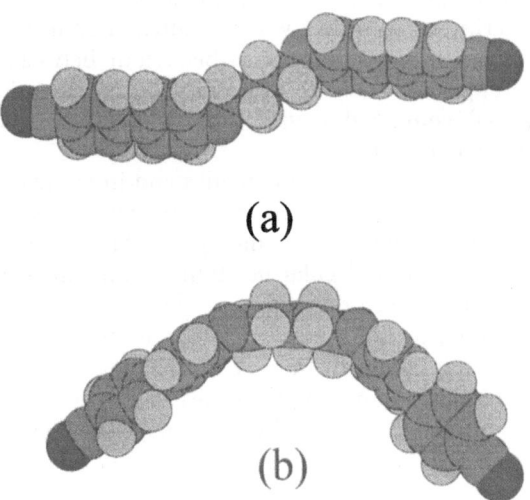

Fig. 4a,b. The molecular shape of: **a** an even; **b** an odd membered liquid crystal dimer with the spacer in the all-*trans* conformation

isotropic entropy than would be expected for a monomer. For odd membered dimers, however, the difference in free energy between the bent and linear conformers is such that the orientational order of the nematic phase is insufficient to convert bent conformers into linear conformers. Hence, the orientational order is not enhanced and a smaller nematic-isotropic entropy would be expected. Models have been developed based on a molecular field theory to describe this synergy between conformational and orientational order which are able to predict most of the properties of nematic liquid crystal dimers [3].

Thus on varying the length and parity of the flexible spacer in dimers the nematic-isotropic transition temperature exhibits a dramatic odd-even effect which attenuates as the length of the spacer is increased while the nematic-isotropic entropy also exhibits the same pronounced alternation but which appears not to attenuate as the spacer length is increased. Such behaviour is also observed for semi-flexible main chain liquid crystal polymers; for example, Figs. 5 and 6 show the dependence of T_{NI} and $\Delta S_{NI}/R$, respectively, on the length of the flexible spacer for the poly{α,ω-[4,4'-(2,2'-dimethylazoxy-phenyl)]alkandioates} [9],

5

As for the dimers, both T_{NI} and $\Delta S_{NI}/R$ exhibit a dramatic dependence on the length and parity of the spacer which is attenuated in the case of T_{NI} but not for $\Delta S_{NI}/R$. This strong similarity in behaviour between the dimers and polymers would appear to indicate that much of the essential physics of the polymers is indeed contained within that of the dimers and hence their suitability as model compounds.

The BCBOn series are exclusively nematics and in many cases it was found that the smectic tendencies of dimers are very much less than those of the analogous conventional monomeric mesogens. This observation has been attributed to the increased molecular flexibility of the dimer [6]. It was clearly of considerable importance, therefore, to determine the influence of the flexible core on smectic behaviour in order to test this suggestion. The definitive study of this issue by Date et al. [37] involved the synthesis and characterisation of eleven homologous series of dimers all belonging to the general family of compounds, the α,ω-bis[4-(4-alkylphenyliminometh-yl)phenoxy]alkanes,

n = 2-12; m = 0-10

6

The acronym used to refer to this family is m.OnO.m in which n refers to the number of methylene groups in the flexible spacer and m that in the

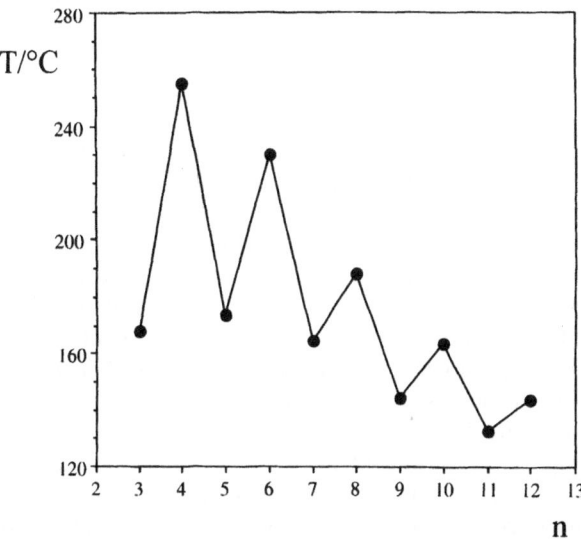

Fig. 5. The dependence of the nematic-isotropic transition temperature, T_{NI}, on the length of the flexible spacer, n, for the poly{α,ω-[4,4'-(2,2'-dimethylazoxyphenyl)]alkandioates} [9]

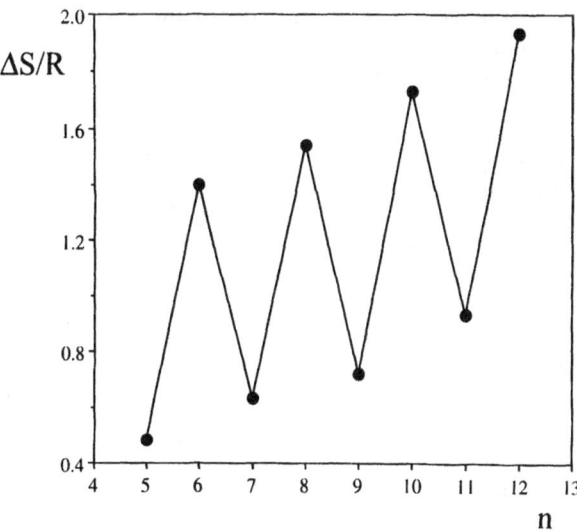

Fig. 6. The dependence of the entropy change associated with the nematic-isotropic transition, $\Delta S_{NI}/R$, on the length of the flexible spacer, n, for the poly{α,ω-[4,4'-(2,2'-dimethylazoxyphenyl)]alkandioates} [9]

terminal alkyl chains. This particular series was chosen for study, at least in part, because they may be considered to be the dimeric analogues of the N-(4-n-alkyloxybenzylidine)-4'-n-alkylanilines,

$$C_nH_{2n+1}O-\!\!\bigcirc\!\!-CH\!\!=\!\!N-\!\!\bigcirc\!\!-C_mH_{2m+1} \qquad 7$$

which are probably the most widely studied series of conventional smectogenic low molar mass liquid crystals. The expectation that the m.OnO.m series would exhibit extensive smectic polymorphism proved correct [37]; members of this family of compounds exhibit smectic A, B, C and F phases and the crystal B, E, G and H phases as well as novel modulated smectic phases (see Sect. 4.3). Rare phase transitions have also been observed, including SmF-SmA and crystal G-isotropic transitions. In addition, this study allowed surprising correlations between structure and phase behaviour to be discovered [37].

Figure 7 shows the dependence of the transition temperatures on the length of the flexible spacer, n, for the 4.OnO.4 series which exhibits both smectic and nematic behaviour [37]. The characteristic, pronounced alternation in the clearing temperatures on varying n can be seen in Fig. 7. The early members of this series (n = 2, 3, 4) exhibit exclusively smectic behaviour but on increasing the spacer length nematic behaviour emerges for n = 5 and 6 while for longer spacers only nematic behaviour is seen. Thus, on increasing the length of the spacer, the tendency is for nematic behaviour to emerge and for smectic phases to be extinguished. This is surprising given the very general observation that increasing the length of alkyl chains in mesogenic structures stabilises smectic phases relative to the nematic phase. Examples of this

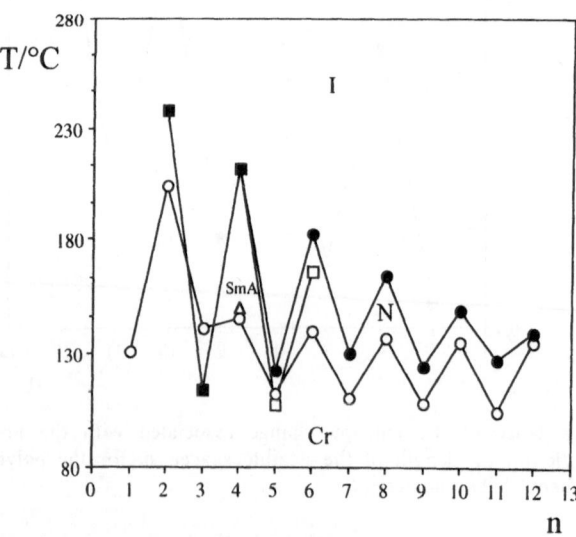

Fig. 7. The dependence of the transition temperatures on the number of methylene groups, n, in the flexible spacer for the 4.OnO.4 series [37]. ■ indicates smectic A-isotropic transitions, ● the nematic-isotropic transitions, ○ the melting points, □ smectic A-nematic transitions and △ smectic B-smectic A transition. *Cr* Crystal; *SmA* smectic A; *N* nematic; *I* isotropic

include increasing the length of the terminal alkyl chain in a conventional low molar mass compound [1] and increasing the spacer length in a side-chain liquid crystal polymer [38]. Furthermore, in this respect the dimers behave quite differently to semi-flexible main chain liquid crystal polymers for which increasing the spacer length promotes smectic behaviour [39].

On examining the wealth of data provided by the m.OnO.m compounds, a simple empirical rule emerged relating the molecular structure to the occurrence of smectic behaviour; specifically, if a symmetric dimer is to exhibit smectic properties then the terminal chain length must be greater than half the spacer length. A molecular interpretation of this empirical rule is discussed in Sect. 4.2.

3.1.2
The Dependence of the Transitional Properties
on the Chemical Nature of the Spacer

The vast majority of liquid crystal dimers contain flexible alkyl spacers. There are, however, a limited number of examples containing chemically different spacers such as oligoethylene oxide [40–44] or siloxane-containing [40, 45–51] chains. For example, Creed et al. [40] compared the transitional behaviour of two sets of dimers in which the chemical nature of the spacer was varied:

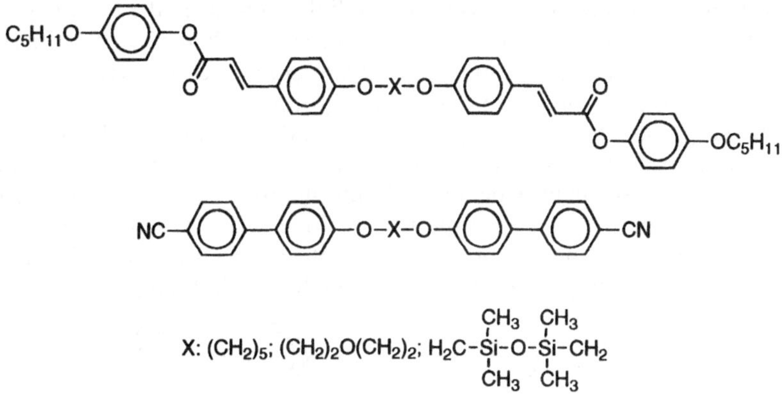

8

For both sets of compounds the nematic-isotropic transition temperature was highest for the pentamethylene chain and lowest for the disiloxane spacer, reflecting the increasing flexibility of the spacer. A surprising feature of dimers containing oligoethylene oxide spacers is that compounds having very long spacers, containing as many as 34 ethylene oxide units, still exhibit liquid crystallinity [42] although this reflects, in part, the low melting points of these compounds.

Jin and co-workers [46, 47] have noted the strong tendency for disiloxane-linked dimers to exhibit smectic rather than nematic behaviour. This is in accord with the behaviour of dimers containing oligosiloxane spacers; for example Aquilera and Bernal [48] have reported the behaviour of

$$H_{2n+1}C_nO-\bigcirc-OOC-\bigcirc-O(CH_2)_3-\overset{\overset{CH_3}{|}}{(SiO)_x}-\overset{\overset{CH_3}{|}}{Si}(CH_2)_3O-\bigcirc-COO-\bigcirc-OC_nH_{2n+1}$$

x = 1, 2, 3, 4; n = 5, 6

9

and with the exception of just one compound (n=6, x=1) these compounds are exclusively smectic. These dimers have clearing temperatures in the proximity of room temperature and considerably lower than expected for their polymethylene analogues. The authors attribute these low clearing temperatures to a combination of enhanced molecular flexibility and the increased steric bulk of the spacer.

Not only does the chemical nature of the spacer effect the transitional behaviour of dimers but so too does the type of link between the spacer and the mesogenic groups. For example, Jin and co-workers [32, 52] showed that exchanging an ether linkage for an ester group in the dimers

$$\bigcirc-\bigcirc-CO.O-\bigcirc-X(CH_2)_nX-\bigcirc-O.OC-\bigcirc-\bigcirc \quad \textbf{10}$$

tends to promote smectic behaviour and similar observations have been made in low molar mass systems on varying the link connecting the terminal chain to the mesogenic core [1]. This preference for smectic behaviour for the ester-linked materials is attributed to enhanced lateral interactions arising from the contribution of the dipole on the carbonyl group.

More surprising, however, is the observation that the magnitude of the odd-even effects exhibited by both the nematic-isotropic transition temperature and entropy are strongly dependent on the nature of this link. This is best illustrated by comparing the properties of three series of dimers containing cyanobiphenyl as the mesogenic group and attached to the spacer via an ether link [18], a carbonate group [16, 53] or a methylene link [16, 54]:

$$NC-\bigcirc-\bigcirc-X(CH_2)_nX-\bigcirc-\bigcirc-CN$$

11

X = -; O; OCO.O

The dependence of the nematic-isotropic transition temperatures and entropies on the length of the flexible spacer for these three series are shown in Figs. 8 and 9, respectively. In order to make meaningful comparisons between these series, in Figs. 8 and 9 the length of the spacer is taken to be the total number of atoms connecting the two mesogenic groups. The nematic-isotropic transition temperatures are higher for the ether than for the methylene linked dimers and this difference is considerably more pronounced for the odd members of the series [16, 54]. This in turns gives rise to an enhancement in the magnitude of the odd-even effect exhibited by the nematic-isotropic

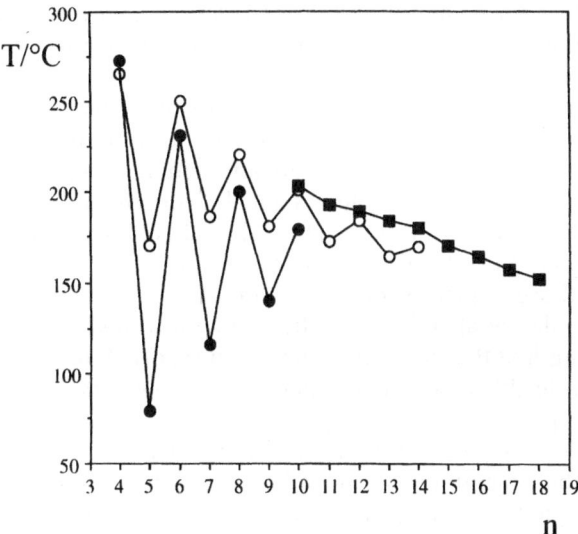

Fig. 8. The dependence of the nematic-isotropic transition temperature, T_{NI}, on the total number of atoms, n, in the spacer for the cyanobiphenyl dimers linked via ether (O) [18], alkyl (●) [16] and carbonate groups (■) [16, 53]

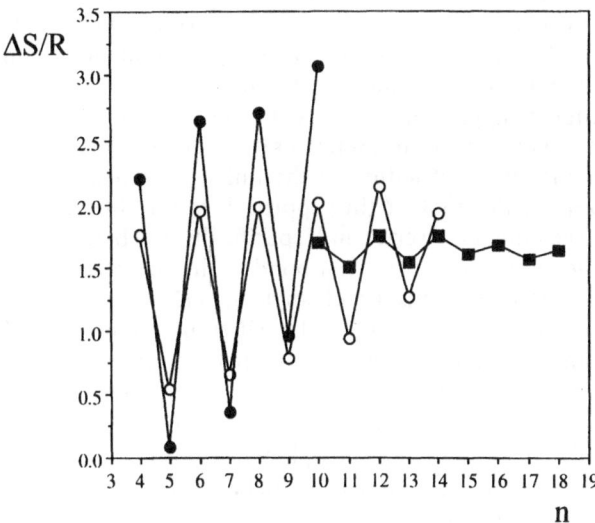

Fig. 9. The dependence of the entropy change associated with the nematic-isotropic transition on the total number of atoms, n, in the spacer for the cyanobiphenyl dimers linked via ether (O) [18], alkyl (●) [16] and carbonate groups (■) [16, 53]

transition temperatures for the methylene linked series (see Fig. 8). Similarly, the odd-even effect seen for the nematic-isotropic entropies is also more pronounced for the methylene than the ether linked series (see Fig. 9);

specifically the even members of the methylene linked series exhibit higher values of $\Delta S/R$ than their ether linked counterparts while this trend is reversed for the odd members.

Quite different behaviour is observed if we now compare the ether and carbonate linked series. The nematic-isotropic transition temperatures for the carbonates decrease essentially without alternation as the spacer length is increased while the entropies show a much reduced odd-even effect when compared to the ether linked series [16, 53]. It should be noted, however, that the nematic-isotropic entropies for the carbonate series are still several times larger than those typically observed for conventional low molar mass mesogens. Similar weak odd-even effects in the transitional properties on varying the length of the spacer have also been reported for other dimer series [55] and semi-flexible main chain liquid crystal polymers [56, 57] containing carbonate groups.

Thus the nature of the link between the spacer and the mesogenic groups can control the magnitude of the odd-even effect exhibited by both the nematic-isotropic transition temperature and entropy. The differences in the nematic-isotropic transition temperatures between these series could, to some extent, be interpreted in terms of the change in the polarisability anisotropy of the mesogenic groups arising form the change in the linking group. For example, the polarisability anisotropy of the mesogens would be expected to increase on passing from the methylene to the ether linked dimers due to the conjugation of the oxygen with the phenyl ring. Such explanations are commonly invoked to account for the relative transition temperatures of conventional low molar mass mesogens [1].

A quite different approach, however, has been adopted by Luckhurst and his co-workers in an attempt to understand the dependence of the transitional properties on the chemical nature of the linkage between the mesogens and spacer [58, 59]. At the root of this approach is the difference in molecular geometry between the three series and, specifically, the bond angle between the para axis of the mesogen and the first bond in the spacer; for example, for the methylene linked dimers this bond angle is 113.5° while for ether linked dimers this angle is 126.4°. Thus, the all-*trans* conformation of an odd membered dimer is more elongated for an ether than a methylene linked dimer and this greater anisotropy should give rise to higher nematic-isotropic transition temperatures. To test this suggestion Luckhurst and co-workers [58, 59] have calculated the nematic-isotropic temperatures for the methylene and ether linked dimers using a molecular field theory [60] in which the only difference between the calculations for the two series was this bond angle. The predictions of this theory were in good agreement with the data shown in Fig. 8 and so it does indeed appear that geometrical factors alone can account for the differences in the nematic-isotropic transition temperatures of these dimers. This approach also predicts the enhancement in the alternation in the nematic-isotropic entropies observed for the methylene linked dimers. Abe et al. [53] have interpreted the unusual behaviour of the carbonate linked materials in terms of the geometry of the linking group and how this affects both the relative orientations of the mesogenic groups and hence, also the conformational distribution found in the nematic phase.

An intriguing possibility arises from the model developed by Luckhurst et al. [61, 62] in which the dimers can adopt just two conformations, one linear and one bent, namely that systems containing high concentrations of bent conformers in the isotropic phase, i.e. dimers containing short odd membered spacers, are predicted to exhibit a nematic-nematic transition. The nematic-isotropic entropy for the member of the methylene linked series containing a pentane spacer is very small (see Fig. 9), suggesting that this transition is approaching second order in nature [54]. Theory suggests that following such a second order transition a biaxial nematic phase should be observed. Experimentally, however, this compound exhibits a smectic phase which would obscure the nematic-nematic transition and so this possibility has still to be realised.

3.1.3
The Dependence of the Transitional Properties on the Molecular Structure of the Mesogenic Group

A wide range of mesogenic groups have been incorporated into dimeric structures and the dependence of the transitional properties on the structure of the mesogenic group parallels that observed for conventional low molar mass mesogens [1]. For example, increasing the length of the core by inserting short unsaturated linkages between phenyl rings enhances the clearing temperature [63] while laterally substituting the core causes the clearing temperature to fall [64]. These investigations have also included examples of metal containing dimers [65].

3.1.4
The Dependence of the Transitional Properties on the Terminal Groups

The most common terminal substituents in dimers are alkyl chains and we return to the m.OnO.m series to illustrate the effects on the transition temperatures and phase behaviour of increasing the length of the terminal chains. Figure 10 shows the dependence of the transition temperatures on m for the m.O4O.m series [37]. The nematic-isotropic transition temperature increases markedly on going from 0.O4O.0 to 1.O4O.1, possibly resulting from the change in shape of the mesogenic unit. Subsequent increases in the length of the terminal chains cause the clearing temperatures to fall, showing initially an alternation which attenuates with increasing m. This behaviour is typical of a series of conventional mesogens having high clearing points [66]. As m increases the smectic phase stability increases as would be anticipated and this will be discussed in Sect. 4.2. Thus, the effects on the transition temperatures and phase behaviour of increasing the length of terminal chains in symmetric calamitic liquid crystal dimers are in accord with those observed for conventional low molar mass mesogens.

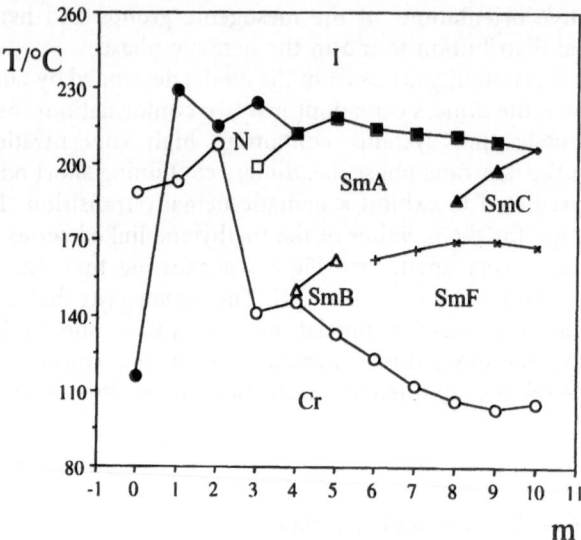

Fig. 10. The dependence of the transition temperatures on the number of carbon atoms, m, in the terminal chains for the m.O4O.m series [37]. Melting points are denoted by O, ● indicates nematic-isotropic transitions, ■ smectic A-isotropic transitions, ◊ the smectic C-isotropic transition, □ the smectic A-nematic transition, Δ smectic B-smectic A transitions, + smectic F-smectic A transitions, × smectic F-smectic C transitions and ▲ smectic C-smectic A transitions. *Cr* Crystal; *SmA* smectic A; *N* nematic; *SmB* smectic B; *I* isotropic; *SmF* smectic F; *SmC* smectic C

Similarly, the effects of a range of terminal substituents on the clearing temperatures of dimers also mirror those observed for conventional low molar mass mesogens [29–32, 34, 46, 67]. For example, Jin et al. [29] established the efficiency of the terminal group in enhancing nematic behaviour for the dimer series

$$X-\!\!\bigcirc\!\!-O.OC-\!\!\bigcirc\!\!-O(CH_2)_{10}O-\!\!\bigcirc\!\!-CO.O-\!\!\bigcirc\!\!-X$$

to be

$$H < CH_3 < Cl < NO_2 \approx CHO < CN < C_6H_5 \qquad\qquad \textbf{12}$$

and this is in good agreement with that found for conventional low molar mass nematogens [1].

3.2
Non-Symmetric Calamitic Liquid Crystal Dimers

As we have seen, non-symmetric dimers consist of molecules containing two differing mesogenic units linked via a flexible spacer (see Fig. 1). In the majority of cases the differing mesogenic groups have been chosen such that they exhibit a specific molecular interaction [68–75]. This normally involves connecting an electron deficient moiety such as cyanobiphenyl via a flexible spacer to an electron rich group such as pyrene [71]. This approach was stimulated, in part, by the behaviour of binary mixtures of conventional low molar mass liquid crystals in which one component consisted of molecules containing electron rich mesogenic units while the other consisted of molecules having electron deficient moieties. For many years it has been known that such mixtures exhibit clearing temperatures higher than the composition weighted average of those of the individual components and show a greater tendency to form smectic phases [76–83]; this includes mixtures of liquid crystal dimers and semi-flexible main chain liquid crystal polymers [84]. Both these observations are attributed to a specific interaction between the unlike cores being more favourable than the geometric mean of the interactions between the like cores. This specific interaction has also been used to induce liquid crystallinity into mixtures of mesogenic and non-mesogenic components [85–89] and to manipulate the phase behaviour of side-group liquid crystal copolymers and blends of homopolymers [90–100]. The smectic phases exhibited by all these various types of mixtures and polymeric systems have conventional structures but the intriguing question was to what extent would the behaviour of such mixtures and the structure of the phases they exhibit be affected if the unlike mesogenic units were covalently linked in the same molecule? Thus many series of non-symmetric dimers have now been characterised and, as we shall see, this resulted in the discovery of a novel family of smectic phases [68]. Most recently, non-symmetric dimers have been reported in which one of the two mesogenic units is assembled via hydrogen bonding [101]

The most extensively characterised examples of non-symmetric dimers are the α-(4-cyanobiphenyl-4'-yloxy)-ω-(4-n-alkylanilinebenzylidene-4'-oxy)al-kanes,

$$NC-\bigcirc-\bigcirc-O(CH_2)_nO-\bigcirc-CH=N-\bigcirc-C_mH_{2m+1} \qquad 13$$

and the acronym used to refer to this family is CBOnO.m where n refers to the number of methylene groups in the spacer and m the length of the terminal alkyl chain [68, 69]. This particular family of materials was chosen, in part, because, as we have seen, the properties of the parent symmetric dimers, ie the BCBOn and m.OnO.m series, are known [18, 37].

Figure 11 shows the dependence of the transition temperatures on the number of methylene units in the spacer, n, for the CBOnO.10 series [69]. Immediately apparent is that the clearing temperatures show a dramatic alternation which attenuates quite rapidly on increasing n. A dramatic alternation is also exhibited by the nematic-isotropic entropies but which is not attenuated on increasing n (see Fig. 12). As we noted earlier this dependence of the nematic-isotropic transition temperatures and associated entropy changes on varying the length and parity of the flexible spacer is

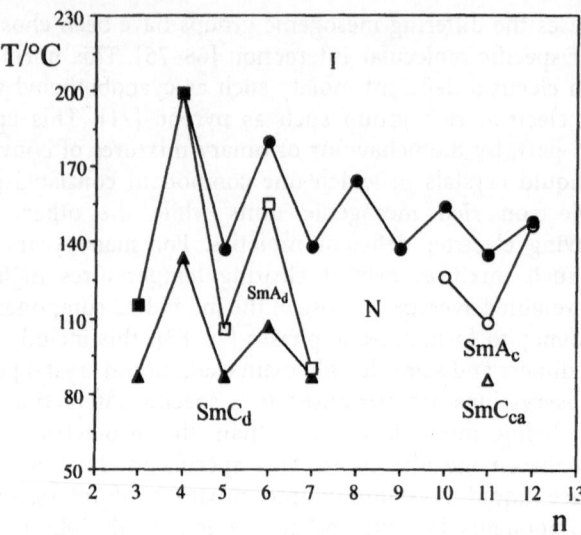

Fig. 11. The dependence of the transition temperatures on the number of methylene groups, n, in the flexible alkyl spacer for the CBOnO.10 series.[69] ■ indicates interdigitated smectic A-isotropic transitions, ● nematic-isotropic transitions, ▲ interdigitated smectic A-interdigitated smectic C transitions, □ interdigitated smectic A-nematic transitions, ○ intercalated smectic A-nematic transitions and △ intercalated smectic A-intercalated smectic C transitions. The melting points have been omitted for the sake of clarity. *SmA_d* Interdigitated smectic A phase; *SmC_d* interdigitated smectic C phase; *SmA_c* intercalated smectic A phase; *SmC_{ca}* intercalated alternating smectic C phase; *N* nematic; *I* isotropic

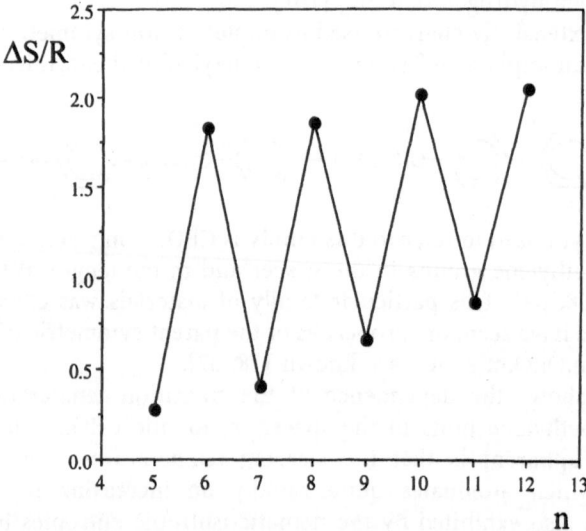

Fig. 12. The dependence of the entropy change associated with the nematic-isotropic transition on the number of methylene units in the flexible spacer, n, for the CBOnO.10 series [69]

characteristic behaviour for liquid crystal dimers. In contrast to such expected behaviour, however, the thermal stability of the smectic A phase is very unusual. CBO3O.10 and CBO4O.10 exhibit smectic A-isotropic transitions while the fifth, sixth and seventh homologues show a smectic A-nematic transition. The thermal stability of the smectic A phase, either T_{SmAN} or T_{SmAI}, exhibits a very pronounced odd-even effect and the underlying trend in the temperatures over these five homologues is a decreasing one. Indeed, the stability of the smectic phase decreases to such an extent that the eighth and ninth members are solely nematogenic. Surprisingly, smectic properties re-emerge subsequently with the decyl homologue and there is an increasing trend in the thermal stability of the phase over the final three homologues. For symmetric dimers we have seen that smectic tendencies simply decrease with increasing the length of the spacer [37] and we will return to a detailed discussion of this most unusual behaviour shown by non-symmetric dimers in Sect. 4.4.

We have noted that interest in these non-symmetric dimers was in part stimulated by mixtures of conventional low molar mass mesogens containing unlike cores and that such mixtures tend to exhibit enhanced clearing temperatures. To see if such behaviour was also observed for the non-symmetric dimers we must compare the clearing temperatures of the non-symmetric dimers with the average clearing temperature of the analogous parent symmetric dimers, ie the m.OnO.m and BCBOn series. It should be stressed that in some instances the nature of the clearing transition varies between the non-symmetric and symmetric dimers and, hence, unlike transitions are being compared. For the majority of liquid crystal series, however, the smectic-isotropic transition temperatures can be obtained from an extrapolation of the nematic-isotropic transition temperatures and this justifies the comparisons of the clearing temperatures. In order to investigate the relative magnitudes of the difference in the clearing temperature of the non-symmetric dimer, T_{AB}, from the mean of those for the symmetric dimers (T_A and T_B), a scaled deviation, ΔT_{SC} can be defined as [69]

$$\Delta T_{SC} = \frac{2T_{AB} - (T_A + T_B)}{(T_A + T_B)} \qquad (1)$$

Figure 13 shows the dependence of ΔT_{SC} on n for three CBOnO.m series containing either an ethyl, hexyl or decyl terminal chain. This reveals that the clearing temperature is not always greater than the average clearing temperature of the symmetric dimers and in fact depends on the length of both the spacer and terminal chain. The general trend is for the deviation in clearing temperature to increase as the ratio m/n decreases although the relative magnitudes of these deviations are typically very small.

To begin to understand such behaviour at least qualitatively we should first consider the case of an ideal binary mixture of nematogens for which $\Delta T_{SC} = 0$. Within the framework of a molecular field theory developed to predict phase diagrams for binary mixtures of nematogens, three intermolecular energy parameters, ε_{AA}, ε_{BB} and the mixed parameter ε_{AB}, must be defined

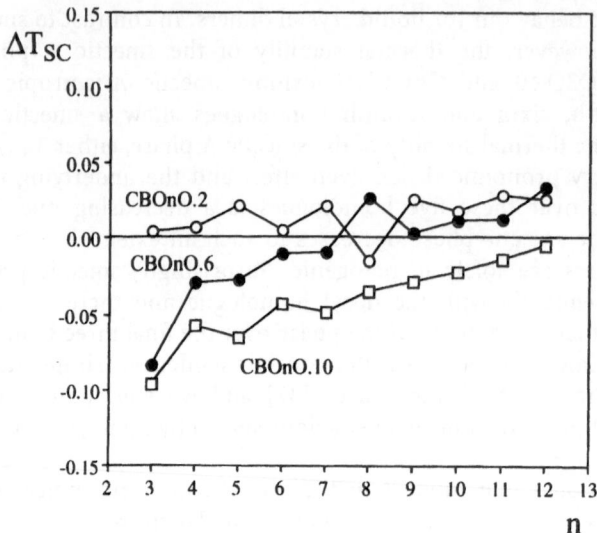

Fig. 13. The dependence of ΔT_{SC} on the length of the flexible alkyl spacer, n, for the CBOnO.m series [69]

[102]. The parameters ε_{AA} and ε_{BB} for the interactions between like species are proportional to the nematic-isotropic transition temperatures of the pure nematogens. If ε_{AB}, the interaction parameter for unlike species, is assumed to be the geometric mean of ε_{AA} and ε_{BB}, then the transition temperature of the mixture is simply the weighted average of those of the components, ie a linear dependence of T_{NI} on composition is predicted and $\Delta T_{SC} = 0$. Such behaviour is often found experimentally. If we allow for a positive deviation in ε_{AB} from the geometric mean approximation, then a curve lying above the straight line is observed, ie $\Delta T_{SC} > 0$, whereas a negative deviation in ε_{AB} results in a curve lying below the straight line, ie $\Delta T_{SC} < 0$. Furthermore, quite small deviations of ε_{AB}, for example ±1.85%, are readily discernible in the phase diagram as curved phase boundaries [102, 103].

The relatively small values of ΔT_{SC} exhibited by the non-symmetric dimers suggest, therefore, that the deviation from the geometric mean approximation is also small, provided of course the dimers approximate to the rigid, cylindrically symmetric particles assumed in the theory [102]. The deviation away from the geometric mean approximation implies a specific interaction between the unlike groups. The precise nature of this interaction is unclear but is normally assumed to be a charge transfer interaction. This assumption has been questioned and more recently it has been suggested that it is in fact an electrostatic quadrupolar interaction between groups with quadrupole moments which are opposite in sign [75]. ΔT_{SC} is most positive for the shortest terminal chain, suggesting that increasing terminal chain length tends to dilute the specific interaction between the unlike groups.

This interpretation of the deviation in clearing temperatures of non-symmetric dimers from the mean of those of the corresponding symmetric dimers should account for the behaviour of all non-symmetric dimers which exhibit a specific interaction between the two unlike mesogenic units. The reason for this is quite simple – the description of the phase behaviour did not refer to the type of mesogenic units in the non-symmetric dimer nor did it consider the nature of the mixed interaction. Similar behaviour was indeed found for the NABOnO.4 series [70],

$$O_2N-\langle O \rangle-N{=}N-\langle O \rangle-O(CH_2)_nO-\langle O \rangle-CH{=}N-\langle O \rangle-C_4H_9 \quad \textbf{14}$$

for which ΔT_{SC} is initially relatively large and negative, becoming more positive as n is increased. If the non-symmetric dimer does not contain a terminal chain then ΔT_{SC} would be predicted to be small and positive for all spacer lengths and such behaviour is seen for the α-(4-cyanobiphenyl-4'-yloxy)-ω-(1-pyreniminebenzylidene-4'-oxy)alkanes[71],

$$NC-\langle O \rangle-\langle O \rangle-O(CH_2)_nO-\langle O \rangle-CH{=}N- \quad \textbf{15}$$

whose clearing temperatures can be compared with those of the BCBOn series [18] and the α,ω-bis(1-pyreniminebenzylidene-4'-oxy)alkanes [104]. Indeed, it is found that ΔT_{SC} is positive and relatively insensitive to changes in n. It appears, therefore, that the terminal chains act to dilute the specific interaction between the cores and this effect is greatest when the m/n ratio is large. We will return to this theme when considering the structure of smectic phases exhibited by these materials in Sect. 4.4.

We now turn our attention to the effects of varying the terminal chain length on the transition temperatures and phase behaviour for a series of non-symmetric dimers. Figure 14 shows the dependence of the transition temperatures on the number of carbon atoms, m, in the terminal chain for the CBO6O.m series [68]. The marked increase in the nematic-isotropic transition temperature on replacing the p-proton by a methyl group is similar to that seen in Fig. 10 for the m.O4O.m series and possibly results from a change in shape of the mesogenic unit. Subsequent increases in the length of the terminal alkyl chain cause the nematic-isotropic transition to fall with a small alternation between odd and even members of the series in a manner comparable to that observed for conventional monomeric nematics [64] and symmetric dimers – see Fig. 10 [37]. In contrast to such expected behaviour, however, the smectic A-nematic transition temperatures show a very unusual dependence on the length of the terminal chain. For conventional mesogens [1] with a single semi-rigid core as well as for symmetric dimeric liquid crystals – see Fig. 10 [37] – the smectic A-nematic transition temperature

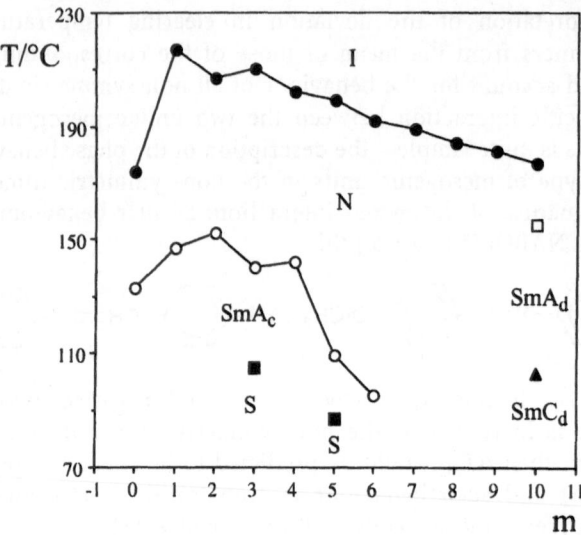

Fig. 14. The dependence of the transition temperatures on the number of carbon atoms, m, in the terminal alkyl chain for the CBO6O.m series [68]. ● denotes nematic-isotropic transitions, ○ intercalated smectic A-nematic transitions, ■ smectic-intercalated smectic A transitions, ▲ interdigitated smectic A-interdigitated smectic C transitions and □ interdigitated smectic A-nematic transitions. The melting points have been omitted for the sake of clarity. SmA_d Interdigitated smectic A phase; SmC_d interdigitated smectic C phase; SmA_c intercalated smectic A phase; S unidentified smectic phase; N nematic; I isotropic

simply increases with increasing chain length. This increase is observed for the first three members of the CBO6O.m series but then the smectic A-nematic transition temperature is found to fall quite rapidly for the next four members. This rapid decrease presumably continues, for the heptyl, octyl and nonyl members do not possess smectic phases. The smectic A phase then reappears dramatically with the decyl member of the series; indeed the smectic A-nematic transition temperature for the decyl member is higher than for any other homologue. This behaviour is unique to non-symmetric dimers and a molecular interpretation will be described in Sect. 4.4.

3.3
Symmetric Discotic Liquid Crystal Dimers

There are relatively few reports of liquid crystal dimers containing discotic mesogenic units [105–112] and this has been due in large part to the difficulties in preparing monofunctionalised discotic precursors. Without such monofunctional precursors the preparation of discotic dimers often involves laborious separation procedures involving several isomers of the discotic core. In recent years, however, there have been many advances in the synthesis of triphenylene-based compounds [113–117] and it is expected that these methodologies will be exploited in the synthesis of liquid crystal discotic dimers [106]. The triphenylene-based dimers that have been reported tend to exhibit glassy behaviour; for example, Boden et al. [106] have shown that the dimer

16

exhibits a monotropic discotic hexagonal columnar (D_h) phase which can be supercooled to form a glassy D_h phase. The clearing temperature of this compound is very similar to that of the analogous discotic monomer, prompting the authors to speculate that the stability of the D_h phase can be largely attributed to the interactions between the aromatic cores while the alkyl chains simply fill space providing that the spacers can bridge the columns. Thus, the authors suggest that the pronounced alternations observed for calamitic dimers will not be observed for discotic dimers. This suggestion has still to be tested.

Discotic dimers have a strong tendency to exhibit columnar phases but an example of a nematogenic discotic dimer has been reported by Praefcke et al. which was based on two discotic multialkylnyl units [111],

17

Conoscopic studies suggest that this compound in fact exhibits a biaxial nematic phase although these experiments indicated only a small optical biaxiality. Indeed, more recently, NMR spectroscopy using selectively deuterated materials has shown that for other materials the assignment of a biaxial nematic phase on the basis of conoscopy alone is not necessarily reliable [118]. This is discussed further in Sect. 3.4.

3.4
Non-Symmetric Discotic-Calamitic Liquid Crystal Dimers

The existence of the thermotropic biaxial nematic phase was theoretically predicted almost thirty years ago [119] but such a phase has yet to be unambiguously identified. Indeed, as we saw in the previous section, a biaxial nematic phase has been claimed for a class of discotic dimers [111]. Theory predicts that mixtures of rod-like and disc-like molecules should exhibit the biaxial nematic phase [120] but experimentally such systems phase separate [121]. This experimental difficulty was overcome by simulating the behaviour of a model system and a biaxial nematic phase was indeed obtained [122]. One way by which the problem of phase separation could be solved experimentally was suggested by Fletcher and Luckhurst and involved covalently linking a rod-like and disc-like unit via a flexible spacer yielding a non-symmetric discotic-calamitic liquid crystal dimer [123]:

These materials did not exhibit liquid crystalline behaviour with just one exception, the hexyl member, which exhibited a strongly monotropic nematic phase. Furthermore, the clearing temperatures of these non-symmetric dimers are considerably lower than the average values of the clearing temperatures for the discotic [112] and calamitic [18] parent symmetric dimers. The authors suggest that this reflects the extreme difficulty in packing the disc-like and rod-like units simultaneously and by increasing the spacer length this problem may, at least in part, be relieved. The equimolar mixtures of these compounds with 2,4,7-trinitro-9-fluorenone (TNF) do exhibit monotropic nematic behaviour, presumably driven by the specific interaction between the electron-rich disc units and the electron-accepting TNF molecules [124]. The monotropic nature of these phases, however, prevented any detailed study of the molecular organisation within the phases and hence it remains to be established whether molecules containing rod-like and disc-like units do exhibit the elusive biaxial nematic phase.

3.5
Laterally Linked Symmetric Liquid Crystal Dimers

All the calamitic dimers discussed so far have had one common structural feature – the mesogenic units have been linked via terminal positions. It is also possible, however, to link the mesogenic units via lateral positions by a flexible spacer [125–128]; for example, Fig. 15 shows the dependence of the transition temperatures on the length of the flexible spacer for the laterally linked dimer series, the α,ω-bis[2,5-bis(4-n-octyloxy-benzoyloxy)-benzamido]alkanes [125],

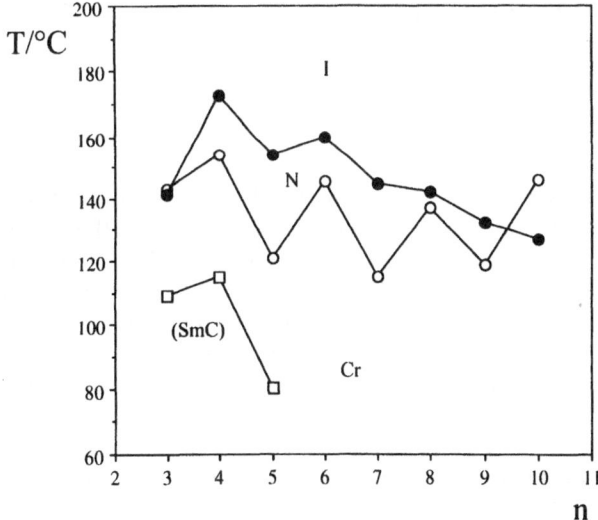

19

All the members of this series (n = 3–12) exhibit nematic behaviour and, in addition, the propyl, butyl and pentyl homologues exhibit a smectic C phase. The nematic-isotropic transition temperatures initially exhibit a large odd-even effect on increasing the spacer length similar to that seen for terminally linked dimers, but this quickly attenuates. It is unclear whether this attenuation is a result of the geometry of the linking group (see Sect. 3.1.2) or whether it results from the change in molecular topology. It is interesting to

Fig. 15. The dependence of the transition temperatures on the length of the flexible spacer, n, for the laterally linked dimer series, the α,ω-bis[2,5-bis(4-n-octyloxy-benzoyloxy)-benz-amido]alkanes [125]. Melting points are denoted by $\bigcirc$, $\bullet$ indicates nematic-isotropic transitions and $\square$ smectic C-nematic transitions. N Nematic; I isotropic; Cr crystal; SmC smectic C phase

note, however, that increasing the spacer length promotes nematic behaviour rather than smectic behaviour in a similar fashion as seen for symmetric terminally linked dimers (see Sect. 3.1.1). As we noted earlier, this is surprising given the very general observation that increasing the length of alkyl chains in mesogenic structures stabilises smectic phases relative to the nematic phase.

Non-symmetric dimers have been prepared in which one mesogenic unit is linked via a terminal position while the other is attached in a lateral position [125, 127]; for example, Weissflog et al. [125] synthesised

20

and this exhibits smectic A and nematic phases. There are insufficient numbers of examples of this molecular architecture, however, to comment upon structure-property correlations.

3.6
Chiral Liquid Crystal Dimers

Chirality in liquid crystalline systems has become an increasingly important area for both technological and fundamental reasons [129] and a wide range of chiral liquid crystal dimers have been reported. These have included both symmetric [50, 74, 130–138] and non-symmetric dimers [74, 139–144] and the chiral centre has been placed either in the terminal chains or in the spacer.

The interest in chiral dimers was stimulated to a large extent by the expectation that placing the chiral centre in the spacer, at least for even dimers, should increase its orientational order compared to that for a chiral centre located in a terminal chain and this in turn should result in an enhancement of the form chirality of the phase. This suggestion has still to be extensively investigated but the limited data available indicates that dimers having chiral spacers actually exhibit ferroelectric smectic C phases with low values of spontaneous polarisation [140].

Given that the transitional behaviour of dimers depends strongly on the length and parity of the spacer it would seem reasonable to assume that the form chirality of the chiral phase should also depend critically on the parity of the spacer. To test this hypothesis, Blatch et al. [74] studied two series of chiral dimers, one symmetric *(S)2 MB.OnO.(S)2MB,*

CH₃CH₂CHCH₂—⟨⟩—N=CH—⟨⟩—O(CH₂)ₙO—⟨⟩—CH=N—⟨⟩—CH₂CHCH₂CH₃ **21**

and the other non-symmetric, CBOnO.*(S)2MB,*

NC—⟨⟩—⟨⟩—O(CH₂)ₙO—⟨⟩—CH=N—⟨⟩—CH₂CHCH₂CH₃   **22**

For the CBOnO.*(S)2MB* series with $n = 7$ and 9 a blue phase was observed but not for $n = 6$ and 8; thus, the chiral properties of these materials do indeed exhibit an odd-even effect as expected. This was rationalised in terms of the smaller pitch for the odd relative to the even membered dimers which arises from the smaller twist elastic constant of odd dimers and is related to their lower orientational order. Surprisingly, the helical twisting power of the dimers in a common monomeric nematic solvent appears to depend solely on the nature of the chiral group, the 2-methylbutyl chiral centre, and not on its environment. Thus similar helical twisting powers are observed for both odd and even membered dimers. We will return to the nature of the phases exhibited by some of these chiral dimers in Sect. 4.4.

4
Smectic Behaviour of Liquid Crystal Dimers

4.1
Smectic Phase Formation in Conventional Low Molar Mass Mesogens

Before we consider the smectic behaviour of liquid crystal dimers it is instructive to review briefly why conventional low molar mass mesogens exhibit smectic phases. The formation of a smectic phase may be thought of as a microphase separation in which the mesogenic cores form one region while the alkyl chains constitute another. There are two possible driving forces behind this separation: energetically if the mean of the mesogenic unit-mesogenic unit and chain-chain interactions is more favourable than the mixed mesogenic unit-chain interaction then phase separation will occur, or entropically the interaction between a core and a chain acts to order the chain and hence is unfavourable. The very general observation for monomeric liquid crystals, therefore, is that increasing the chain length in a given homologous series promotes smectic behaviour [1]. There are, however, a small number of exceptions to this rule, most notably the *trans-4-(trans-4-n-*alkylcyclohexyl)-cyclohexylcarbonitriles [145],

NC—⟨⟩—⟨⟩—CₙH₂ₙ₊₁   **23**

for which smectic behaviour is observed for $n = 2$–5 while the heptyl homologue exhibits only a nematic phase. This suggests that the presence of strong lateral interactions between the molecules promotes smectic behaviour even for short terminal chain lengths while packing considerations or the dilution effect of the terminal chain become increasingly important as the chain length is increased. It is not known whether smectic behaviour would reappear for longer terminal chains.

This view that strong lateral interactions may be important in the formation of smectic phases is supported by the behaviour of sexiphenyl,

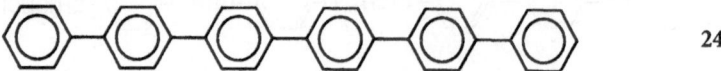   24

which does not possess terminal chains but exhibits smectic behaviour [146]. Furthermore, a computer simulation study of a system of particles interacting via the Gay-Berne potential has shown that the introduction of lateral interactions does indeed promote smectic behaviour [147]. The final driving force for the formation of smectic phases is thought to be largely a steric one giving rise to the so-called filled smectic phases in which the molecules simply pack in the most efficient manner [148].

4.2
Conventional Smectic Phases

It is easy to imagine three quite different smectic A structures composed of symmetric calamitic liquid crystal dimers. To do this we simply place the mesogenic units into smectic layers and then consider the different ways in which the spacers and terminal chains can be added (see Fig. 16). In the first of these the spacers and terminal chains are added in a random fashion (Fig. 16a) such that differing parts of the molecules are overlapping and this is termed an intercalated smectic phase; in the second, the spacers and terminal chains essentially phase separate such that the mesogenic units in one layer are all attached to mesogenic units in the same adjacent layer giving rise to a conventional monolayer smectic phase (Fig. 16b); finally the molecules may adopt horseshoe-like conformations (Fig. 16c). On entropic grounds alone, the intercalated arrangement in which the spacers and terminal chains are randomly mixed should be the most favourable. Energetically, the phase consisting of molecules in horseshoe-like conformations would be expected to be the least favoured but it is difficult to predict which of the other two would be most favoured. We noted in Sect. 4.1 that lateral forces between the mesogenic units are thought to play an important role in the formation of smectic phases and these are identical in the structures shown in Fig. 16a,b. Experimentally, however, it should be straightforward to distinguish between these three possibilities on the basis of the ratio of the smectic layer periodicity to the estimated all-*trans* molecular length: for structures shown in Fig. 16a,c

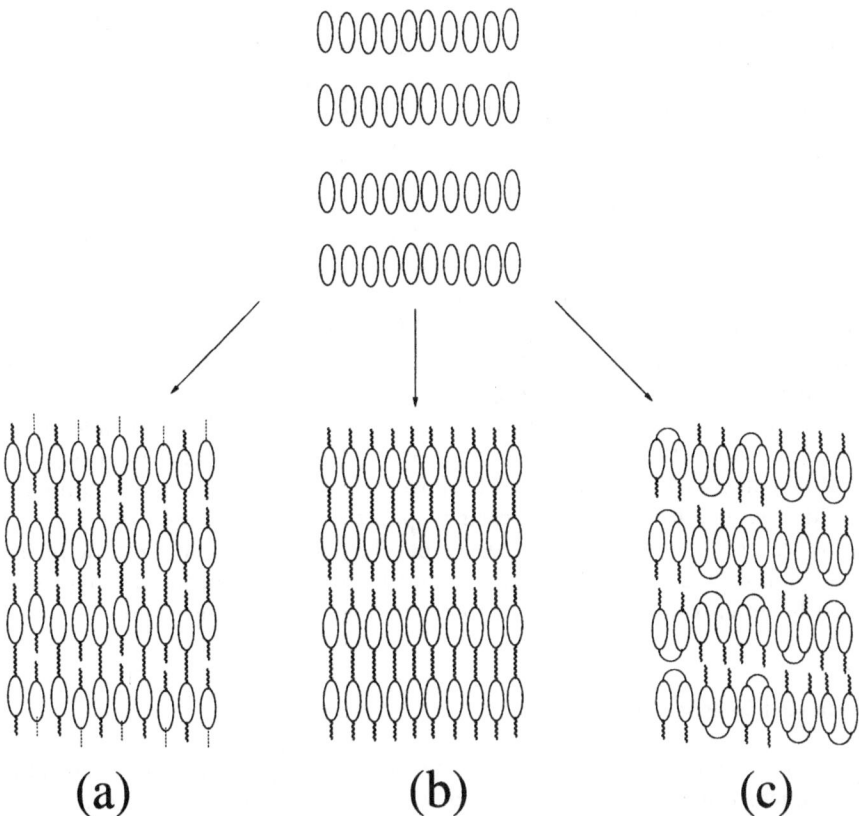

Fig. 16a–c. Sketches showing the formation by symmetric dimers of: **a** an intercalated; **b** a monolayer smectic A phase; **c** a smectic A phase composed of horseshoe-like molecules

this ratio is approximately 0.5 while for that shown as Fig. 16b it is approximately 1.

We saw in Sect. 3.1.1 that for a symmetric dimer containing terminal alkyl chains to exhibit smectic behaviour then the terminal chain length must be greater than half the spacer length. This empirical rule effectively eliminates the possibility that the dimers form an intercalated structure simply because the terminal chains can only be accommodated within such a structure if the total length of the two terminal chains is equal to or less than the length of the spacer (see Fig. 16a). This view is supported by X-ray diffraction studies which reveal that for the overwhelming majority of symmetric dimers the smectic phases have a monolayer structure (see Fig. 16b).

Symmetric dimers appear therefore to have a strong tendency to exhibit monolayer smectic phases even though the mesogenic unit-mesogenic unit interactions in both the monolayer and intercalated smectic phases are identical and furthermore, on entropic grounds the intercalated structure should be favoured. This strongly suggests that the relative stabilities of the two arrangements must rest with the nature of the interaction between the spacer and the terminal alkyl chains. The observation of only the monolayer structure implies that the terminal chain-spacer interaction is an unfavourable one and so destabilises the intercalated arrangement. It would appear, therefore, that the favourable entropic term driving the intercalated arrangement is offset by an unfavourable enthalpic term arising from the interaction between the spacers and the terminal chain.

This discussion has made no reference to the type of mesogenic unit present in the dimer and so it would be reasonable to assume that these observations should hold true for all dimeric liquid crystals possessing terminal alkyl chains. For the vast majority of symmetric dimers containing terminal chains monolayer phases are indeed observed if the terminal chain length exceeds half the length of the spacer (see, for example, [35, 149, 150]) and there are very few exceptions to this rule (see for example [151]).

Finally, we have noted already in Sect. 3.1.1 that for semi-flexible main chain liquid-crystalline polymers, increasing the spacer length increases the smectic tendencies of the polymer [39], whereas the reverse is true for dimeric liquid crystals, for which increasing spacer length decreases the smectic tendencies of the dimer. This suggests that the driving force for smectic phase formation differs for the two systems and for the polymers the driving force must be an entropic one in order to disentangle the polymer chains.

4.3
Modulated Smectic Phases

It is widely believed that frustrated smectic phases which include re-entrant, incommensurate and modulated smectic phases arise as a result of a competition between different characteristic length scales [152]. Such phases are most commonly observed for mesogens possessing a terminal polar group, for example, the cyano group[153], for which the competing periodicities are the molecular length and the length of anti-parallel molecular pairs. Thus packing the molecules with the periodicity corresponding to the molecular length fills space efficiently but gives rise to unfavourable dipolar interactions between neighbouring molecules. On the other hand, the dipolar energy can be minimised by allowing the molecules to form anti-parallel dimers but these fill space much less efficiently. These competing periodicities can both be accommodated in a number of structures including the modulated phases such as the Sm $\tilde{A}$ and Sm $\tilde{C}$ phases [153, 154]. The packing frustration required to drive the formation of these phases can also arise from purely steric effects; for example, if the mesogenic core is bulkier than the terminal chains then this will create a packing stress within a layer which can be relieved by the system adopting a modulated structure [155].

Aside from exhibiting a rich range of conventional smectic phases, members of the m.OnO.m family also exhibit novel modulated hexatic phases [155] in which the tilted hexatic smectic monolayers have a periodic modulation in their smectic structure, analogous to the Sm $\tilde{C}$ ribbon phase. Two such phases have been identified and have been termed the Sm1 and Sm2 phases. The Sm1 phase is exhibited by members of the m.OnO.m family with m = 10, 12 and 14 and n = 9 and 11 [155] and has been obtained on cooling the isotropic, smectic A and smectic C phases. In the Sm1 phase the mesogenic groups are tilted symmetrically with respect to the layer normal and the modulation is purely displacive along the b-axis. In essence, the Sm1 phase can be considered as the modulated SmF or SmI phase in which the tilt direction of the director with respect to the local hexagonal lattice has yet to be determined. A sketch of this molecular arrangement is shown in Fig. 17 although in a more realistic representation of the Sm1 phase the domain boundaries would be somewhat less well-defined. All the dimers which exhibit this phase possess odd membered spacers and thus in Fig. 17 the molecules are assumed to have a bent molecular shape (see Fig. 4). It is not apparent, however, what the competing length scales are which give rise to the phase and indeed we have seen in Sect. 4.2, that symmetric dimers form monolayer smectic structures in which the layer periodicity is approximately equal to the molecular length. It is important to note that the Sm1 phase is shown only by odd membered dimers which strongly suggests that molecular shape plays an important role in the formation of the phase given that this is the most striking difference between odd and even membered dimers (see Fig. 4). Thus, the bent odd dimers may interlock giving rise to the periodicity L' with $L<L'<2L$ which competes with the monolayer periodicity L (see Fig. 17). The view that molecular shape plays an important role in the formation of modulated phases by dimers is supported by the observation of the Sm $\tilde{C}$ phase for odd members of the chiral non-symmetric series [139],

$$C_{10}H_{21}O-\bigcirc-CO.O-\bigcirc-CO.O(CH_2)_nO-\bigcirc-CO.O-\bigcirc-CO.O-\bigcirc-CO.O-\overset{CH_3}{\underset{*}{CH}}-C_6H_{13}$$

25

In contrast the even members of this chiral series do not exhibit modulated smectic phases.

On cooling the Sm1 phase for 10.090.10 a second modulated hexatic phase is observed, the Sm2 phase [155]. This is a monotropic phase and its optical texture is indistinguishable from the preceding Sm1 phase. In addition, the phase transition could not be observed using differential scanning calorimetry.

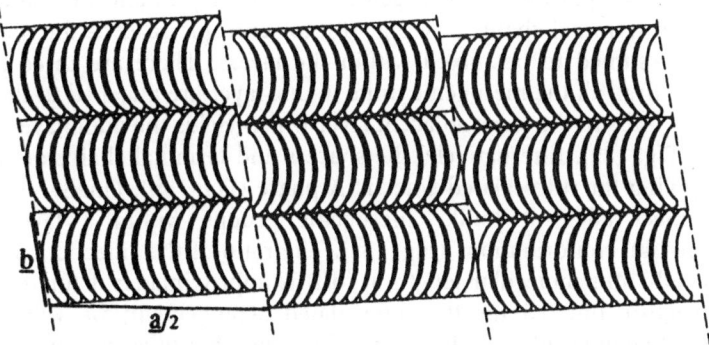

Fig. 17. A sketch of the Sm1 phase composed of bent molecules [155]

Surprisingly, the two phases appear to coexist over a relatively wide temperature range although this may be a kinetic effect. Further speculation on the nature of the Sm2 phase must await detailed structural studies which have been prevented by the monotropic nature of the only known example of this phase [155].

Another class of frustrated smectic phases are the incommensurate smectic A phases in which the competing periodicities coexist along the layer normal. The first example of such a phase was observed for a non-symmetric dimer, KI5 [144]

26

In this phase the larger periodicity appears to correspond to the molecular length while the shorter period may indicate a coexisting intercalated arrangement of the dimers[143, 144].

4.4
Intercalated Smectic Phases

We now return to consider the unique dependence of the occurrence of smectic phases on molecular structure seen for the CBOnO.m series in Sect. 3.2. Specifically, for the CBOnO.10 series [69] smectic behaviour was observed for n = 3–7 and 10–12 but not for n = 8 and 9 (see Fig. 11) while for the CBO6O.m series [68] smectic behaviour is observed for m = 3–6 and 10–12 but not for m = 7–9 (see Fig. 14). This behaviour is unique to non-symmetric dimers and in order to interpret it we must first consider the possible molecular arrangements within a smectic A phase composed of non-symmetric dimers. We can construct the possible smectic structures as in Sect. 4.2 by first placing the mesogenic units into layers and then adding the spacers and terminal chains (see Fig. 18); for entropic reasons the mesogenic units are assumed to be mixed within the layers. There are three possible structures; first, all the mesogenic units in one layer are attached to the unlike mesogenic units in the same adjacent layer which yields the monolayer smectic A phase (Fig. 18a). Alternatively, the spacers and terminal chains are allowed to mix, giving the intercalated smectic A phase (Fig. 18b). Finally, the spacer may connect adjacent unlike mesogenic groups within a layer giving horseshoe-like conformations (Fig. 18c). The ratio of the smectic periodicity, d, to the all-*trans* molecular length, l, (d/l) is approximately 1 for the mono-layer structure but 0.5 for the intercalated arrangement; we will see that horseshoe-like molecules can be arranged in a number of ways giving differing values of d/l.

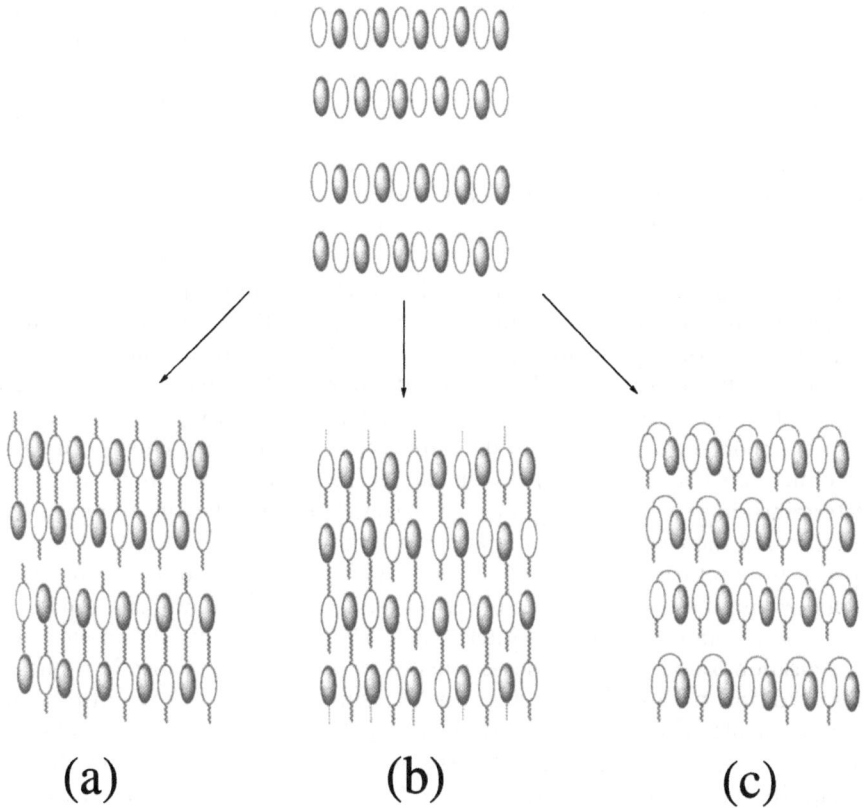

Fig. 18a–c. Sketches showing the formation by non-symmetric dimers of: **a** a monolayer and **b** an intercalated smectic A phase from an equimolar mixture of differing mesogenic units; **c** the smectic A phase composed of horseshoe-like molecules

For the CBOnO.10 series (see Fig. 11) the ratio d/l for the smectic A phase shows a dramatic dependence on n, and for n = 10–12, d/l is approximately 0.5 [69]. This strongly suggests that these homologues exhibit the intercalated smectic A phase in which differing parts of the molecules overlap (Fig. 18b). The notation used to designate smectic phases was extended to allow for this new structural variant such that a subscript c denotes intercalated by analogy to the use of a subscript d to denote interdigitated [69].

The driving force for the SmA_c phase is normally attributed to the specific interaction between the unlike cores [68–70] described in Sect. 3.2 to account for deviations in the clearing temperatures when compared to the weighted averages of the parent symmetric dimers. In addition, mixing the unlike mesogenic units within a layer rather than letting them phase separate is entropically favourable. The interactions between the mesogenic groups are, however, identical in the monolayer and intercalated structures (see Fig. 18a,b) and hence the specific interactions between the unlike mesogenic units cannot solely account for the formation of the intercalated structure. Entropically the most favourable arrangement would be one in which the spacer connects pairs of neighbouring unlike mesogenic groups essentially randomly throughout the structure giving the SmA$_c$ phase [75]. We have already seen that the SmA$_c$ phase is not normally exhibited by symmetric dimers and that this is thought to imply that the mixing of the terminal chains and spacers is unfavourable (see Sect. 4.2). Presumably the reduction in the number of terminal chains to just one for the non-symmetric dimers reduces this unfavourable enthalpic term and this can now be offset by the gain in entropy on mixing the spacers and terminal chains. It should be noted that the sketch of the phase shown in Fig. 18b implies that the structure has ferroelectric ordering but this is removed by a random arrangement of such domains at the macroscopic level. Indeed, attempts to detect a macroscopic polarisation for several typical SmA$_c$ phases have failed [156].

It is important to note that in the SmA$_c$ phase the terminal chains must be accommodated in the space between the layers of mesogenic units which is determined largely by the length of the spacer. Thus for intercalated phases to be observed the terminal chain must be equal to or shorter in length than the spacer. If we return to the CBOnO.10 series and consider the local organisation found within the smectic phases observed for n = 3–7 it is now quite apparent that these cannot have an intercalated structure. Thus the molecular organisation found in these phases must differ to that in the smectic A phase observed for n = 10–12 and this view is supported by X-ray diffraction measurements of the smectic periodicities [69]. For n = 3–7 the d/l ratio is approximately 1.8 which does not correspond to any of the structures shown in Fig. 18. Instead it is indicative of an interdigitated smectic A, SmA$_d$, phase in which like parts of the molecules overlap (see Fig. 19). This molecular arrangement is stabilised by the electrostatic interactions between the polar and polarisable cyanobiphenyl groups while the smectic phase results from the molecular inhomogeneity arising from the long terminal alkyl chains – see Sect. 4.1. The question arises, however, why do these dimers form the SmA$_d$ phase rather than the monolayer SmA phase? The answer to this rests with the large difference in steric bulk of the two terminal substituents, i.e. the cyano group and a decyl chain. If these dimers were to pack into a monolayer arrangement (Fig. 18a) then this mismatch would create a significant volume fraction of voids. Thus the dipolar interaction between the cyanobiphenyl groups intervenes and modifies the monolayer arrangement creating the SmA$_d$ phase. The apparent voids in this structure are presumably filled by the flexibility of the terminal chains.

The structure of the smectic A phase, therefore, is governed by the relative lengths of the spacer and terminal chains. For short spacers, n = 3–7, the CBOnO.10 series exhibits the SmA$_d$ phase while for long spacers, n = 10–12, the SmA$_c$ phase is observed. The disappearance of smectic behaviour for intermediate chain lengths (see Fig. 11) implies that neither smectic modification is favourable and hence nematic behaviour results. There is a strong similarity here to re-entrant nematic behaviour which is also driven by two different length scales.

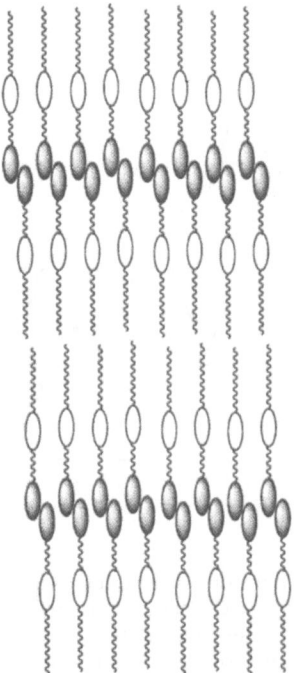

Fig. 19. A sketch of the interdigitated smectic A phase composed of non-symmetric dimers

If we now consider the CBO6O.m series for which smectic behaviour is observed for m = 3–6 and 10–12 but not for m = 7–9 (Fig. 14), then within the framework established for the CBOnO.10 series, we would expect the short terminal chain lengths to exhibit the SmA_c phase while for the long terminal chain lengths, the SmA_d phase should be favoured. This view is indeed supported by X-ray diffraction studies [68]. For this series the cross-over in structure is a result of insufficient space between the layers to accommodate the terminal chains.

There is a remaining difficulty with this interpretation of the novel behaviour of non-symmetric dimers in terms of the intercalated structure composed of dimers having long spacers but short terminal chains which is best illustrated using the CBOnO.2 series [68] (see Fig. 20). For this series SmA_c phases are exhibited for dimers with n=4–12. The SmA_c-N transition temperature shows a pronounced odd-even effect as the spacer length is increased in which the even members have the higher values. This reflects, in part, the ease with which an even or odd membered dimer may be incorporated into the intercalated structure. The underlying trend in the SmA_c-N transition temperature for both odd and even members is an increasing one with increasing n and this is presumably a measure of the ability of the terminal ethyl chain to be located in the space between the intercalated layers. Thus CBO12O.2 exhibits a SmA_c phase which persists to

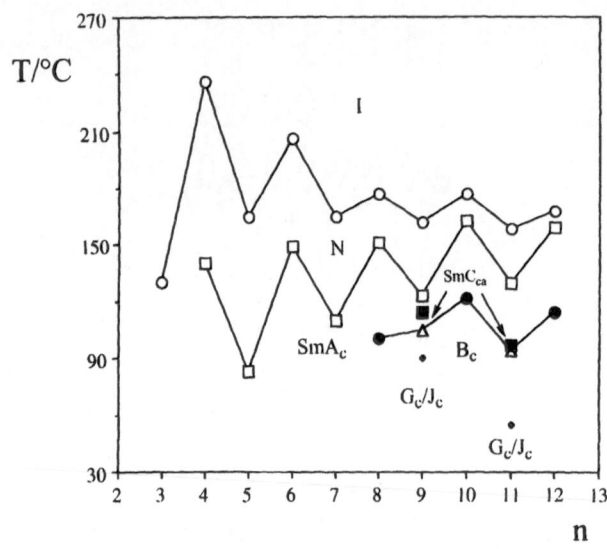

Fig. 20. The dependence of the transition temperatures on the number of methylene groups, n, in the flexible alkyl spacer for the CB.OnO.2 series [69]. The nematic-isotropic transition is denoted by ○, □ indicates the intercalated smectic A-nematic transition, ● the intercalated crystal B-intercalated smectic A transition, ■ the intercalated alternating smectic C-intercalated smectic A transition, △ the intercalated crystal B-intercalated alternating smectic C transition and ◇ the intercalated crystal G/J-intercalated crystal B transition. The melting points have been omitted for the sake of clarity. SmA_c Intercalated smectic A phase; SmC_{ca} intercalated alternating smectic C phase; N nematic; B_c intercalated crystal B phase; G_c/J_c intercalated crystal G/J phase; I isotropic

higher temperatures than that shown by CBO4O2. Figure 21 sketches the SmA_c phase exhibited by CBO12O.2 and there appears to be a considerable concentration of voids within the structure. By contrast, space is filled more efficiently in the intercalated structure exhibited by CBO4O.2. The problem of efficient packing evident in Fig. 21 may be overcome to some extent by reducing the translational order within the phase. In doing so, however, the overlap between the unlike mesogenic units would be reduced and remember it is this interaction which is thought to help stabilise the phase.

This question of the difficulty of filling space within the SmA_c phase led to the suggestion that the molecules might adopt horseshoe-like conformations (see Fig. 18c) [75]. Such structures could be achieved through the introduction of gauche defects into the spacer and the energetic cost in the creation of these horseshoe-like conformations is presumably regained from the favourable interaction between the unlike mesogenic groups. These horseshoe structures can then be packed into smectic arrangements for which d/l = 0.5 or 1.8 (see Fig. 22).

Although the molecular organisation within smectic phases composed of non-symmetric dimers remains ambiguous, there is a growing body of evidence in favour of the intercalated arrangement (see Fig. 18b) as opposed to

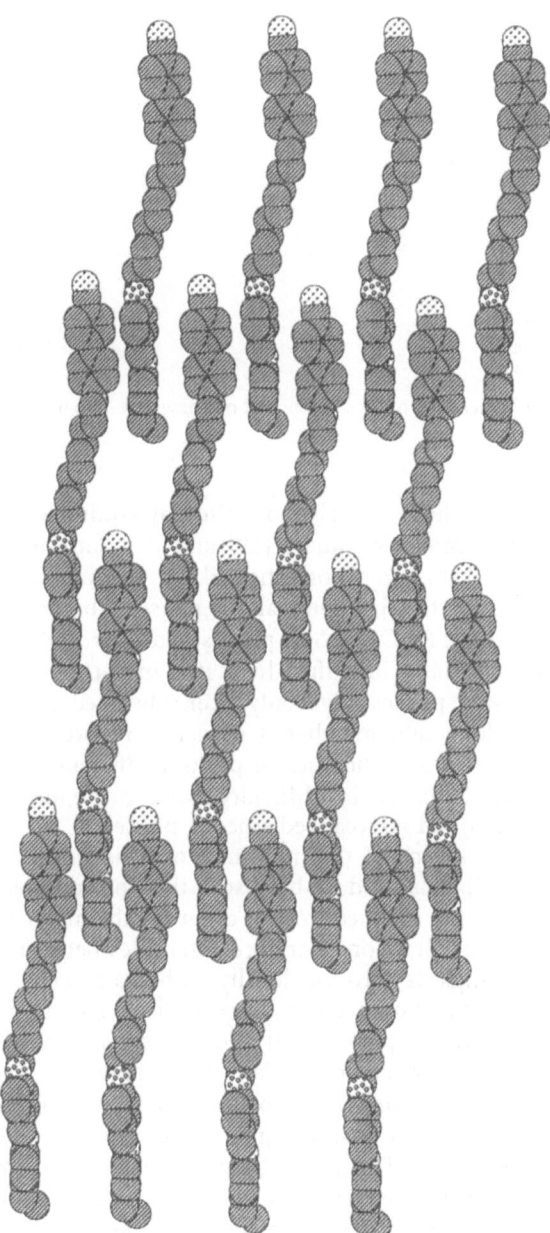

Fig. 21. A sketch of the intercalated smectic A phase exhibited by CBO12O.2

those containing horseshoe-like conformations (see Fig. 22). For example, the SmA_c phase has considerably higher twist viscosities in comparison to those of the interdigitated smectic phases at comparable temperatures [157]. In

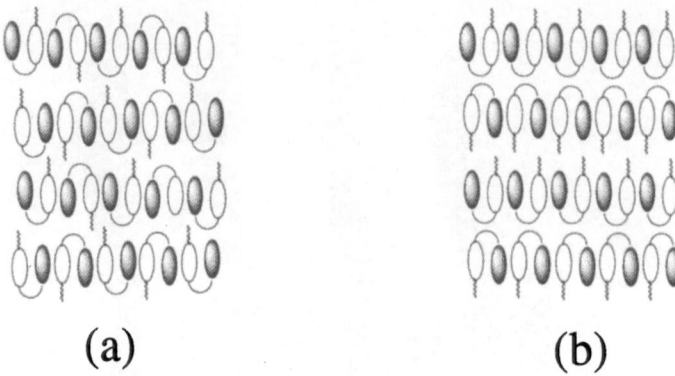

Fig. 22a,b. Sketches of smectic A phases composed of horseshoe-like molecules in which: **a** d/l = 0.5; **b** d/l = 1.8

addition, intercalated phases are also exhibited by equimolar mixtures of the appropriate parent symmetric dimers [69] and it is far from clear why the symmetric dimers should adopt horseshoe-like conformations.

The SmA_c phase was the first intercalated phase to be discovered [68] but examples of intercalated SmC and SmI phases and intercalated crystal B and J phases have since also been identified [69]. It is important to note, however, that tilted intercalated phases have only been observed for non-symmetric dimers containing long odd membered spacers and presumably such bent molecules experience great difficulties in packing efficiently into the intercalated network. Indeed, this packing difficulty may provide the driving force for the formation of the tilted intercalated smectic phases.

Figure 23 shows a sketch of the molecular organisation found within the intercalated smectic C phase and we have seen that for odd membered dimers the mesogenic groups are inclined to each other for the majority of the spacer conformations. The tilt direction, therefore, alternates between the layers and thus the global tilt angle is zero but locally within a layer is non-zero. This alternation of the tilt direction between layers is also a structural feature of the antiferroelectric smectic C* phase [129] and of a smectic phase exhibited by certain semi-flexible main chain liquid crystal polymers [11–15]. It was decided to extend the notation used to describe smectic phases to include reference to this structural feature and a subscript a now denotes an alternating tilt direction. Thus, SmC_{ca} represents the intercalated and alternating SmC phase. The requirement for the observation of this alternation in the tilt direction is simply a correlation of the mesogenic groups. For the polymers this correlation is provided by the flexible spacer and the polymeric nature of the system. For the antiferroelectric smectic C* phase, it is provided through dipolar forces between the chiral molecules while for the non-symmetric dimers the correlation may originate from the specific interaction between the unlike mesogenic groups. A possible model to describe the SmA_c-SmC_{ca} transition involves a biasing of the distribution of the tilt directions of

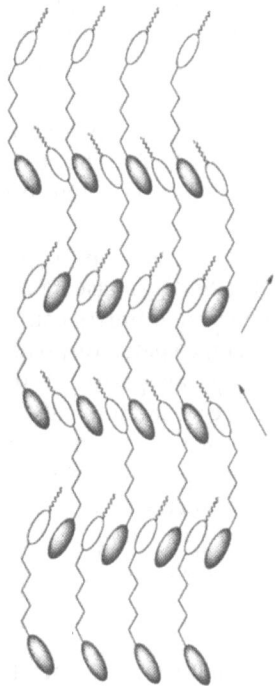

Fig. 23. A sketch of the intercalated alternating smectic C phase

the mesogenic units resulting in a long range correlation of the tilt angle [69]. This model is essentially that described by Wulf to account for the SmA-SmC transition in conventional mesogens [158].

The proposed structure of the SmC_{ca} phase shown in Fig. 23 has recently been supported by the results of a study using electron spin resonance spectroscopy [159]. Specifically, it is clear that within the SmC_{ca} phase there are two distinct directors with tilt directions differing by 180°. The tilt angle within a layer is reported to have a maximum value of 18° which is consistent with the bent geometry of an odd membered dimer in which the mesogenic groups make an angle of 15° with the spacer.

Intercalated smectic phases have also been observed for symmetric liquid crystal dimers. Watanabe and his co-workers [13, 160–163] have described the properties of the α,ω-bis(4,4'-butoxybiphenyl-carbonyloxy) alkanes,

$$C_4H_9O-\bigcirc\!\!-\!\!\bigcirc\!\!-CO.O(CH_2)_nO.OC-\bigcirc\!\!-\!\!\bigcirc\!\!-OC_4H_9 \qquad 27$$

The even members of this series exhibit the intercalated smectic A phase while the odd members show only the intercalated alternating smectic C phase. In this respect the dimers are behaving in an identical fashion to the analogous polymers for which even membered spacers give rise to smectic A phases while odd members give the alternating smectic C phase [11–15]. It is not clear why these particular symmetric dimers exhibit intercalated phases while the vast majority do not [37]. It has been suggested that a specific dipolar interaction may exist between the ester groups of the spacer and the ether groups of the terminal chain which might offset the apparently unfavourable interaction between the spacers and terminal chains. The strong dependence of the phase type on the parity of the spacer has been interpreted in terms of molecular shape and the ease at which the more linear even members are accommodated within the intercalated smectic A phase. By comparison an odd membered dimer packs efficiently into the SmC$_{ca}$ phase. We have seen, however, for the CBOnO.m series that odd-membered dimers can also exhibit the intercalated smectic A phase. Thus, the average spatial dispositions of the mesogenic units which are largely governed by the parity of the spacer do not, in fact, exclusively determine the nature of the intercalated phase formed by dimers.

An intercalated smectic C phase in which the tilt direction does not alternate between the layers has apparently been reported by Suzuki et al. [134, 137] for two chiral symmetric dimer series:

$$C_8H_{17}O-\bigcirc-\bigcirc-CO.O-\bigcirc-CO.O-\overset{R}{\underset{\bullet}{CH}}-(CH_2)_n-\overset{R}{\underset{\bullet}{CH}}-O.OC-\bigcirc-O.OC-\bigcirc-\bigcirc-OC_8H_{17}$$

R=CF$_3$, n=7-12; R=CH$_3$, n=8, 10-12

28

For the shorter spacer lengths even membered dimers exhibit the ferroelectric smectic C phase but for odd membered dimers the antiferroelectric smectic C phase is observed. As the spacer length is increased both odd and even membered dimers exhibit the antiferroelectric smectic C phase in which the tilt direction alternates between the layers. Although the authors sketch these structures as being intercalated there is no X-ray data supporting these assignments and there remains the possibility that they are conventional monolayer arrangements. The authors note that there are two principle forces responsible for the stabilisation of the antiferroelectric structure over the ferroelectric arrangement, namely conformational and electrostatic interactions. Thus, for the shorter spacers the molecular shape plays the dominant role such that only the odd members exhibit the antiferroelectric phase. As the spacer length increases, however, the electrostatic interactions become increasingly more important, presumably as a result of the increased molecular flexibility, and antiferroelectric behaviour can be stabilised for both even and odd membered spacers.

5
References

1. Gray GW (1979) In: Luckhurst GR, Gray GW (eds) The molecular physics of liquid crystals, chap 1. Academic Press, London
2. Demus D (1989) Liq Cryst 5: 75
3. Imrie CT, Luckhurst GR (1998) In: Demus D, Goodby JW, Gray GW, Spiess HW, Vill V (eds) Handbook of liquid crystals, vol 2B. Low molecular weight liquid crystals, chap 10. Wiley-VCH, Weinheim

4. Vorländer D (1927) Z Phys Chem 126: 449
5. Rault J, Liebert L, Strzelecki L (1975) Bull Soc Chem Fr 1175
6. Griffin AC, Britt TR (1981) J Am Chem Soc 103: 4957
7. Ober CK, Jin J-I, Lenz RW (1984) Adv Polym Sci 59: 103
8. Attard GS (1993) Trends Polym Sci 1: 79
9. Blumstein A, Thomas O (1982) Macromolecules 15: 1264
10. Ungar G, Percec V, Zuber M (1992) Macromolecules 25: 75
11. Watanabe J, Hayashi M (1989) Macromolecules 22: 4083
12. Watanabe J, Hayashi M (1988) Macromolecules 21: 278
13. Watanabe J, Kinoshita SJ (1992) J Phys II France 2: 1237
14. Watanabe J, Hayashi M, Morita M, Niori T (1994) Mol Cryst Liq Cryst 254: 221
15. Tokita M, Osada K, Watanabe J (1998) Liq Cryst 24: 477
16. Luckhurst GR (1995) Macromol Symp 96: 1
17. Imrie CT, Karasz FE, Attard GS (1993) Macromolecules 26: 3803
18. Emsley JW, Luckhurst GR, Shilstone GN, Sage I (1984) Mol Cryst Liq Cryst Lett 102: 223
19. Abe A, Furuya H (1988) Polym Bull 19: 403
20. Abe A, Nam SY (1995) Macromolecules 28: 90
21. Abe A, Furuya H, Shimizu RN, Nam SY (1995) Macromolecules 28: 96
22. Dunmur DA, Wilson MR (1988) J Chem Soc, Faraday Trans II 84: 1109
23. Maeda Y, Furuya H, Abe A (1996) Liq Cryst 21: 365
24. Emsley JW, Luckhurst GR, Shilstone GN (1984) Mol Phys 53: 1023
25. Emsley JW, Luckhurst GR, Timimi BA (1985) Chem Phys Lett 114: 19
26. Ionescu D, Luckhurst GR, de Silva DS (1997) Liq Cryst 23: 6
27. Heeks SK, Luckhurst GR (1993) J Chem Soc, Faraday Trans 89: 3289
28. Malpezzi L, Brückner S, Galbiati E, Zerbi G, Luckhurst GR (1991) Mol Cryst Liq Cryst 195: 179
29. Jin J-I, Chung Y-S, Lenz RW, Ober C (1983) Bull Korean Chem Soc 4: 143
30. Jin J-I (1995) Mol Cryst Liq Cryst 267: 249
31. Jin J-I, Park J-H (1984) Mol Cryst Liq Cryst 110: 293
32. Jin J-I, Seong C-M, Jo B-W (1985) Bull Korean Chem Soc 5: 40
33. Jin J-I, Kang J-S, Jo B-W, Lenz RW (1983) Bull Korean Chem Soc 4: 176
34. Jin J-I (1983) Polymer Preprints (Jpn) 32: 935
35. Aquilera C, Ahmad S, Bartulin J, Muller HJ (1988) Mol Cryst Liq Cryst 162B: 277
36. Buglione JA, Roviello A, Sirigu A (1984) Mol Cryst Liq Cryst 106: 169
37. Date RW, Imrie CT, Luckhurst GR, Seddon JM (1992) Liq Cryst 12: 203
38. Imrie CT (1996) In: Salamone JC (ed) Polymeric materials encyclopedia. CRC Press, Boca Raton, 5: 3770
39. Finkelmann H (1987) In: Gray GW (ed) Thermotropic liquid crystals. Critical reports on applied chemistry, vol 22, chap 6. Wiley, New York
40. Creed D, Gross JRD, Sullivan SL, Griffin AC, Hoyle CE (1987) Mol Cryst Liq Cryst 149: 185
41. Chien JCW, Zhou R, Lillya CP (1987) Macromolecules 20: 2341
42. Sledzinska I, Bialecka-Florjanczyk E, Orzesko A (1996) Eur Polym J 32: 1345
43. Hoshino H, Jin J-I, Lenz RW (1984) J Appl Polym Sci 29: 547
44. Vora RA, Teckchandani VR (1991) Mol Cryst Liq Cryst 209: 285
45. Ibn-Elhaj M, Skoulios A, Guillon D, Newton J, Hodge P, Coles HJ (1995) Macromolecules 19: 373
46. Jo B-W, Lim T-K, Jin J-I (1988) Mol Cryst Liq Cryst Inc Nonlin Opt 157: 57
47. Jo B-W, Choi J-K, Bang M-S, Chung B-Y, Jin J-I (1992) Chem Mater 4: 1405
48. Aquilera C, Bernal L (1984) Polym Bull 12: 383
49. Rivas BL, Hisgen B, Schmidt HW, Ringsdorf H (1987) Bul Soc Chil Quim 32: 143
50. Poths H, Wischerhoff E, Zentel R, Schonfeld A, Henn G, Kremer F (1995) Liq Cryst 18: 811
51. Hohmuth A, Schiewe B, Heinemann S, Kresse H (1997) Liq Cryst 22: 211
52. Jin J-I, Choi E-J, Ryu S-C, Lenz RW (1986) Polym J 18: 63
53. Abe A, Furuya H, Nam SY, Okamoto S (1995) Acta Polym 46: 437

54. Barnes PJ, Douglass AG, Heeks SK, Luckhurst GR (1993) Liq Cryst 13: 603
55. Carfagna C, Iannelli P, Roviello A, Sirigu A (1985) J Thermal Anal 30: 1317
56. Roviello A, Sirigu A (1979) Eur Polym J 15: 423
57. Roviello A, Sirigu A (1982) Makromol Chem 183: 895
58. Emerson APJ, Luckhurst GR (1991) Liq Cryst 10: 861
59. Ferrarini A, Luckhurst GR, Nordio PL, Roskilly SJ (1994) J Chem Phys 100: 1460
60. Luckhurst GR (1985) In: Chapoy LL (ed) Recent advances in liquid crystalline polymers, chap 7. Elsevier, London and New York
61. Ferrarini A, Luckhurst GR, Nordio PL, Roskilly SJ (1993) Chem Phys Lett 214: 409
62. Ferrarini A, Luckhurst GR, Nordio PL, Roskilly SJ (1996) Liq Cryst 21: 373
63. Griffin AC, Vaidya SR, Hung RSL, Gorman S (1985) Mol Cryst Liq Cryst Lett 1: 131
64. Imrie CT (1989) Liq Cryst 6: 391
65. Bruce DW, Hall MD (1994) Mol Cryst Liq Cryst 250: 373
66. Imrie CT, Taylor L (1989) Liq Cryst 6: 1
67. Jin J-I, Chung Y-S, Kang J-S, Lenz RW (1982) Mol Cryst Liq Cryst Lett 82: 261
68. Hogan JL, Imrie CT, Luckhurst GR (1988) Liq Cryst 3: 645
69. Attard GS, Date RW, Imrie CT, Luckhurst GR, Roskilly SJ, Seddon JM, Taylor L (1994) Liq Cryst 16: 529
70. Attard GS, Garnett S, Hickman CG, Imrie CT, Taylor L (1990) Liq Cryst 7: 495
71. Attard GS, Imrie CT, Karasz FE (1992) Chem Mater 4: 1246
72. Griffin AC, Vaidya SR (1988) Liq Cryst 3: 1275
73. Ikeda T, Miyamoto T, Kurihara S, Tsukada M, Tazuke S (1990) Mol Cryst Liq Cryst 182B: 357
74. Blatch AE, Fletcher ID, Luckhurst GR (1997) J Mater Chem 7: 9
75. Blatch AE, Fletcher ID, Luckhurst GR (1995) Liq Cryst 18: 801
76. Pelzl G, Novak M, Weissflog W, Demus D (1987) Cryst Res Technol 22: 125
77. Sharma NK, Pelzl G, Demus D, Weißflog W (1980) Z Phys Chem 261: 579
78. Schneider F, Sharma NK (1981) Z Naturforsch 36a: 1086
79. Demus D, Pelzl G, Sharma NK, Weissflog W (1981) Mol Cryst Liq Cryst 76: 241
80. Sadowska KW, Zywocinski A, Stecki J, Dabrowski R (1982) J Phys (Paris) 43: 1673
81. Boy A, Adomenas P (1983) Mol Cryst Liq Cryst 95: 59
82. Srikanta BS, Madhusudana NV (1983) Mol Cryst Liq Cryst 99: 203
83. Pelzl G, Demus D, Sackmann H (1968) Z Phys Chem 238: 22
84. Ujiie S, Uchino H, Iimura K (1995) Chem Lett 195
85. Schroeder JP, Schroeder DC (1968) J Org Chem 591
86. Araya K, Matsunaga Y (1980) Bull Chem Soc Jpn 53: 3079
87. Araya K, Matsunaga Y (1981) Mol Cryst Liq Cryst 67: 153
88. Araya K, Matsunaga Y (1981) Bull Chem Soc Jpn 54: 2430
89. Homura N, Matsunaga Y, Suzuki M (1985) Mol Cryst Liq Cryst 131: 273
90. Imrie CT (1995) Trends Polym Sci 3: 22
91. Portugall M, Ringsdorf H, Zentel R (1982) Makromol Chem 183: 2311
92. Meredith GR, Vandusen JG, Williams DJ (1983) In: Williams DJ (ed) Nonlinear optical properties of organic and polymeric materials. ACS Symposium Ser No 233, chap 5
93. Griffin AC, Bhatti AM, Hung RSL (1988) In: Prasad PN, Ulrich DR (eds) Nonlinear optical and electroactive polymers. Plenum, New York, p 375
94. Imrie CT, Karasz FE, Attard GS (1991) Liq Cryst 9: 47
95. Schleeh T, Imrie CT, Rice DM, Karasz FE, Attard GS (1993) J Polym Sci Polym Chem Ed 31: 1859
96. Kosaka Y, Kato T, Uryu T (1994) Macromolecules 27: 2658
97. Kosaka Y, Uryu T (1995) Macromolecules 28: 870
98. Imrie CT, Paterson BJA (1994) Macromolecules 27: 6673
99. Imrie CT, Attard GS, Karasz FE (1996) Macromolecules 29: 1031
100. Craig AA, Imrie CT (1997) Polymer 38: 4951
101. Wallage MJ, Imrie CT (1997) J Mater Chem 7: 1163
102. Humphries RL, James PG, Luckhurst GR (1971) Symp Faraday Soc 5: 107

103. Humphries RL, Luckhurst GR (1973) Chem Phys Lett 23: 567
104. Attard GS, Imrie CT (1992) Liq Cryst 11: 785
105. Lillya CP, Murthy YLN (1985) Mol Cryst Liq Cryst Lett 2: 121
106. Boden N, Bushby RJ, Cammidge AN, Martin PS (1995) J Mater Chem 5: 1857
107. Adam D, Schuhmacher P, Simmerer J, Häussling L, Paulus W, Siemensmeyer K, Etzbach K-H, Ringsdorf H, Haarer D (1995) Adv Mater 7: 276
108. Kranig W, Hüser B, Spiess H-W, Kreuder W, Ringsdorf H, Zimmermann H (1990) Adv Mater 2: 36
109. Zamir S, Poupko R, Luz Z, Hüser B, Boeffel C, Zimmermann H (1994) J Am Chem Soc 116: 1973
110. Haarer D, Simmerer J, Adam D, Schuhmacher P, Paulus W, Etzbach K-H, Siemensmeyer K, Ringsdorf H (1996) Mol Cryst Liq Cryst 283: 63
111. Praefcke K, Kohne B, Singer D, Demus D, Pelzl G, Diele S (1990) Liq Cryst 7: 589
112. Praefcke K, Kohne B, Gündogan B, Singer D, Demus D, Diele S, Pelzl G, Bakowsky U (1991) Mol Cryst Liq Cryst 198: 393
113. Boden N, Borner RC, Bushby RJ, Cammidge AN, Jesudason MV (1993) Liq Cryst 15: 851
114. Boden N, Bushby RJ, Cammidge AN (1994) J Chem Soc, Chem Commun 465
115. Boden N, Bushby RJ, Cammidge AN (1995) J Am Chem Soc 117: 924
116. Boden N, Bushby RJ, Cammidge AN, Headdock G (1995) J Mater Chem 5: 2275
117. Stewart D, McHattie GS, Imrie CT (1998) J Mater Chem 8: 47
118. Hughes JR, Kothe G, Luckhurst GR, Malthete J, Neubert ME, Shenouda I, Timimi BA, Tittelbach M (1997) J Chem Phys 107: 9252
119. Freiser MJ (1970) Phys Rev Lett 24: 1041
120. Alben R (1973) J Chem 59: 4299
121. Pratibhar R, Madhusudana NV (1985) Mol Cryst Liq Cryst Lett 1: 111
122. Hashim R, Luckhurst GR, Prata F, Romano S (1993) Liq Cryst 15: 283
123. Fletcher ID, Luckhurst GR (1995) Liq Cryst 18: 175
124. Praefcke K, Singer D, Kohne B, Ebert M, Liebman A, Wendorff JH (1991) Liq Cryst 10: 147
125. Weissflog W, Demus D, Diele S, Nitschke P, Wedler W (1989) Liq Cryst 5: 111
126. Andersch J, Tschierske C (1996) Liq Cryst 21: 51
127. Andersch J, Tschierske C, Diele S, Lose D (1996) J Mater Chem 6: 1297
128. Surendranath V, Lokanath NK, Sridhar MA, Prasad JS (1998) Liq Cryst 24: 361
129. Goodby JW (1991) J Mater Chem 1: 307
130. Shiraishi K, Kato K, Sugiyama K (1990) Chem Lett 971
131. Barberá J, Omenat A, Serrano JL (1989) Mol Cryst Liq Cryst 166: 167
132. Barberá J, Omenat A, Serrano JL, Sierra T (1989) Liq Cryst 5: 1775
133. Yoshizawa A, Matsuzawa K, Nishiyama I (1995) J Mater Chem 5: 2131
134. Suzuki Y, Isozaki T, Kusumoto T, Hiyama T (1995) Chem Lett 719
135. Yoshizawa A, Soeda Y, Nishiyama I (1995) J Mater Chem 5: 675
136. Yoshizawa A, Nishiyama I (1995) Mol Cryst Liq Cryst 260: 403
137. Suzuki Y-I, Isozaki T, Hashimoto S, Kusumoto T, Hiyama T, Takanishi Y, Takezoe H, Fukuda A (1996) J Mater Chem 6: 753
138. Marcelis ATM, Koudijs A, Sudhölterr EJR (1996) J Mater Chem 6: 1469
139. Faye V, Babeau A, Placin F, Nguyen HT, Barois P, Laux V, Isaert N (1996) Liq Cryst 21: 485
140. Marcos M, Omenat A, Serrano JL (1993) Liq Cryst 13: 843
141. Marcelis ATM, Koudijs A, Sudhölterr EJR (1994) Recl Trav Chim Pays-Bas 113: 524
142. Marcelis ATM, Koudijs A, Sudhölterr EJR (1995) Liq Cryst 18: 843
143. Hardouin F, Achard MF, Jin J-I, Yun Y-K (1995) J Phys II Fr 5: 927
144. Hardouin F, Achard MF, Jin J-I, Shin J-W, Yun Y-K (1994) J Phys II Fr 4: 627
145. Brownsey GJ, Leadbetter AJ (1981) J Phys Lett, Paris 42: 135
146. Irvine PA, Wu DC, Flory PJ (1984) J Chem Soc, Faraday Trans 1 80: 1795
147. Luckhurst GR, Stephens RA, Phippen RW (1990) Liq Cryst 8: 451
148. Pelzl G, Humke A, Diele S, Demus D, Weissflog W (1990) Liq Cryst 7: 115

149. Jin J-I, Chung B-Y, Park J-H (1991) Bull Korean Chem Soc 12: 583
150. Rozhanskii IL, Tomita I, Endo T (1996) Liq Cryst 21: 631
151. Jin J-I, Oh H-T, Park J-H (1986) J Chem Soc Perkin Trans 2 343
152. Prost J, Barois P (1983) J Chim Phys 80: 65
153. Hardouin F, Levelut AM, Achard MF, Sigaud G (1983) J Chem Phys 80: 53
154. Hardouin F (1986) Physica 140A: 359
155. Date RW, Luckhurst GR, Shuman M, Seddon JM (1995) J Phys II Fr 5: 587
156. Carboni C, Farrand FJ, Luckhurst GR, de Silva D (unpubl. data)
157. Le Masurier PJ (1996) PhD thesis, University of Southampton
158. Wulf A (1978) Mol Cryst Liq Cryst 47: 225
159. Le Masurier PJ, Luckhurst GR (unpubl. data)
160. Watanabe J, Komura H, Niiori T (1993) Liq Cryst 13: 455
161. Takanishi Y, Takezoe H, Fukuda A, Komura H, Watanabe J (1992) J Mater Chem 2: 71
162. Niori T, Adachi S, Watanabe J (1995) Liq Cryst 19: 139
163. Watanabe J, Hayashi M (1998) Macromolecules 22: 4083

Metallomesogens

Bertrand Donnio, Duncan W. Bruce

Institut für Makromolekulare Chemie, Albert-Ludwigs Universität,
Hermann-Staudinger-Haus, Stefan-Meier-Straße 31, D-79104 Freiburg,
Germany and School of Chemistry, University of Exeter,
Stocker Road, Exeter EX4 4QD, UK
bdonnio@fmf.uni-freiburg.de and d.bruce@exeter.ac.uk

Metallomesogens are metal complexes which exhibit liquid-crystalline properties, forming the same type of mesophases as found in purely organic materials. Indeed, nematic, smectic and columnar mesophases have all been observed, as well as, in a few cases, the more elusive cubic mesophases. The complexes may have a covalent (neutral) or an ionic character, and a large number of low molar mass thermotropic metallomesogens have been described as well as, to a lesser extent, examples of lyotropic complexes. Furthermore, in the last few years, increased attention has also been devoted to the study of metallomesogenic polymers [1], with main-chain, side-chain, and cross linked examples having been obtained. Metal-lomesogens are becoming a very important class of mesogenic materials, as new properties may be expected on the introduction of metals into a liquid-crystalline material. In addition, they offer wider possibilities for structural variations than simple organic materials as, for example, several types of coordination geometry can be envisaged through the metal and by the use of polydentate ligands. This may in turn lead to new mesophases or new types of molecular organisations, which could eventually form the basis for new effects and devices.

Keywords: Metallomesogens, liquid crystals, inorganic chemistry, mesomorphism

Structure and Bonding, Vol. 95
© Springer Verlag Berlin Heidelberg 1999

1
An Historical Perspective

It is commonly accepted that the first report on metallomesogens was by Vorländer [2] who reported, in 1923, a number of mercury-based mesomorphic materials [3], such as the Schiff base complexes of diarylmercury(II) (**1**), which showed smectic phases. In fact, the first report on metal-containing liquid crystals can be traced back as far as the 1850s with Heintz's work on magnesium tetradecanoate [4]. However, it was not until 1976, some 50 years after Vorländer's mercury complexes, that a new series of well-characterised metallomesogens appeared in the literature. These were some ferrocenyl Schiff base derivatives (**2**), synthesised by Malthête and Billard [5], and they showed nematic phases which were either enantiotropic (alkoxy derivatives) or monotropic (alkyl derivatives). Then, in 1977, Giroud and Mueller-Westerhoff published a study of mesomorphic dithiolene complexes of nickel(II) [6] (**3**, M=Ni), later followed by a study of the platinum(II) and palladium(II) congeners [7] (**3**, M=Pt and Pd). Both the nickel and platinum complexes showed smectic and nematic phases, with melting points at moderate temperatures, but with some decomposition before entering the isotropic liquid. Curiously, the palladium complexes were not mesomorphic, although the rationalisation of this behaviour by citing strong metal–metal interactions which inhibited mesophase formation is questionable, since it does not explain the occurrence of mesomorphism in the platinum complexes, which also possess similar metal–metal interactions as evidence by crystallographic data found in related complexes. These two pieces of work are now accepted as being the beginning of the systematic research into metal-containing liquid crystals.

Now, the field has properly expanded, and many different structures have been realised. However, it is not our intention to write a comprehensive review of the area, since several reviews [8] and a self-contained book [9] already exist. Rather, we would emphasise what are, in our view, some of the more significant developments of the last 10 years, paying particular attention to the relationship between the shape and the type of a complex, and the mesomorphism it shows. This will be discussed in relation both to the metal centre and to the ligands. Furthermore, metals can induce mesomorphism when incorporated in non-mesomorphic materials. Or, when the ligands are themselves mesomorphic, the introduction of a metal centre can have dramatic effects on the thermal behaviour as it can suppress or change completely the liquid-crystalline properties of the ligands. These effects, as well as the previously controversial rôle of the metal–metal interactions in promoting (or destroying) mesophase formation, will also be discussed.

The structures of the first metallomesogens were based mainly on the rod-like and/or disc-like shapes typically found in organic mesogens. This architecture led usually to high transition temperatures and decomposition was often observed, making the materials difficult to handle for physical measurements. In order to obtain stable complexes with accessible meso-phases, significant efforts have been made to reduce the transition temper-atures on the one hand, but also, on the other hand, to increase the chemical stability of the complexes, by increasing the strength of the metal-ligand bonds. In order to achieve these necessary requirements, the structures of the ligands have been modified, and a wider range of metals has been more systematically investigated, such as, for example, those giving rise to high coordination geometries [10]. New concepts such as the lowering of the symmetry, the diminution of the lateral interactions, the use of several aliphatic terminal chains, and the use of correlated structures have been successfully used and will be discussed here. Those modifications have led to new molecular structures, showing that the design of new materials was not simply limited to single rods or discs, and interesting relationships between the structure of the metallomesogens and the structure of the mesophases have been found.

2
Compounds with Reduced Symmetry

2.1
Unsymmetric Compounds

Reducing symmetry by building unsymmetrical materials in order to decrease the melting point is an approach adopted by several groups. This concept lies with a simple principle: if the molecular symmetry was to be reduced, then the molecules ought to pack less favourably in the solid state, hence reducing the melting points and hopefully the clearing temperatures. This approach has proved quite productive as evidence by the following examples.

Maitlis et al. [11] reported the non-centrosymmetric trans-(η^2-alkene)(4'-alkoxy-4-stilbazole)-dichloroplatinum(II) complexes, 4, for which the possibility existed of varying independently the alkoxy chain length of the stilbazole (n) and the alkyl chain length of the alkene (m). This was an early example of such a systematic variation in metallomesogens.

However, this approach is slightly limited by the fact that many metal complexes lacking a centre of symmetry unfortunately show strong tendencies to disproportionate (Eq. 1) and thus can be difficult to isolate. In this case, an inert metal centre (such as Pt) was needed to avoid such disproportionation.

$$2[MLL'X_2] \rightleftharpoons [ML_2X_2] + [M(L')_2X_2]. \qquad (1)$$

The complexes exhibited S_A phases at very accessible temperatures (between 50 and 100 °C) when $(m+n) \geq 8$, which were monotropic for $8 \leq (m+n) \leq 13$, and enantiotropic for $(m+n) \geq 14$; T_{Crys-S_A} typically varies between 50 and 70 °C and T_{S_A}-I is in the range of 90–100 °C. From this collection of complexes, it was observed that the simultaneous increase of both m and n decreased the stability of the crystal phase (reflected by the low melting point) and increased the clearing temperature thus, increasing the range of existence of the smectic phase. Some of the complexes with $m = 0$ were also mesomorphic ($n \geq 8$), showing S_A phases, although at higher temperatures. The comparison with the mesomorphism of the *trans*-bis(4′-dodecyloxy-4-stilbazole)dichloroplatinum(II) [8a], which showed a S_C phase at temperatures greater than 280 °C (with extensive decomposition), indicated that this effect had to be due to the lower symmetry in the molecules, resulting from the substitution of one alkoxystilbazole by a terminal alkene. X-Ray experiments confirmed the S_A structure, with layer spacings smaller than the full extended length of the molecules, suggesting either that the S_A layers were interdigitated or that the tails were folded back or disordered. Mixing experiments on these complexes were also carried out, and resulted in the enhancement of the S_A phase range, essentially due to the depression of the melting points. In some mixtures, temperature ranges of as much as 50 °C were obtained.

Related, non-centrosymmetric metal complexes having the general formula *cis*-[MCl(CO)$_2$L] have also been investigated (5: M = Ir [12], Rh [13]; 6: M = Ir, Rh [14, 15].

$$H_{2n+1}C_nO-\!\!\!\!\bigcirc\!\!\!\!-X \quad \text{CO} \qquad 5: X = N$$
$$N-M-CO \qquad 6: X = CH$$
$$Cl$$

The mesomorphism of compounds **5** was less extensive than compounds **6**. Indeed, while the imine complexes of both Ir and Rh showed enantiotropic mesophases (N and S_A) from the octyloxy homologue onward, the stilbazole complexes gave an enantiotropic nematic phase at $n = 5$ and S_A phase at $n = 7$ for the Rh complexes, and at $n = 6$ and $n = 8$ respectively, for the Ir complexes. In all cases, the metal has been found to promote mesomorphism since the imines are non-mesomorphic and the S_B and E phases of the free stilbazole have been replaced by N and S_A phases. For compounds **5**, the Ir complexes have lower transition temperatures than the Rh, while the opposite is true for compounds **6**, although the general mesomorphism is not metal-dependent. Curiously, the more anisotropic complexes *trans*-[RhCl(CO)L$_2$] (with an imino-linked pyridine) were non-mesomorphic, decomposing above 300 °C. These results were in agreement with those found for the non-centrosymmetric Pt complexes above, showing that the reduction in symmetry can lead to materials with lower melting points.

Espinet has recently described the thermal behaviour of several unsymmetrical organoisonitrile complexes of gold having the general formula [AuX(CNR)]. All the complexes, **7**, formed S_A phases between 100 and 180 °C, with one iodo derivative showing a monotropic S_A phase [16]. The biphenylisonitrile derivatives, **8**, all showed S_A phases, but at higher temperatures (120–250 °C), and decomposition was found to be a problem [17]. In general, the melting points followed the series Cl<Br<I, on account of the polarisability, but the variation of the clearing points was not so easily interpreted. Lateral monofluorination of complexes **7**, in the *ortho* (3-F) or *meta* (2-F) position relatively to the alkoxy chain, has also been investigated

[18]. The 3-fluoro (7: X = Cl, Br and I) derivatives were mesomorphic, showing a S_A phase, although at much higher temperatures than in the non-fluorinated systems. The 2-fluoro derivatives (7: X = Cl) are very weakly mesomorphic in comparison, but a narrow nematic range was seen for the hexyloxy homologue and a S_A for the others. Such promotion of the S_A phase or the N phase by selective substitution is consistent with other observations made on metallomesogenic systems [19]. The transition temperatures were found to decrease in the order 3-F > 2-F and in the order Cl > Br > I, the latter in accordance with the decrease in polarity of the Au–X bond, as above for complexes 7 and 8. Interestingly, the transition temperatures of the non-fluorinated derivatives 7 are intermediate between the 3-fluoro and 2-fluoro derivatives. Finally, thermally stable perhalophenylgold(I) isocyanide complexes (9–11) have been prepared and their liquid-crystalline behaviour reported [20]. They showed a nematic phase at short chain-lengths, N and S_A mesophases for intermediate chain-length and S_A phases only for longer chain-length. Melting points ranged between 70 and 100 °C for average chain-lengths and between 120 and 140 °C for short chain-lengths, the clearing points being greater than 150 °C. As observed by the authors, the transition temperatures decrease as follows: p-$C_6F_4Br \geq C_6F_5 > o$-C_6F_4Br for $n \leq 6$ and p-$C_6F_4Br \geq o$-$C_6F_4Br > C_6F_5$ for $n \geq 8$, which are explained in terms of polarisation and polarisability arguments. The transition temperatures have been lowered, the mesomorphic range diminished, and the nematic phase enhanced when compared to the mesomorphic properties of complexes, 8.

7: y = 0, X = Cl, Br, I
8: y = 0, X = Cl, Br, I

X—Au—CN—[—⟨ ⟩—]$_y$—⟨ ⟩—OC$_n$H$_{2n+1}$

y = 2:
9: X = C_6F_5
10: X = o-C_6F_4Br
11: X = p-C_6F_4Br

Takahashi previously reported some structurally related gold systems, 12a and 12b [21]. Complexes 12a showed S_A phases at elevated temperatures (170–270 °C), while the metal-free ligand showed nematic and S_A phases, depending on the chain length. Addition of an alkoxy chain in the *ortho* position, relatively to the isonitrile group, 12b, depressed considerably the transition temperatures (90–190 °C), and a nematic phase was observed in addition to the S_A phase. Here, complexation induced mesomorphism since the ligand itself was not liquid crystal.

Cl—Au—CN—⟨ ⟩—C(=O)—O—⟨ ⟩—OC$_n$H$_{2n+1}$ 12a: X = H
          | 12b: X = OC$_m$H$_{2m+1}$
          X

H$_{2n+1}$C$_n$—≡—Au—CN—⟨ ⟩—Y—⟨ ⟩—OC$_{10}$H$_{21}$ 13a: Y = COO, R = OMe, OEt, OPr
              | 13b: Y = OOC, R = Et, OMe, OEt
              R

$$H_{2n+1}C_nO \text{—} \langle \bigcirc \rangle \text{—} Y \text{—} \langle \bigcirc \rangle \text{≡——} Au\text{—}CN\text{—}C_8H_{17}$$

14a: Y = COO
14b: Y = OOC

$$C_nH_{2n+1}O \text{—} \langle \bigcirc \rangle \text{—} C \overset{S}{\underset{S}{\langle}} \overset{CH_3}{\underset{CH_3}{Au}}$$ **15**

Takahashi [22] and Espinet [23] both reported liquid-crystalline isonitrile gold(I) acetylide complexes, with the general formula [R—NC—Au—≡—R′]. When both the ligands contained aromatic rings, the complexes were mesomorphic, showing mainly S_A phases, but due to the elevated temperatures, decomposition occurred before clearing. In order to reduce the transition temperatures, and hopefully to avoid the decomposition of the complexes, Takahashi designed and synthesised new alkynylgold(I) isonitrile complexes which consisted of either an aromatic isonitrile and an aliphatic acetylene, **13**, or an aliphatic isonitrile and an aromatic acetylene, **14**. Compound **13a** showed S_A and N phases, the type of phase depending on the length of the side chain. Melting points increased with increasing n, while clearing points decreased, thus, narrowing the temperature range of the mesophase. Compounds **13b** ($n = 4$) showed only nematic phases, with a fairly large enantiotropic domain for the methoxy derivative (Crys·78·N· 108·I). It is worth recalling that these complexes are amongst the first gold complexes to show a nematic phase, and also to possess very low transition temperatures; decomposition was less systematically observed. Other low-melting gold complexes (**15**) were dithiobenzoate derivatives of dimethyl-gold(III) which showed S_A phases below 100 °C [24]. All the compounds **14** formed nematic phases, in the temperature range 80–160 °C. They appeared more thermally stable than compounds **13**. Complexes **14b** usually exhibit higher transition temperatures than complexes **14a**. Branching of the isonitrile octyl chain depressed the transition temperatures, but in some cases, simply suppressed the mesomorphism. So, a careful design of the ligand led to stable mesomorphic gold complexes, and decomposition does not occur below 160 °C.

Takahashi also recently reported the first mesomorphic gold(I) carbene complexes, **16** [25]. All the complexes showed S_A phases (monotropic or enantiotropic depending on Y and Z), those with enantiotropic phases having a temperature range of around 20 °C. The transition temperatures have been reduced when compared to the gold-isonitrile complexes **12**, this being attributed to the large lateral substituents which is thought to disrupt the strong intermolecular interactions, probably Au—Au interactions.

$$H_{2m+1}C_mO \text{—} \langle \bigcirc \rangle \text{—} Y \text{—} \langle \bigcirc \rangle \text{—} N \overset{H}{\underset{Z}{\langle}} =Au\text{—}X$$

X = Cl, I
16: Y = COO, OOC, —≡—
Z = OC_nH_{2n+1}, NHC_nH_{2n+1}

2.2
Compounds with Mixed Ligands

A variation of the above idea has consisted in the development of non-symmetrical 'twin' structures, that is metal complexes made of two different, usually bidentate ligands (such as Schiff bases or β-diketones). Here, the unsymmetric shape of the mononuclear complex [MLL'] is thought to disrupt the molecular arrangements more than the 'H-shaped' mononuclear [ML$_2$] and dinuclear [M$_2$X$_2$L$_2$] complexes, favouring less ordered mesophases, and at lower temperatures.

Ghedini reported a number of mononuclear cyclopalladated 2-phenylpy-rimidine complexes, some of them, 17 and 18, showing mesomorphic properties [26]. These compounds were obtained by bridge splitting reactions from the dinuclear (μ-chloropyrimidine)palladium(II) complexes with various acetylacetone and dinitrogen ligands.

17: Crys • 83 • (S$_A$ • 68) • I

18: Crys • 146 • N • 158 • I

Thus, only the acetylacetonate (acac), 17, and 2,2'-bipyridine, 18, derivatives were mesomorphic, showing a monotropic S$_A$ phase, and an enantiotropic nematic phase respectively, and at much lower temperature than the mesophases of the dinuclear parent complex (Crys·177·S$_A$·219·I) [27]. Complex 18 is a further rare example of ionic materials showing a nematic phase. Interestingly, related complexes with PF$_6$ or SbF$_6$ anions were non-mesomorphic.

Other structurally related complexes have also been prepared by cleavage of dinuclear ortho-palladated complexes with acac (19 [28], 20 [29], 21 [30]), with various chelating O,O monoanionic ligands [31], or with cyclopentadienyl groups (Cp) [30, 32]. It appears that in most cases, and for the appropriate chain length, the mononuclear complexes usually give rise to a nematic phase, and/or a S$_A$ phase, and at lower temperatures than the dinuclear precursors.

Furthermore, the mesomorphic behaviour of some complexes, such as **20**, can be very comparable with that of the free ligand (the azoxy ligand of **20** possess a nematic phase between 80 and 126 °C). The cyclopentadienyl cyclopalladated complexes mainly show nematic phases, although with wider temperature ranges and at lower temperatures than the acetylacetonate complexes.

19: $n = 6$: Crys • 84 • S_A • 118 • N • 125 • I
$n = 10$: Crys • 74 • S_A • 128 • I

20: Crys • 90 • N • 105 • I

21: $n = 4$, $m = 7$: Crys • 155 • S_A • 225 • N • 249 • I
$n = 7$, $m = 4$: Crys • 162 • S_A • 241 • N • 261 • I

Espinet and Serrano also reported metallomesogens based on monomeric, β-diketonate palladium complexes derived from ortho-palladated imines, **22**. These complexes, which are 'K-shaped', are interesting since they combine low transition temperatures, and the possibility to generate ferroelectric materials. Indeed, complex **22**, with R^2, R^3, $R^4 = OC_{10}H_{21}$, and $R^1 = (S)$-2-methylheptyloxy, is amongst the first examples of metallomesogens to show ferroelectric behaviour (Crys·114·(S_C^*·111) S_A·118·I) [33]. The number and position of the chiral chain has been systematically varied, and a large number of mixtures have been studied [34], in order to design very effective ferroelectric materials [35]. In general, the spontaneous polarisation was increased when the chiral centre was at R^1 and R^2 rather than at R^3 and R^4, or when the number of chiral centres was raised. Mixtures also gave very promising results for the use of metallomesogens in optical devices. The achiral complex **22**, with R^1, R^2, R^3, $R^4 = OC_{10}H_{21}$, also showed, in addition to thermotropic mesophases (Crys·80·S_C·150·S_A·154·N·155·I), lyotropic mesophases (enantiotropic nematic, chiral nematic and monotropic lamellar phases) in mixtures with apolar organic solvents (linear alkanes, limonene) [36].

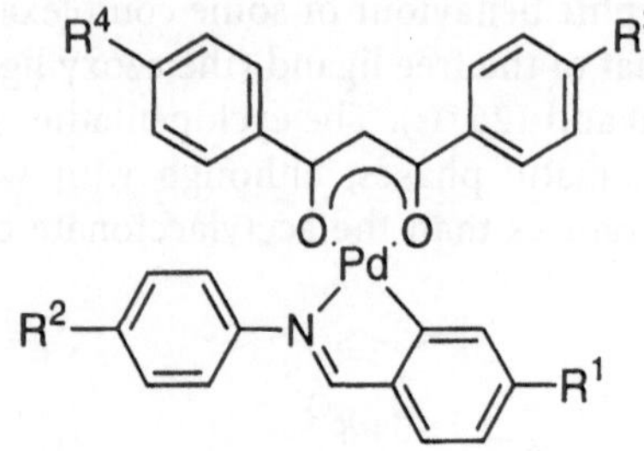

Other complexes with two distinct, elongated Schiff base ligands, with the same general formula [LML′], have also been described. Complexes **23a** (with $m = 12$), exhibit S_A and/or nematic phase as a function of the chain length n [37]. All the compounds of this series cleared at about 140 °C, despite the large difference in the chain length. This behaviour may be attributed to the existence of a mixture of two isomers for all the complexes, since the aniline groups of the two ligands can be *trans* (as shown below) or *cis* to each other. ¹H-NMR analysis showed that both the isomers, in a *cis*-to-*trans* ratio of 1:5, are present in all the complexes **23a**. In contrast, complexes **23b**, have been obtained exclusively in the *trans* conformation [38]. They all showed nematic phases, enantiotropic only for the two derivatives $n = 1$ and $m = 12$ and 14, and one complex displaying a monotropic S_C phase ($n = 8$, $m = 12$). The clearing temperatures increase with n but decrease with m.

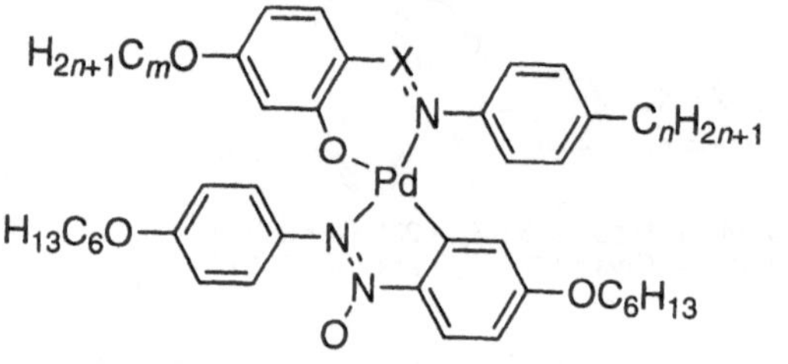

23a: X = CH
23b: X = N

2.3
Roof-Shaped Compounds

Dinuclear *ortho*-metallated complexes consist of two ligands joined through a central bridge (Fig. 1). They have been widely studied, and several types of mesogenic, metal complexes derived from azobenzene, benzylideneaniline, benzalazine (imine) and phenylpyrimidine derivatives, with a wide variety of intermetallic bridges, such halogens, thiocyanates, carboxylates, have been reported [8, 9].

In principal, the complexes exist as a mixture of two isomers (*anti* and *syn*, Fig. 1), their proportion in the mixture depending on both the type of ligands and the intermetallic bridge and being evaluated by ¹H NMR. In general, when X=Cl, Br, I, SCN, the complexes are planar or slightly bent, and show mesomorphic properties. However, when X=OAc, the complex adopts the original 'roof' or 'open-book' molecular shape, and none of the palladium

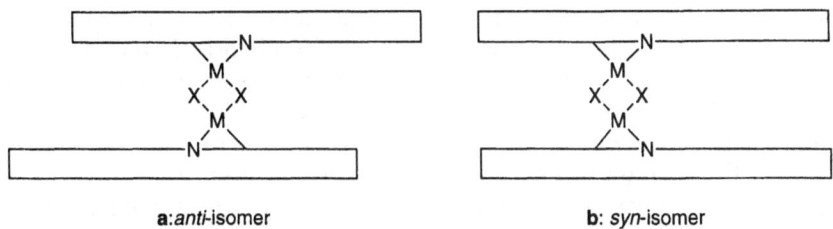

a:_anti_-isomer   **b:** _syn_-isomer

Fig. 1. Schematic diagram of the _syn_ and _anti_ isomers possible for di-μ-halo-bridged _ortho_-palladated complexes

complexes derived from azobenzene, benzylideneaniline and phenylpyrimidine was liquid crystalline. The substitution of the acetate by the chiral carboxylate, 2-chloropropionate, generated a new complex, **24**, which showed a S_A^* phase [28b, 39]. The mesogenic behaviour of **24** was attributed to the dipole moment introduced by the C–Cl bond, which increased the molecular interactions sufficiently to compensate the unfavourable steric effects. Furthermore, chiral nematic phases were obtained when one of the chloropropionato bridges was replaced by a thiolato group. The mixed-bridge complexes thus produced, **25**, exclusively as the pure _syn_ isomer, represent the first metallomesogens to display the cholesteric phase. The platinum congeners of **24** and **25** showed a tendency to display more ordered mesophases (S_A phase in place of N or N*), with higher transition temperatures, due to the higher polarisability of Pt [40].

24: Crys • 151 • S_A^* • 193 • I

25: $n = 6$:Crys • 140 • N* • 157 • I
$n = 10$:Crys • 109 • S_A^* • 115 • N* • 138 • I
$n = 18$:Crys • 92 • S_A^* • 128 • N* • 132 • I

In contrast, dinuclear, acetato-bridged palladium complexes derived from benzalazine ligands, **26** ($m = 1$), showed S_C ($n \geq 7$) and N ($n = 6-8$) phases, although, over much narrower temperature ranges than the chloro-, bromo- and thiocyanato-related complexes [41]. They melt at around 100–130 °C and clear at 150 °C. The existence of a mixture of _syn/anti_ isomers is thought to be responsible for the decrease in the transition temperatures, and thus of the liquid-crystalline properties. Interestingly, the mixture is composed of 25% of the _syn_ isomer and 75% of a racemic mixture of the two enantiomers of the _anti_ isomer. An important systematic study has also been carried out, consisting of the variation in the length of the carboxylato bridge (26: $n = 10$, $m = 0$–17) [42]. A nematic phase was seen for almost all the complexes (except for $m = 0,1$), and a S_C phase observed for the shortest ($m = 0$–3) or longest carboxylate chains ($m \geq 10$), the mesomorphic temperature range being smaller for middle-m values. This complex behaviour was explained in terms of the position of the carboxylate chain relatively to the spine of the 'open-book' complex. Short carboxylate chains do not affect dramatically the anisotropy of the molecule, nor sterically disrupt the molecular packing. However, when m increases the chains reduce the lateral interactions and the anisotropy of the molecules (decrease in the length-to-width ratio) thus favouring the nematic phase. When m becomes longer, the chains stretch along the main molecular axis, and unfavourable effects are reduced, allowing the complexes to have both types of liquid-crystalline molecular organisation.

26

Finally, the first example of a ferroelectric metallomesogen was obtained with a chiral derivative of **26** [43]. A derivative was synthesised by ligand exchange where the carboxylate was (R)-2-chloropropionate (instead of C_mH_{2m+1}), and a mixture was produced whose composition was evaluated to be *anti*-ΛR,R: 34%; *anti*-ΔR,R: 34%; *syn*-R,R: 32% by ^{1}H NMR. The resulting mixture showed a S_C^* phase (Crys·102·S_C^*·119·S_A^*·149·I) which was found to switch at a speed of 330 ms at a square wave voltage of ±17 V and 0.5 Hz and a cell thickness of 11 mm.

3
Metallomesogens Possessing Lateral Chains

The introduction of a lateral aliphatic chain in organic liquid crystals has proved to be a very good way of producing low-melting, organic materials, and of promoting nematic phases. Many studies have been performed on the 1,4-bis(4-alkoxybenzoyloxy)-2-alkylbenzenes, **27** [44], which showed that the increase in the length of the lateral alkyl chain was associated with a substantial decrease in the clearing temperature.

27

Another interesting result was the strong tendency of the materials to give nematic phases and to suppress completely the smectic phases. This latter fact was attributed to the lateral chain acting as a perturbation and reducing the lateral interactions between the molecules, thus decreasing the segregation between the aromatic and aliphatic parts of the molecule. Similar work, but with two lateral chains, has also been carried out and shown also to generate low melting nematogens [45].

Many square planar complexes have the general formula $trans$-$[ML_2X_2]$, where L is an organic ligand containing the rigid aromatic part (nitrile, isonitrile and alkynyl ligands) and X is a co-ligand (often a halide [46, 47], or a tertiary phosphine [48, 49]. The substitution of the co-ligands by an amphiphilic ligand, such as an alkylcarboxylate, has yielded interesting results when applied to the $trans$-bis(4'-alkoxy-4-stilbazole)dichloropalladium(II) complexes. In the complex **28** (X=Cl, $n = 12$), the transition temperature was greater than 280 °C and extensive decomposition occurred before reaching the clearing point, making the identification of the mesophase difficult [8a]. However, when the chlorine atoms were replaced by two alkylcarboxylate groups [50] (**28**: $X = C_mH_{2m+1}CO_2$, $m = 5$–7), the complexes formed an enantiotropic nematic phase which cleared at around 150–170 °C. A substantial decrease in the melting point ($\Delta T = 130$ °C) was, therefore, easily achieved in this way.

28: X = Cl, $OOCC_mH_{2m+1}$

A similar approach has been used for metallomesogens derived from Schiff base ligands. This type of complex represents an important class of metallomesogens, which has been widely studied by several groups [8, 9]. Upon complexation, the two salicylaldiminato-based ligands arrange themselves in an antiparallel fashion giving rise to 'fused-twin' metallomesogens, by analogy with organic liquid crystals. In general, Ni(II), Cu(II), VO(IV), Pd(II), Pt(II) and FeCl(III) complexes of N-4'-alkoxy- and N-4'-alkylphenyl-4-alkoxysalicylaldimine usually show a strong tendency to promote mainly smectic phases (A and C), while complexes of N-4'-alkoxyphenyl-4-benzoyloxy salicylaldimine show mainly smectic C and nematic phases, the T_{M-N} (M=Crys, S_C) being around 160–200 °C, with T_{N-I} often greater than 200 °C. In order to obtain nematic phases at lower temperatures, the N-phenyl moieties were replaced by N-alkyl groups. Considerable attention has thus been focused on *N*-alkylated metal complexes of the type **29**, in which both *n* and *m* were varied independently. Several metal ions were used, and in a few case, the position of anchoring was changed (that is the ligand were derived from 2,4-dihydroxy- or 2,5-dihydroxybenzaldehyde, **29** and **30** respectively). Several series of complexes **29** (abbreviated *n*M*m*) were prepared (10Cu*m* and 10Ni*m* [51], *n*Cu3 and 6Cu*m* [52], 10VO*m* [53], *n*Cu1 [54], 14Cu*m* [55], *n*Ni3 and *n*VO3 [56], 6Cu*m*, 10Cu*m*, *n*Cu8 and *n*Cu13 [57] and 7FeCl12 [58]).

29

Almost all of the complexes show a nematic phase, in addition to a S_C phase for a few of them. The complexes *n*Ni3, 10Ni*m* and *n*VO3, all display a nematic phase only. In contrast, the Fe(III) complex has a S_A phase (Crys·85·S_A·151·I). The high number of copper complexes synthesised allowed a rather accurate view of the mesomorphic behaviour as a function of *n*, *m* and T, and their behaviour was rationalised (Scheme 1). The nematic phase was seen for all the complexes ranging from $1 \leq m \leq 14$ and $1 \leq n \leq 18$, and the S_C phases were only observed for long alkoxy substituents ($n \geq 10$ when *m*=1) and for both large *n* and *m* ($m \geq 13$). At constant *m*, T_{cl} decreases smoothly with increasing *n*, and T_{M-I} (M=Crys or S_C) is almost constant for *m*=8 and 13, but decreases quite rapidly for *m*=3. At constant *n*, T_{cl} follows the same behaviour, while T_{M-I} starts by decreasing rapidly and then plateaus or even slightly increases. Thus, the mesophase temperature range decreases quite substantially with both *n* and *m*, varying from ΔT=40–50 °C to ΔT=1 °C. The most accessible nematic phases (between 100 and 140 °C) were found for average values of *n* and *m* (between 6 and 10). Furthermore, the effect of the metal on the mesomorphism can be explained qualitatively by the geometry of the coordination which should be square planar for the copper, square pyramidal for the vanadyl and an intermediate situation between tetrahedral and square planar for the nickel complexes (distorted square planar). In general, the nematic thermal stability follows the order $NiL_2 > CuL_2 > VOL_2$.

The incorporation of lateral chains in the series of complexes **30** had dramatic effects on the mesomorphism [59]. Indeed, almost all the complexes displayed only monotropic mesophases, essentially nematic, with a few only showing S_C phases. The only enantiotropic mesophases were obtained for *m*=1 for both Ni and Cu complexes (Crys·239·S_C·267·N·281·I and Crys·188·S_C·243·N·267·I, respectively).

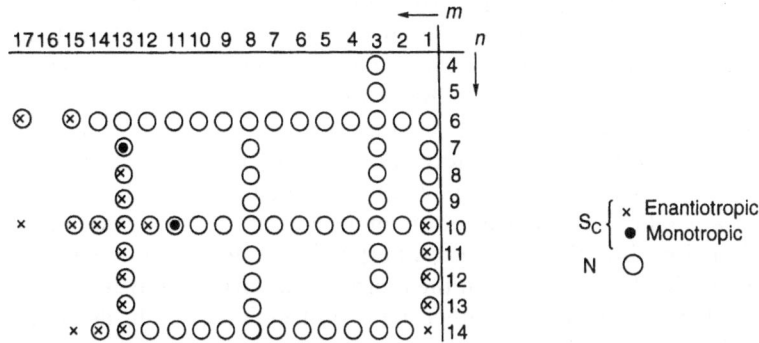

Scheme 1. Grid to show the dependence of mesomorphism upon the chain lengths in salicylaldimato complexes of copper(II)

$$H_{21}C_{10}O \quad\cdots\quad C_mH_{2m+1} \quad\cdots\quad H_{2m+1}C_m \quad\cdots\quad OC_{10}H_{21}$$

30

Thus, nematic phases have been observed for most of the N-alkylimine complexes. The type of mesophase obtained is directly linked to the length of the alkyl attached to the nitrogen atom. For the complexes 29, when $m > 1$, the alkyl groups probably lie parallel to the direction of the main molecular axis, and as such, prevent the molecular interactions responsible for the formation of the S_C phase (Fig. 2b). Further, with the loss of segregation between the aromatic and aliphatic parts of the molecules, the nematic phase is greatly promoted. When the substituent is a methyl group, the interactions between the aromatic parts of the molecules remain almost unaffected, and, due to the segregation of the aliphatic and aromatic parts, the smectic phase is formed (Fig. 2a). Interestingly, when m becomes longer ($m \geq 11$), the S_C reappears, and this corresponds to the crossing of the aromatic/aliphatic interface by the alkyl

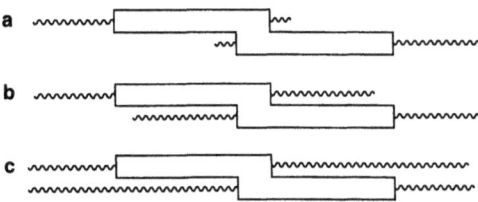

Fig. 2a–c. Schematic diagram of different lateral chain lengths in 4-substituted salicylaldimato complexes

chain; as a consequence, the segregation is increased leading to layered structures (Fig. 2c). X-Ray measurements suggested two different types of molecular organisation for the S_C phase for compounds nCu1, and that there was no evidence of a local coupling of the molecules in pairs or in ribbons, implying that metal–metal interactions are not essential to drive liquid crystallinity, and that the complexes behave exactly as organic mesogens.

The overall structure of compounds 30 is such that the alkyl substituents prevent a parallel arrangement of the molecules in the mesophase on the one hand, but also, since the chains grow perpendicular to the molecular main axis (Fig. 3a,b), the distance between the molecules is considerably increased. Both of these chain effects resulted either in the destruction or in the decrease of the thermodynamic stability of the mesophase.

From X-ray diffraction experiments carried out on a few of these complexes [60, 61], it was also deduced that the copper complexes of the series 29 possessed a negative anisotropy of the magnetic susceptibility, while it was positive for all the other complexes 29 and 30. Chiral nematic phases (and blue phases in some examples) have also be obtained by replacing the linear alkyl group by a chiral chain in the complexes 29 [62].

Structurally related β-diketonato metal complexes with lateral chains have also been synthesised. The first complex of this type, 31a, was not mesomorphic [63], nor was the analogous copper complex, 31b [64], essentially due to the small anisotropy of the complexes. However, more 'rod-like' complexes, such as 32, showed monotropic nematic phases [65]. Only a few members have been synthesised, and thus it is rather difficult to rationalise the thermal behaviour of these compounds. Anyway, melting points were found to decrease very steeply with rising m, suggesting again that the lateral chains disrupt the efficient molecular packing of the complexes in the crystal state. The crystal structures of two homologues (32: n=3 and m=1 and 8) have been investigated, and indicated that the molecules as a whole are not strictly planar. The benzene rings of the substituent groups are slightly tilted off-plane relatively to the plane containing the metal chelate ring (ca 12° and 21° for the two compounds respectively). Furthermore, the molecules are organised into layers in the crystal state, with no intermolecular contacts. The thermal stability of these complexes being small above 150 °C, a binary mixture between two of them has been investigated (A: 32: n=3, m=7 and B: 32: n=3, m=8), and an enantiotropic nematic phase was found between 30 and 60 mol% of B, with an eutectic point at 54 mol% of B.

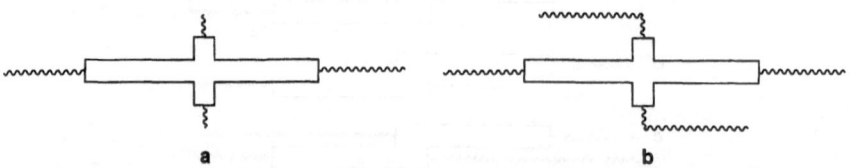

Fig. 3a,b. Schematic diagram of different lateral chain lengths in 5-substituted salicylaldimato complexes

31a: $m = 1$, M = Pd, R = OC_8H_{17}
31b: $m = 1$, M = Cu, R = C_8H_{17}
32: M = Cu, R = $C_6H_{10}C_nH_{2n+1}$
33: M = Cu, R = $C_6H_4OC_nH_{2n+1}$

Another series of more rigid, copper-based β-diketonato complexes, **33**, have also been prepared and showed interesting liquid-crystalline behaviour, in which subtle structural changes led to mesophases of different nature. Indeed, the mesomorphism of complexes **33** included columnar and nematic phases [66], the type and the thermodynamic stability of the mesophases being greatly influenced by the length of the lateral chains. The ligand itself was mesomorphic, showing a crystal E and a S_A phase. The rectangular columnar phase (Col$_r$) was found for all homologues with $m=1$, although when m was raised, the columnar phase disappeared at the expense of an enantiotropic nematic phase (**33**: $m=2$ and $n=6$, 8, 10, 11, 12, 14, $m=3$, $n=12$, and $m=4$ and $n=4$, 8) and a monotropic nematic phase for longer m and n pairs. Furthermore, re-entrant nematic phases, between two crystalline phases, were also observed for the compounds in the series $m=2$. T_{M-N} and T_{N-I} decreased very rapidly with increasing m, as observed for the N-alkylimine complexes, **29**. Ohta and co-workers suggested that, owing to the small size of the lateral methyl groups, the complexes would be able to form dimers with an overall discoid shape, suitable to generate a columnar phase [66b]. However, this model generates a large unoccupied volume, which cannot be realistic in the mesophase. Another hypothesis would be to consider the breaking of the crystalline lamellae into finite ribbons of molecular clusters (Fig. 4a), in such a way that the diffraction pattern could be indexed into a 2-D rectangular lattice. Further increase in the length of the lateral chains would decrease (or suppress) the lateral interactions between molecules (Fig. 4b), thus leading to the formation of the nematic phase. When the lateral chains are even more bulky, the intermolecular interactions would be reduced to a greater extent, and the nematic phase would be less thermodynamically stable. Yet again, this is another example that confirms that the shape of the molecule itself does not necessarily accounts for the nature of the mesophase, but rather it is the nature and shape of the molecular aggregate.

Copper complexes derived from enaminoketone ligands having two lateral chains attached on the nitrogen atom (**34**) or on the carbonyl group (**35**), were found to behave in a similar way to the complexes above, that is, in most cases the nematic phase was promoted over the smectic phases [67]. In the first series (**34a**), the nematic phase is present for all the complexes ($m=3$–18), in addition to a monotropic and an enantiotropic S_C phase for the longer homologues ($m=15$ and 18 respectively). The ligand itself showed these two phases too, but the S_C phase appeared enantiotropic from $m=10$ onward. T_{N-I} decreased for all homologues, while T_{Crys-N} firstly decreased, then rose for

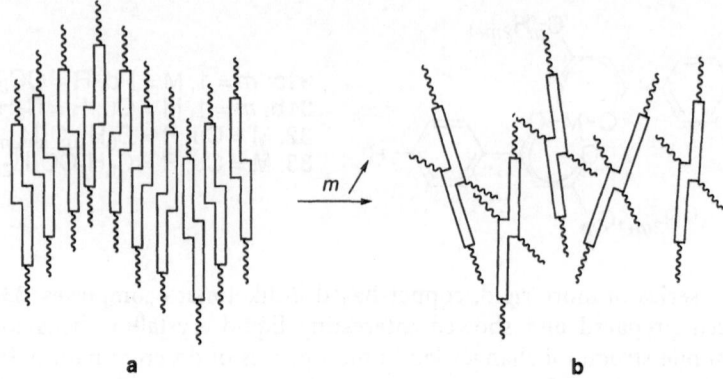

Fig. 4a,b. Schematic diagram showing effect of lateral chains on the mesomorphism of copper β-diketonate complexes

$m \geq 11$. The authors attributed this phenomenon to non-planarity of the alkyl chain relative to the chelating ring (slight distortion of the chain). Substitution by terminal alkylamino groups as the terminal chain significantly depressed clearing and melting points. For compounds, **34b**, layered structures were favoured, while in the other series, **34c**, only nematic phases were observed, this difference being attributed to the stronger possibility of H-bonding formation in series **34b** (through the NH groups) than in series **34c**. In a second part of this study, the lateral chains were attached to the carbonyl groups, **35**. The complexes showed smectic and nematic phases, the smectic phases being largely promoted over the nematic phase (for the azo compounds **35a**). It can be deduced from this comparative study, that when the phenyl substituents are attached to the amino groups, **35**, smectic phases are favoured, while when attached to the ketone groups, **34**, nematic phases are promoted.

34a: X = O
34b: X = NH
34c: X = N(CH₃)

35a: X = N=N
35b: X = N=N(O)

These results are in agreement with some previous work carried out on similar, structurally related enaminoketone complexes (**36, 37**), and suggest similar explanations [68]. Only enantiotropic nematic phases were formed for complexes **36**, which were destabilised as m increased, while complexes **37** showed nematic and S_A phases.

36

37

An interesting piece of work, carried out with salicylaldimato complexes of Cu(II), **38**, containing four aromatic rings per ligands [69], has shown that the nematic phase was largely promoted over smectic phases when the alkoxy chains were added at the centre of the molecule (**38**: position 3), despite the considerable length of the complexes. An evident decrease in the transition temperatures could also be observed when compared to non-substituted parents [70], or to the free ligand, if the lateral chains contain a sufficient number of carbon atoms: the longer the lateral chain, the wider the nematic temperature range. The grafting of an additional chain in the terminal benzoate ring resulted in the decrease in the thermal stability of the nematic phase. When two lateral chains were grafted to the central core of the copper complexes (**38**: positions 2 and 3), liquid-crystalline properties were preserved, and fairly wide nematic domains were also observed [71]. The clearing points, as well as the temperature range, were also very similar to the free-ligand. It seems that the key to enlarging the range of the nematic phase was due to the grafting of lateral chains in the centre of a rigid, rod-like core, which resulted in the decrease of lateral interactions.

38

Rourke has recently demonstrated [72] that di-cyclopalladated Schiff base complexes, with β-diketones of various chain-length as co-ligands, **39**, were mesomorphic. They showed nematic phases, with a decrease in the clearing points as m was raised (melting points were reduced, when compared to complexes **39** with $m=1$, by around 100 °C), although at the expense of a large mesophase temperature range.

39

4
Polycatenar Compounds

4.1
Tetracatenar and Hexacatenar Complexes

In most cases, the shape of a mesogen determines the symmetries of the resulting thermotropic mesophases. For example, ellipsoidal molecules give rise to nematic and lamellar phases [73] (N, S_A, S_B, S_C,...), whereas discoid molecules display columnar mesomorphism [74]. This led to an inappropriate mesophase classification based on the structure of the mesogen itself rather than the structure of the molecular aggregate. This was especially misleading particularly in the light of Demus' review [75] containing nearly 50 types of thermotropic liquid crystals with 'conventional and unconventional molecular structures', which fulfil neither the ellipsoid nor discoid criterion. Amongst many classes of fascinating liquid crystals, such as the carbohydrate [76], dimeric [77] and twin [78] liquid crystals, the polycatenar mesogens [79] were found to constitute a class of particular interest.

Polycatenar liquid crystals are molecules that have a long aromatic core with several paraffinic chains grafted onto the terminal aromatic rings. As such, they can be seen as a 'missing link', having an aspect ratio intermediate between that of the classical ellipsoidal and discoid mesogens. Their remarkable mesomorphism includes nematic, S_C, cubic (Cub) and columnar (Col) mesophases. In particular, some tetracatenar systems possess all of the mesophases listed above, as the chain length is varied. Polycatenar molecules can be regarded as being amphiphilic [80] in character in the sense that segregation of the aromatic and aliphatic parts tends to occur in the mesophase, resulting in the variation of the curvature of the paraffinic-aromatic interface. Increasing the number of carbon atoms in the aliphatic chains will increase the discrepancy between the core and the aliphatic chains (the ratio between the mean chain area and the core area increases), leading to the destabilisation of the N and S_C phases in favour of the mesophases with curved interfaces such as the columnar and the cubic phases.

There are, in fact, rather few such compounds. We have recently reported the thermal behaviour of several series of polycatenar metal complexes, 40–45, and observed very interesting relationships between the structure of the metal complexes and their thermal behaviour [81].

Tetracatenar
40: $R^1 = R^4 = H$; $R^2 = R^3 = OC_nH_{2n+1}$
41: $R^1 = R^3 = H$; $R^2 = R^4 = OC_nH_{2n+1}$
42: $R^1 = R^3 = OC_nH_{2n+1}$; $R^2 = R^4 = H$
Hexacatenar
43: $R^1 = H$; $R^2 = R^3 = R^4 = OC_nH_{2n+1}$
44: $R^1 = R^2 = R^3 = OC_nH_{2n+1}$; $R^4 = H$
45: $R^1 = R^3 = R^4 = OC_nH_{2n+1}$; $R^2 = H$

Of those investigated, the tetracatenar complexes of Pd and Pt with 3,4-disubstituted stilbazoles, 40, one complex 41 (M=Pd, n=14), and the hexacatenar complexes of Pd with 3,4,5- and 2,3,4-trisubstituted ligands (43 and 44, respectively) were mesomorphic. The two other series of complexes

(42 and 45) were not mesomorphic but melted to the isotropic liquid at moderate-to-low temperatures. The tetracatenar palladium complexes, 40, showed a S_C phase (n=4–12) and a hexagonal columnar mesophase (n=13, 14, 18). Such a sudden change from smectic to columnar behaviour as a function of chain-length is not so common in polycatenar mesogens, and it is often the case that the two mesophases co-exist in one or more single compounds in a sort of 'cross-over' behaviour. This is often accompanied by the appearance of a cubic phase. The transition temperatures (still relatively high), and the mesophase temperature range decrease with increasing n. Nevertheless, the mesophase are more accessible than the related trans-bis(4'-alkoxy-4-stilbazole)dichloropalladium(II) complexes 28 (X=Cl) showing this to be an effective strategy for reduction in transition temperatures; the complexes were also rather stable at elevated temperatures. The platinum complexes, 40, behaved in a very similar manner, although they did not display a columnar mesophase for higher values of n. As expected, the hexacatenar complexes, 43, and 44, showed columnar mesomorphism close to room temperature. Complexes 43 displayed only one mesophase (for $n \geq 8$), probably Col_h, from optical textures, while complexes 44 displayed two columnar mesophases. In the latter, the transition between the two mesophases was not detected by DSC, suggesting minimum structural differences between them. The transition temperatures of the hexacatenar complexes were much lower than the tetracatenar complexes, and the mesomorphic domains much larger (ΔT=50–80 °C).

Looking at the number and the position of the alkoxy substituents, a relationship between the molecular structure and the mesomorphic properties can be emphasised. While substituents at the para position seem fairly important to promote mesomorphism, the columnar behaviour is clearly connected to substitution at the meta position. Substitution at the ortho position seemed to disfavour the induction of calamitic mesomorphism, as the lateral interactions and the molecular segregation are reduced, but does not disfavour columnar mesophases, providing that the space is properly filled by aliphatic substituents.

A two-step model can account for the formation of the columnar phases, starting from layered structures. Previous studies showed that both tetracatenar and hexacatenar molecules are arranged in crystalline planes (Fig. 5a), with segregation of the aliphatic chains and aromatic cores, the interactions between adjacent planes being weak. When heated (and n increased), these interactions are even more weakened allowing the planes to undulate (Fig. 5b), and creating discrete clusters of molecules, and free volumes. The void thus created will be filled by the sliding of adjacent 'layers' (Fig. 5c), leading to a columnar type of molecular organisation. This hypothesis suggested by Guillon and co-workers [82] is also supported by the general high enthalpies of transition suggesting a mechanism of transformation containing several steps.

4.2
Fused-Polycatenar Mesogens

Fused-polycatenar metallomesogens is the name given to metallomesogens resulting from metal-complexation to polycatenar ligands, by analogy to

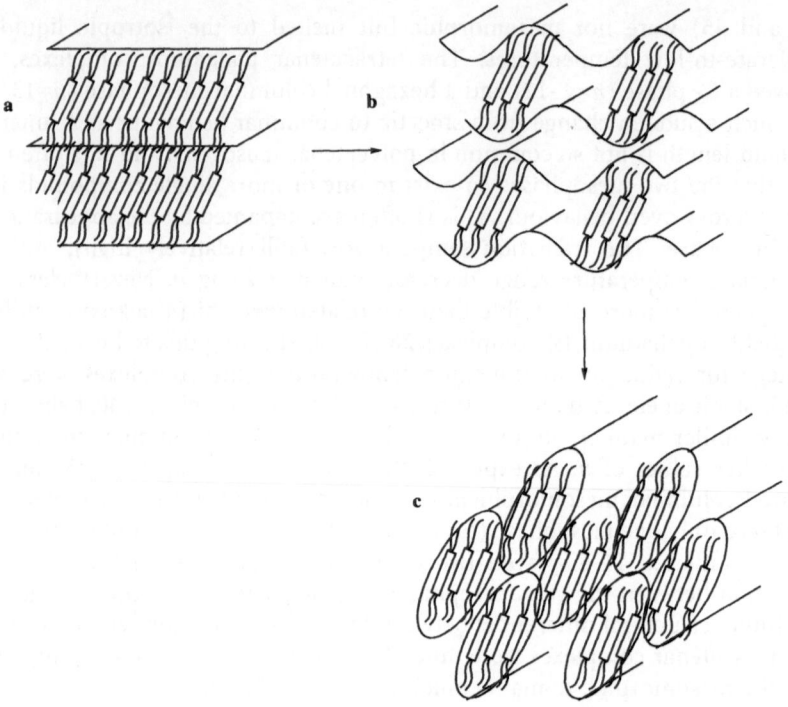

Fig. 5a–c. Schematic diagram to show the transition from the smectic C to the columnar phase

Siamese-twins mesogens. Courtieu and co-workers [83] have investigated the effects of such substitution on some copper complexes based on elongated salicylaldimines, **46**. The position of the additional chain was varied, as was its length, which could be varied independently. When the chain was in the 2-position, the nematic phase was monotropic for $n=m=4$, 8 and 16 and enantiotropic for $n=m=12$), although over a range of only a few degrees. However, when the chain was attached at the extremity of the molecules (3-position), only S_C phases were observed, and the nematic phase seen for the ligand was suppressed upon complexation at the expense of the S_C phase. These series of compounds do not strictly behave as true polycatenar mesogens in the sense that no columnar mesomorphism was observed, despite their relatively low aspect ratio.

Krowczynski and co-workers [84] have recently studied the mesomorphism of some 'lath-like' copper complexes of enaminoketones, **47**. It was thought that this kind of structure would fulfil the necessary requirements to generate materials displaying biaxial nematic phases [85]. Four types of such complex were prepared, all of them differing by the number and position of the alkoxy chains. In the first series, **47a**, the complex showed a S_C phase at rather elevated temperature (Crys·164·S_C·203·I). When an extra chain was added, **47b**, the melting and clearing point dropped substantially, and the mesomorphic range considerably reduced ($\Delta T \leq 10$ °C). **47b** showed enantiotropic S_C and N phases (Crys·74·S_C·84·N·85·I), although the nematic phase existed over a very narrow temperature range (around 1 °C). In contrast, **47c** behaved rather differently, giving rise to monotropic S_C and S_A phases, Crys·108(S_C·73·S_A·80)·I. Furthermore, the melting points were raised by almost 30 °C, when compared to its isomeric complex, **47b**. Finally, the octasubstituted complex, **47d**, was not mesomorphic and simply melted at 105 °C. Thus, in this system, the substitution pattern influences the mesomorphism in a very unusual manner, which can be the consequence of the relative position of the two aromatic rings of each ligand with the chelating system: the aniline rings lie perpendicular to the rest of the complex, and so the chains attached to the two phenyls lie in two different perpendicular planes. Conoscopic investigations revealed the nematic phase to be uniaxial.

47a: $R^1 = R^2 = H$
47b: $R^1 = H$, $R^2 = OC_8H_{17}$
47c: $R^1 = OC_8H_{17}$, $R^2 = H$
47d: $R^1 = R^2 = OC_8H_{17}$

Praefcke and co-workers reported a series of disc-shaped metallomesogens, the dinuclear *ortho*-palladated benzalimine complexes, **48**, which are the first example of metal complexes showing the nematic phase of disc-like molecules, N_D [86]. The flat, dinuclear halogeno- and thiocyanato-bridged complexes (**48**: $m=n=6$) exhibited a monotropic N_D phase, whereas the acetato-bridged complex was not mesomorphic, a consequence of its book-shaped structure. The structurally related chloro-bridged platinum complex also showed the N_D phase on cooling, although, it melted at much higher temperature than the palladium homologue [87]. Interestingly, unlike the halogeno-bridged palladium complexes, the platinum complex exists as an isomeric mixture *syn/anti* 17:83, and attempts to separate the two isomers were unsuccessful because of decomposition processes. Thus, in this study, the number of peripheral chains is crucial in determining the type of mesophase observed, since related *ortho*-palladated imine complexes with four alkoxy chains exhibited exclusively calamitic mesomorphism, mostly S_A phases [88].

X = Cl, M = Pd: Crys • 79 (• N • 43) • I
X =Br, M = Pd: Crys • 73 (• N • 43) • I
X = I, M = Pd: Crys • 97 (• N • 43) • I
48: X = SCN, M = Pd: Crys • 96 (• N • 43) • I
X =OAc, M = Pd: Crys • 56 • I
X = Cl, M = Pt: Crys • 92 (• N • 54) • I

All these complexes form charge transfer systems when doped with strong electron acceptors such as 2,4,7-trinitrofluorenone (TNF) [89]. Thus, columnar hexagonal phases are induced in all the halogeno-bridged complexes-TNF binary mixtures, the iodo-bridged palladium complex showing an additional N_D phase at higher TNF concentrations, although an N_D phase alone was stabilised for the thiocyanato-bridged complex-TNF system. The authors explained these differences in mesophases 'induction' or 'stabilisation' in terms of structural differences caused by the nature of the bridging group, X, different in length, space fillings (steric effects), and electronic effects [90].

Praefcke and co-workers also carried out an important piece of work on large, macroheterocyclic, tetrametalorganyls derived from polycatenar bis(i-mine) ligands, **49** and **50** [86, 91]. Almost all the complexes were mesomorphic, showing mainly oblique columnar mesophases (an ill-defined mesophase was observed for **49a** and **50a**).

M = Pd, Pt
X = **a**: OAc, **b**: Cl, **c**: Br, **d**: I, **e**: SCN, **f**: N₃

Complexes **49** melted below 70 °C, but cleared at very high temperatures (270–300 °C), with extensive decomposition in the isotropic liquid. Complexes **50**, differing by the spacer unit, and for which the ligand itself showed a nematic phase, melted at much higher temperatures than complexes **49** (between 54 (**50e**) and 125 °C (**50b**) higher), and their clearing were significantly lower, being in the range 240–260 °C. Thus, the transition temperatures, and then the mesophase temperature ranges, are very much affected by the choice of spacer (1,4-phenylene and 4,4'-stilbenylene), but the nature of the thermotropic mesophases does not seem to depend on the metal, nor on the rigid spacer, but rather depends on the type of bridging group. However, a recent study showed that the number of peripheral chains (12 or 8 chains), their position (2,3, and/ or 4) and length (n=8–18), also influenced the nature of the columnar phases [92].

In addition, these complexes, except **49a** and **50a**, form lyotropic columnar (oblique) and nematic phases when dissolved in linear, apolar organic solvents (alkanes) over wide temperature and concentration ranges. Interestingly, for some of them, **49b–c**, an unexpected transition between two lyotropic nematic phases has been observed, for which a model has recently be proposed [93]. As for **48**, formation of lyotropic nematic and columnar mesophases is also extended by π–π interactions with electron-acceptors, such as TNF, in apolar solvents (pentadecane). Induction of chiral nematic phases by charge transfer interactions, in a ternary mixture (**49b**/alkane/TAPA; TAPA is 2-(2,4,5,7-tetranitro-9-fluorenylideneaminooxy)-propionic acid and is used (and is available commercially) enantiomerically pure), has recently been demonstrated for the first time [94], and opens new perspective for producing chiral nematic phase of disc-like compounds.

The dipalladium organyls **51**, derived from **49** (M=Pd, X=Cl) by ligand exchange reaction between the bridging group and acetylacetonate, are not mesomorphic in their pure state, but form mesomorphic charge transfer systems with the electron acceptor TNF [95]. The identity of the induced mesophase is still unknown, but seems very likely to be columnar [96]. Furthermore, lyotropic nematic phases were obtained in the ternary mixture **51**/TNF/linear alkane (the binary mixture **51**/alkane did not yield mesomorphism). The nematic phase in this system is though to have a columnar nature, namely a nematic columnar phase.

51

5
Cubic Phases in Thermotropic Metallomesogens

Cubic phases have long been known in the field of amphiphilic liquid crystals (surfactants [97] and lipids [98, 99], and have been studied extensively. For example, it has recently been proved that lyotropic, cubic phases play an important rôle in the control of biological processes such as membrane fusion, fat digestion and the ultra-structural organisation of cells [100]. They have also been observed in conventional calamitic liquid crystals but still remain rather rare [101]. Both lyotropic and thermotropic cubic phases show a very well-developed, long-range, three-dimensional order, but are, however, highly orientationally disordered as evidenced by the presence of diffuse reflections in the wide angle regions, corresponding to the liquid-like, disordered state of the hydrocarbons chains. They have relatively large lattice parameters and contain a large number of molecules within the unit cell. These phases appear transparent and optically isotropic under a polarising microscope, owing to their cubic symmetry.

Silver(I) complexes of 4′-alkoxy-4-stilbazoles, 52, have proved to be metal-containing liquid crystal systems displaying a remarkable mesomorphism. Derivatives with $X=BF_4$, NO_3, CF_3SO_3 and $C_mH_{2m+1}OSO_3$ ($m=8, 10, 12, 14$) have been synthesised, and all were mesomorphic. Complexes 52 with $Y=Z=H$ showed S_C and S_A phases, with a nematic phase at short length being found for the triflate and alkylsulphate complexes; the BF_4 and NO_3 derivatives [102] were thermally unstable at the high clearing temperatures (around 250 °C), while the triflate salts were more stable. However, the alkylsulphate derivatives differed significantly by their much lower melting and clearing temperatures.

The dodecylsulphate [103] complexes (52: $X=C_{12}H_{25}OSO_3$, $Y=Z=H$) were found to exhibit a cubic phase between a S_C and a S_A phase when the alkoxy chain length was hexyloxy or longer. In the octyl- [104] and decylsulphate derivatives (52: $X=C_8H_{17}OSO_3$ and $C_{10}H_{21}OSO_3$), the cubic phase was not present and the complexes showed a typical calamitic mesomorphism (S_A and S_C), in addition to a nematic phase at short chain lengths. However, the isotropic phase reappeared for the tetradecylsulphate [105] derivatives (52: $X=C_{14}H_{29}OSO_3$). Interestingly, the cubic phase was also found in a few derivatives of the 4′-alkoxy-2′-fluoro-4-stilbazole silver(I) dodecylsulphate complexes (52: $X=C_{12}H_{25}OSO_3$, $Y=H$, $Z=F$), but not in its structural isomer ($Y=F$, $Z=H$) [19]. Here the cubic phase appeared below the isotropic liquid and, in one case, below a nematic phase. In a recent part of our studies on silver-containing liquid crystals, we described an X-ray determination of a cubic monodomain [106] shown by some derivatives of the fluorinated and non-fluorinated silver dodecylsulphate systems. These monodomains were easily indexed in the Ia3d cubic space group, similar to some lyotropic and other thermotropic materials; this space group was later confirmed by freeze-fracture electron microscopy (FFEM) [107]. The cubic phase is thought to arise, at least in these systems, from the combination of the effects of the curvature at the aromatic/aliphatic interface, and the relative volumes occupied by the core and chain parts of the molecule. In addition to this very rich mesomorphism, a metastable and previously uncharacterised phase, the S_4 phase, which was considered by many as columnar on the basis of its optical texture, has also been observed (52: $X = C_{12}H_{25}OSO_3$, $Y = Z = H$). By X-ray diffraction, we found that the phase has a tetragonal structure, with the $I4_1/acd$ space group, corresponding to a slight distortion in one direction of the $Ia\bar{3}d$ cubic phase [108].

Complexation of 3,4-dialkoxystilbazole to silver(I) (52: $X = C_{12}H_{25}OSO_3$, $Y = OC_nH_{2n+1}$, $Z = H$) has led to a new series of metal-containing liquid crystals [109]. In principal, the complexes can be identified as tetracatenar, but this classification is confused by the presence of the dodecylsulphate anion which, we assume, acts as a lateral group. Nevertheless, the complexes were found to behave broadly as polycatenar mesogens do, showing cubic ($4 \leq n \leq 10$) and hexagonal columnar ($n \geq 6$) phases. The $Ia\bar{3}d$ symmetry of the cubic phase, and the 2D-hexagonal lattice were identified by means of X-ray diffraction and FFEM [110]. It is important to note that these two techniques were in good agreement, and that the freezing and replication processes did not alter the structural parameters of the mesophases. The transition $Ia\bar{3}d$-to-Col_h has been studied by a combination of X-ray diffraction and dilatometry. There are six complexes in the columnar repeat unit (or three dimers), and the exponential variation of the columnar cross section (with n and T) may be explained by undulations in the columnar structure. The linear relationships between the volumes of the cubic lattice and the repeat unit of the columnar phase, and the proposed undulations, allows a model to be advanced for the transition between the two mesophases. At the transition, the amplitude of the undulations become sufficiently large to allow the coalescence of the columns to form a complicated 3D-network, which further rearranges to generate the two 3D-rod networks, typical of the $Ia\bar{3}d$ structure. This model successfully accounts for the observed epitaxial relationship between the $Ia\bar{3}d$ and Col_h phases, and is consistent with models proposed for related transitions in lyotropic systems [111].

Cubic phases have also been observed in silver thiolates ($AgSC_nH_{2n+1}$) [112]. These complexes possess a rather unusual thermotropic mesomorphism, despite their non-rigid structures. Indeed, in addition to the cubic phase, lamellar (S_A), and columnar phases were observed. This similarity with lyotropic systems is not surprising, since these complexes can be seen as amphiphilic, in the sense that they possess a long hydrophobic tail and a polar head. A single crystal structure determination showed that the molecules are organised in tight layers, the silver atom having a trigonal coordination; that is they are coordinated to three thiolate groups (the chains pointed in the normal directions of the crystalline planes). On heating, short-medium chain length derivatives ($n=4-10$) underwent a transition to a lamellar phase, explained by a change of the coordination geometry of the silver atom, from trigonal to linear. On further heating, the compounds formed a columnar hexagonal phase, although through an intermediate mesophase, a cubic phase. Long chains derivatives ($n=11$) only showed a columnar hexagonal mesophase. The structure and the space group of the cubic phase have not been determined.

6
Metallomesogens Based on High Coordination Number Metal Centres

As mentioned in the introduction, the considerable variety of metal coordination geometries offers good opportunities for building new molecular structures that are not readily attainable in purely organic materials. This may lead to the formation of new phase types, or more precisely, to new ways of molecular organisation as the structural features of the mesophases may depend strongly on the choice of the metal. In addition, new properties may become available. Initially, most of the metallomesogens reported in the literature were restricted to metals with a d^8-d^{10} electronic configuration, and as such, most of the geometries of the metal complexes were limited to linear or square planar. This was convenient as these structural types readily gave rod- and disc-shaped molecules, essential structural requirements for the-rmotropic mesophase formation. However, one recent development has been the design of new systems containing high coordination metals such as Cr, Mo, W, Mn, Re, Ru, Co, and the lanthanides. It was rather interesting to verify whether the geometries were compatible with the formation of thermotropic mesophases, but furthermore, if mesomorphism was to be observed, to discuss the rôle of metal–metal or metal-ligand interactions, as in these cases, the metal is shielded from any interactions by its surrounding ligands. A few groups have recently demonstrated that the use of such molecular geometries can lead to mesomorphic materials, both in lyotropic and thermotropic systems. In the latter, both columnar and lamellar mesophases have been obtained by a subtle, and appropriate combination of metal and ligands. In addition to new geometries, chirality can be induced through other means than the introduction of an optically active, chiral carbon in the organic ligand, owing to the presence of the metal. In principle, the formation of chiral complex having a metal atom as the centre of dissymmetry [113] is possible (C_n and D_n symmetry groups), as, for example, in the octahedral chelation of polydentate ligands. Some selected examples will be discussed here.

6.1
Butadiene Iron Tricarbonyl Complexes

Butadiene iron tricarbonyl complexes represent an interesting class of compounds as they have the double advantage of possessing a potentially chiral rigid unit, with a rather large lateral dipole, making them good candidates for the construction of new ferroelectric liquid crystal materials. In these systems, chirality (here, axial chirality) can be induced by resolving the racemic mixtures of asymmetric complexes.

Metallomesogens containing the iron tricarbonyl fragment, 53, were first reported by Ziminski and Malthête in 1990 [114], who described the thermal behaviour of two series of racemic mixtures. The racemic mixture 53, displayed a S_A and a nematic phase at relatively low temperatures, and over a narrow domain of stability (typically from 5 to 10 °C). Note the relatively small size, and, therefore, the low structural anisotropy, of the ligand. To assess the effect of axial chirality on the mesophases, resolution of one racemic mixture (53: $n=10$) was successfully achieved, and the pure enantiomer was found to exhibit a S_A^* phase and a chiral nematic phase [115]. Both the racemic mixture and the pure enantiomer displayed similar mesophase temperature ranges, with a slight increase in the melting transition for the pure isomer.

53

Two series of polycatenar complexes containing the same metallic fragments were also reported [115]. As expected for polycatenar complexes, nematic (54: $n=6$, 10 and 11), S_C (54: $n=10$ and 11) and Col_h (54: $n=12$–16) phases were obtained. Resolution of the pure enantiomers gave rise to the equivalent chiral phases for the same n (N^* and S_C^*), and Col_h phases (chiral?) for $n \geq 12$, with slight differences in the transition temperatures as compared to the racemic mixtures. Small spontaneous polarisations were measured.

54

Other metallomesogens containing the optically active butadiene iron tricarbonyl moiety, incorporated into chiral nematic mesogens, have recently been reported [116]. Both the diastereoisomers exhibited a chiral nematic and a monotropic smectic phase, the chiral nematic phase being monotropic for one diastereoisomer, and enantiotropic for the other.

The results reported above clearly indicate that the butadiene iron tricarbonyl moiety can be inserted into mesogenic structures, to give, when suitably designed, ferroelectric properties. This opens up a new and fascinating area of research in which metallomesogens can be used in electro-optical devices. However, little information is yet available on the structure of the mesophases.

6.2
π-Arene-Based Metallomesogens

6.2.1
η^6-Benzenechromiumtricarbonyl Derivatives

The (η^6-arene)tricarbonylchromium complexes 55 were reported by De-schenaux et al. [117]. The uncomplexed ligands showed a mesomorphism characteristic of polycatenar mesogens, progressing from N and S_C phases at shorter chain lengths to Col_{ob} and Col_h phases at longer chain lengths.

$$H_{2n+1}C_nO \quad \text{—} \quad O \quad \text{—} \quad N \quad \text{—} \quad Cr(CO)_3 \quad \text{—} \quad N \quad \text{—} \quad O \quad \text{—} \quad OC_nH_{2n+1}, OC_nH_{2n+1}$$

55

Upon complexation, the columnar mesophases were suppressed, the nematic phase was more extended, at least for the first two derivatives, and the S_C phase was present for all homologues. An important observation was the rather rapid decrease in the mesophase temperature range as the alkoxy chain length was increased, the decrease being associated with a huge reduction in the clearing temperature of the complexes, the melting point being almost invariant (around 90 °C). Compared to the structurally related butadiene iron tricarbonyl complexes, the disappearance of the columnar mesophases was rather surprising. On the basis of results from X-ray studies, this was due to a conformation change of the ligand upon complexation. Indeed, the layer spacings of the S_C phases were smaller for the metal-free ligand than for the complex. The metal-free ligand can adopt several conformations, and thus, for example, the smectic layer can contain a mixture of molecules with extended and zigzag conformation; in such a case the apparent length of the molecules is shorter. However, when complexed, the ligand very likely exists in its fully extended conformation, as evidenced by the increase of the interlayer spacings. The conformation of the complexed ligand accounts for an optimisation of the π–π interactions, and thus, of the S_C phase. However, as the temperature was raised the lateral interactions and the segregation decrease, giving rise to the nematic phase. However, the suppression of the columnar mesophase is more surprisingly. It is generally accepted that the S_C-to-Col transition is a consequence of a long-range breaking of the ribbon structure of the S_C to form small cluster, driven by the larger volume of the alkyl chains compared to the volume of the core. However, in this case the volume of the core is enhanced by the $[Cr(CO)_3]$ group so that the volume of the chains at the lengths studied are insufficient to stabilise columnar phases.

It is worth mentioning that another type of tricarbonylchromium complexes derived from cholesteryl 4-alkoxybenzoate ligands have been reported to show broad chiral nematic phases [118]. The complexes are unsymmetrical, and this may the be the reason of the important decrease of both the melting and clearing transition temperatures as compared to the metal-free ligand.

6.2.2
Ferrocenes

Ferrocene-containing liquid crystals represent a large and important class of materials. They have been widely studied (one of the first metallomesogen to be reported was based on ferrocene), and most of the different systems made so far, in addition to other metallocene-containing liquid crystals, have been reviewed [8b]. Therefore, it is not our intention to review them all again, but rather to discuss some of the more recent results.

The relatively low-melting ferrocene systems, **56** and **57**, were described by Imrie and co-workers. Generally, monosubstituted ferrocenes have been found to be poor mesomorphic materials [119], this being essentially due to the bulkiness of the ferrocenyl moiety which disrupts the overall linear shape of the molecule. However, compound **56** showed an enantiotropic nematic phase at accessible temperatures [120], and compound **57** showed a chiral S_C phase below 100 °C, which supercooled to room temperature, the frozen mesophase remaining as such for several weeks [121]. These two examples seem to indicate that the insertion of the large, bulky ferrocene unit into a mesogen can destabilise the crystalline phase, as evidence by the moderate melting points, without significantly disrupting the tendency to form liquid-crystalline phases.

56: Crys • 115 • N • 129 • I

57: Crys • 83 • S_C^* • 99 • I

A very interesting finding was that of a 3,3′-disubstituted-[3]ferrocenophane, a 'U'-shape type of ferrocene complex, **58**, which was found to be mesomorphic [122]. It first proved that a *cisoid* conformation (U-shape) does not prevent mesophase formation, but also that this type of structure can lead to potentially ferroelectric materials, since compound **58** was obtained as a mixture of isomers, some of them being optically active (planar chirality).

58: Crys • 100 • S_C • 139.5 • N • 158 • I

Deschenaux and co-workers have recently reported the first 1,1′,3-trisubstituted ferrocene-containing thermotropic liquid crystals [123]. The first series of materials (**59**: x=1), showed monotropic S_A and S_C phases (n=12, 14), and monotropic S_C phases (n=16, 18), all the complexes melting in the isotropic liquid at ca 100 °C. However, the incorporation of an additional aromatic ring in each of the three pendant side groups yielded the complex **60** (x=2), showing enantiotropic mesophases, although at rather elevated temperatures, Crys·167·S_C·244·S_A·260·I, the complex decomposing when entering the isotropic liquid. These results are of fundamental importance since they showed that the 1,1′,3-trisubstitution pattern considerably enhances the mesomorphic properties over the di- and monosubstitutions [8b]. Indeed, one aromatic ring per pendant group is sufficient to promote liquid crystallinity in these systems on the one hand, and the mesomorphic range is increased on the other hand. A structure-property relationship can be formulated in which the liquid-crystal tendency varies as: 1,1′,3-trisubstitution > 1,3-disubstitution > 1,1′-disubstitution >monosubstitution.

59: x = 0, n = 12, 14, 16, 18
60: x = 1, n = 16

Goodby and co-workers looked at a whole series of mesogens and investigated the effect on the transition temperatures, and more generally, on the mesomorphic properties, caused by the replacement of a 1,4-disubstituted benzene ring (e.g. **62**) with a 1,1′-disubstituted ferrocene unit (e.g. **61**). The ferrocene unit was found to be a poorer component of a mesogen when inserted in place of a benzene ring in a linear liquid-crystalline system [124]. The mesophase stability was dramatically reduced, essentially due to a huge reduction in the clearing point. This reduction of the (virtual) $T_{N/I}$ was attributed to the step-like structure of the molecule, which implies that the alkyl chains cannot be co-planar with the rest of the molecule, and to the rotation of the two cyclopentadiene rings which would reduce the rigidity of the systems, since the molecule can adopt a fully extended 'S'-geometry and an alternative 'U'-shape.

61: Crys • (N • 63) • 107 • I

62: Crys • 138 • S_C • 176 • N • 280 • I

Interestingly, a similar set of studies has been carried out on laterally connected twin liquid crystals, yielding opposite conclusions. Here, the metallomesogens consist of two linear, 4,4″-disubstituted p-terphenyl rigid cores, which are laterally connected, in two different positions, to each other via a 1,1′-disubstituted ferrocene unit (Fc), **63**, **65**. These are compared to the related organic systems, **64**, **66**, in which a 1,4-disubstituted benzene ring is used as a spacer [125]. For these two twin systems, it was found that the mesophase temperature ranges were enhanced in comparison to the related benzoates, this being essentially due to the reduction in the melting points, as the clearing points remained similar or slightly higher than the structurally related terephthalates. Thus, when used as a spacer, the bulky ferrocene unit must disrupt the efficiency of the packing in the crystal state, as evidenced by the lowering of the melting points, and the rather hindered crystallisation on cooling.

63: (X = Fc)
Crys • 86 • S_A • 102 • I

64: (X = Ph)
Crys • 132 • (S_A • 94) • I

65: (X = Fc)
Crys • 120 • S_A • 153 • I

66: (X = Ph)
Crys • 144 • S_A • 149 • I

Finally, a new class of ferrocene-containing liquid crystals has recently been reported, based on persubstituted ferrocene [126]. Peralkylated ferrocenes are more readily oxidised than ferrocene itself, giving rise to stable, paramagnetic ferroceniums. A careful chemical oxidation of the monosubstituted permethylated ferrocene led to the corresponding stable, ferrocenium derivative, **67**. While the parent ferrocene was not mesomorphic (clearing at 154 °C), the ferrocenium species exhibited a monotropic S_A phase. On further cooling, the sample did not crystallise, and a glass transition was observed ($T_g = 37$ °C), a relatively common observation in ionic liquid crystals [127]. A detailed X-ray investigation showed that the layer spacing of the S_A phase equals the length of the complex in its fully extended conformation, suggesting a monolayer structure rather that the expected bilayer. Thus, it has been demonstrated for the first time that electron-transfer is responsible for mesophase induction in the ferrocene-ferrocenium redox system, and this finding should lead to further studies, for which the structure, charge and counter ion can be systematically varied for a better understanding of this behaviour.

67

6.3
Octahedral Complexes as Calamitic Metallomesogens

For many years, the only examples of formally six-coordinate metal complexes forming lamellar mesophases were derivatives of ferrocene, while the vast majority of calamitic metallomesogens were based essentially on two- and four-coordinate complexes of metals from groups 8 and 10. Novel metallomesogens based on high coordination number metal centres were recently reported to show nematic and lamellar mesophases.

Octahedral manganese(I) and rhenium(I) [128] complexes, **68** and **69**, based on the orthometallation of some mesogenic imines, have recently been reported to be mesomorphic. While the ligands showed a rich, polymorphic behaviour, including nematic and smectic phases, both the Mn(I) and Re(I) complexes, **68** and **69**, displayed only the nematic phase over a wide range of temperature (usually between about 120 and 190 °C). The mesomorphic properties of the manganese and rhenium complexes are virtually identical, the rhenium complexes being more thermally stable in the isotropic liquid than the manganese congeners. The high structural anisotropy of the ligands was reflected by their very high clearing temperatures (around 300 °C), but upon complexation to Mn and Re, the clearing temperatures were substantially reduced, the decrease in the transition temperatures being consistent with the reduction of the structural anisotropy due to the insertion of a bulky, metal tetracarbonyl fragment (M(CO)$_4$) in the central core. This also reduces the lateral intermolecular interactions, and the smectic phase becomes disfavoured. Reduction of the structural anisotropy of the ligand had profound effects on the thermal properties of the manganese complexes [129], in that the melting temperature was unaffected, but the nematic phase was destabilised, becoming monotropic. In addition, the molecule has a reduced symmetry, which can also be taken into account for the reduction of the melting.

68

M = Mn, Re **69**

A series of calamitic, octahedral halorheniumtricarbonyl complexes, **70**, derived from diazabutadiene has been reported [130]. Some structural parameters, such as the length of the terminal chains ($n=8$, 10 and 12), and the halide ($X=$Cl, Br and I) have been varied systematically, and their effects on the thermal properties studied in details.

70

The increase of the ligand chain length and of the halide size resulted in the reduction of the transition temperatures, which was quite substantial for the clearing temperatures. Here again, a strong structural anisotropy of the ligand was an essential requirements for driving liquid crystallinity after complexation to the bulky, metallic fragment. The ligands showed S_C and nematic phases and started to decompose before reaching the isotropic liquid. For the same reasons as above, most of the complexes exhibited a nematic phase, but unlike the manganese Schiff base complexes, they cleared, in most cases, at similar or higher temperatures than the free ligands. This can perhaps be explained by the configurational change of the ligand upon complexation (*transoid* to *cisoid* configuration), which results in the formation of a non-negligible dipole moment, in addition to that associated to the Re—X bond, thus, increasing the intermolecular dipolar interactions. This is consistent with the findings with some mesomorphic halorheniumtricarbonyl complexes based on 2,2'-bipyridine ligands, **71** ($m=1$), for which both the complexes and metal-free ligands have rather high transition temperatures [131]. Interestingly, the same bipyridines led to the suggestion of certain guidelines for the construction of liquid crystals based on high coordination number metal centres. Thus, previously it has been found that while the 2,2'-bipyridine ligands derived from **71** ($m=0$) were strongly mesomorphic [132], their metal complexes were not [133] despite the fact that the ligands contained four rings in common with the *ortho*metallated imines, the diazabutadienes, and 1,3-disubstituted ferrocenes. However, in these last three systems, it is approximately the case that they contain a metal core unit *plus* four rings in the ligand. If the [(2,2'-bipyridine)Re(CO)$_3$] unit is considered as the core unit, then it is clear that a bipyridine containing six rings would be necessary to have a core unit plus four rings. Thus, the complexes **71** ($m=1$) were synthesised and found to be mesomorphic [131].

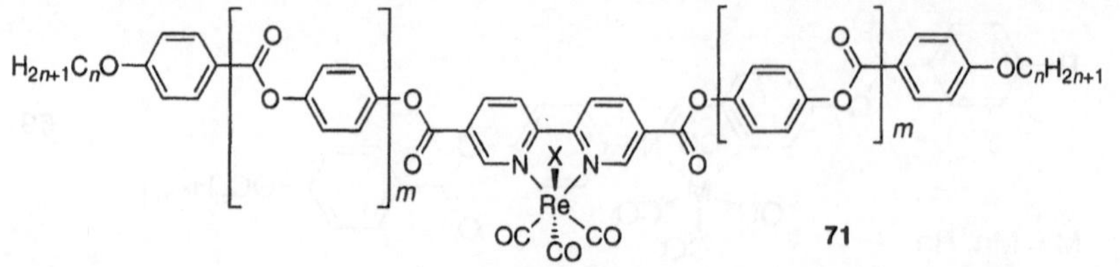

$H_{2n+1}C_nO$— ... —OC_nH_{2n+1}

71

It is worth mentioning the very recent report of new calamitic liquid crystals containing a metal–metal bond, based on the stable $[Ru_2(CO)_4(\mu^2\text{-}\eta^2\text{-}O_2CR)_2L_2]$ 'sawhorse' unit, **72**, [134]. This class of metallomesogens represents a rare example of mesomorphic compounds containing covalent bonds between two metal atoms, and are of considerable interest due their structure and electronic properties, and also to their potential catalytic reactivity [135].

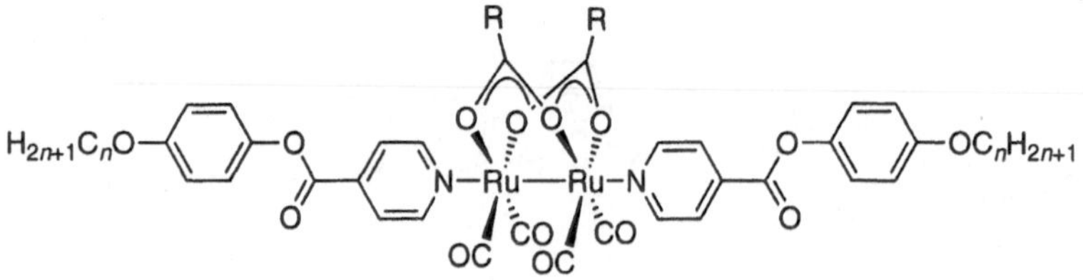

$H_{2n+1}C_nO$— ... —OC_nH_{2n+1}

72

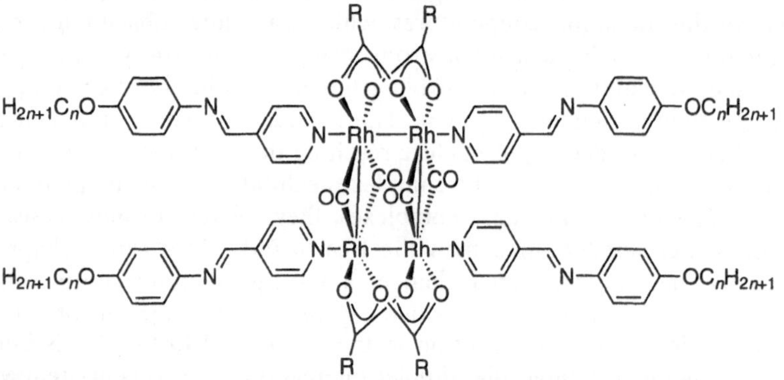

$H_{2n+1}C_nO$— ... —OC_nH_{2n+1}

73

$H_{2n+1}C_nO$— ... —OC_nH_{2n+1}

Despite the bulky central cluster unit, induction of mesomorphism (**72**: R=H, C_6H_5, C_6H_4—CH_3, C_6H_4—OCH_3) was achieved by introducing mesogenic ligands in the axial positions, which led to rod-shape molecules. However, a minimum of two aromatic rings per axial ligand was a necessary condition to overcome the large disruption of the molecular anisotropy. With the exception of the compound functionalised by the smallest substituent (**72**: R=H, $n = 10$) in the bridging positions, which gave an enantiotropic nematic and a monotropic S_C phase, all the other compounds showed a nematic phase only. The stability of the mesophase was found to be dependent upon the type of carboxylato bridges (decreasing order of stability: $C_6H_5 > C_6H_4$—CH_3, $> H$, C_6H_4—OCH_3). It was suggested that the benzene rings of the carboxylato bridges may act as 'tweezers', that is to say they interact with the aromatic rings of the neighbouring pyridine ligand, allowing the complexes to be organised along one direction, and to generate the nematic phase. Substitution in the *para* position by a methyl or a methoxy group resulted in a dramatic decrease in the thermal stability of the nematic phase. This is consistent with a disruption of the π-π interactions between the benzene rings of R and those of the ligands of adjacent complexes, when the benzoato bridges are substituted by methyl or methoxy substituents, essentially due to steric hindrance. As for R=H, the mesomorphism should be the consequence of the appropriate anisotropic structure giving rise to favourable lateral interactions (as evidenced by the observation of the S_C phase). Interestingly, Serrano and co-workers reported, a few years ago, a series of tetrarhodium clusters, 73, having the 'butterfly' unit [136]. Their structure can be seen as a dimer of the 'saw-horse' unit (73: R=Me, CF$_3$), in which the CO groups are transform to CO bridges. Unfortunately, none of them was mesomorphic.

6.4
Columnar Mesophases from Octahedral Complexes

The understanding of the relationship between molecular structure and liquid-crystalline properties led to the development of unusual structures. In a previous attempt to obtain octahedral-based liquid crystals, Giroud-Godquin and Rassat [137] prepared tris(β-diketonato)iron(III) complexes. The viscous phase observed on both heating and cooling was found to be very ordered, as evidenced by strong supercooling of the isotropic liquid. Later, Swager and Zheng [138] prepared several series of tris(β-diketonato) complexes of Fe(III), Mn(III) and Cr(III), using more highly substituted ligands. In these systems, the ligand has either 4, 5 or 6 alkoxy chains, leading to complexes having 12 (74), 15 (75), and 18 (76) alkoxy chains. All the complexes were mesomorphic, displaying columnar mesophases from room temperature to above 100 °C (74), below 100 °C (75) and below 80 °C (76).

The number of chains seems to have an effect on the mesomorphism, sometime in conjunction with the metal. Thus, 75 and 76 bearing 15 and 18 chains, respectively, showed a single Col$_h$ phase with little dependence on the metal, presumably as the metal is well encapsulated within an aliphatic sphere. However, 74 with 12 chains and, therefore, a less dense peripheral coverage by alkyl groups, shows two Col$_h$ phases for M=Fe and Mn, and a Col$_h$ and a Col$_r$ phase for M=Cr. Further, there are significant effects on the mesophase stabilities, e.g. 74 with M=Fe, the mesomorphism is described by Col$_{h1}$·90 Col$_{h2}$·110·I, while for 74 with M=Mn, the mesomorphism is described by Col$_{h1}$·60·Col$_{h2}$·108·I.

M = Fe, Mn, Cr

74: R$_1$ = R$_2$ = OC$_{12}$H$_{25}$, R$_3$ = R$_4$ = H
75: R$_1$ = R$_2$ = R$_3$ = OC$_{12}$H$_{25}$, R$_4$ = H
76: R$_1$ = R$_2$ = R$_3$ = R$_4$ = OC$_{12}$H$_{25}$

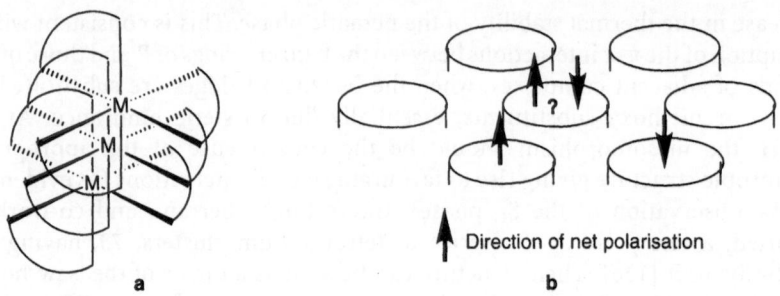

Fig. 6a,b. Schematic diagram to illustrate: **a** the possible packing in the columnar phase of an octahedral mesogen; **b** the origin of net polarisation in the columnar hexagonal phase of a polar mesogen

A possible model for the structure of the columnar phases is sketched below. Since two optical isomers (Δ and Λ) are present in a 50:50 ratio, they must be randomly distributed in a column. Also, it is possible to consider a stacking of the complexes in such a way that a net polarisation is induced in the column direction (Fig. 6a: stacking of a same enantiomer). In columnar phases, the symmetry is such that the net dipole of every column cannot be cancelled out by the one of the neighbouring column, that is a perfectly antiferroelectric arrangement of the columns is impossible; the system is frustrated (Fig. 6b). Thus, a net polarisation must result in the columnar phase whose direction it should be possible to switch.

Lattermann and co-worker have reported a series of mesomorphic metal tricarbonyl complexes, **77**, formed by complexation of the metal fragments $M(CO)_3$ (**77**: $M=Cr(0)$, $Mo(0)$ and $W(0)$) by the non-mesomorphic tridentate azamacrocyclic ligand, i.e. 1,4,7-tri-(3',4'-didecyloxy)benzyl-1,4,7-triazacyclononane [139].

In this system, the ligand is responsible for the formation of the monofacial, octahedral geometry of the complex. The chromium, molybdenum and tungsten (d^6 electronic configuration) complexes, **77**, all exhibit a wide-range, stable, Col_r phase, with no signs of decomposition being observed on reaching the isotropic liquid. While the melting points remain constant for the three complexes (around 62 °C), the clearing points increased steeply from the

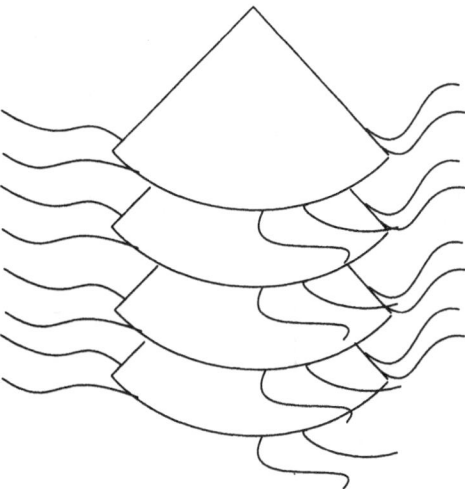

Fig. 7. Schematic diagram to show the columnar stacking present in mesomorphic metal complexes of tri-substituted triazacyclononanes

chromium to the tungsten complexes (T_{cl}: 174, 195 and 227 °C for Cr(0), Mo(0) and W(0), respectively). When coordinated to the metal, the triazacyclononane ring must lose its flexibility and, therefore, the inner part of the complex can be seen as a cone. The formation of the columnar mesophase then presumably results from the assembly of cones on top of one other, the summit being made of the tricarbonyl moieties, and the base of the cone by the ligands (Fig. 7). Thus, a net polarisation must result in the columnar phase, as for the complexes above. These new findings seem to agree with previous work, in which nickel-dinitrate was complexed to the same ligand [140]. The octahedral complex was mesomorphic giving a Col_r mesophase between 140 and 167 °C.

Complexation of branched, dendrimeric (first and second generation, L^1 and L^2, R=3,4-didecyloxybenzyl) amino ligands (derived from tris(2-amino-ethyl)amine) to metal salts such as $CoCl_2$, $NiCl_2$, CuX_2 (where X = Cl, SCN and NO_3) and $ZnCl_2$, allows access to new metallomesogens, with two possible coordination geometries, the trigonal bipyramid (78) and the octahedron (79) [141]. On the basis of IR and UV/Vis spectroscopies, all the complexes, except the Ni complexes, are based on the trigonal-bipyramidal geometry ('azatran' geometry). The melting temperatures of the metallomesogens were all very low, some of them being liquid crystals at room temperature (L^1CoCl_2, L^1CuCl_2, and L^1ZnCl_2), and the other melted at around 50 °C (L^1NiCl_2, $L^1Ni(NO_3)_2$, $L^1Cu(NO_3)_2$, $L^1Cu(SCN)_2$). X-Ray diffraction showed that the Co and Ni complexes have a lamellar mesophase, while all the other have a Col_h phase. The complexation of $CuCl_2$ with the second generation dendrimer, L^2, also gives rise to metallomesogens, with a Col_h phase from room temperature to 140 °C; the metal has the azatran geometry.

This type of stacking of cone-shape molecules has already been proposed for the description of the molecular organisation in the mesophase of some bowlic liquid crystals based on tungsten-oxo calix[4]arenes, **80, 81** [142].

80: R = H Crys • 136 • Col • 320 • I

81: R = $OC_{12}H_{25}$ Crys • 54 • Col_1 • 77 • Col_2 • 267 • I

While the ligands melted to give a transient mesophase, which appeared only on first heating, the tungsten complexes displayed enantiotropic columnar mesophases. The stability of the mesophases upon complexation was attributed to the increase of the rigidity of the calixarene conformation. Furthermore, the 1:1 host-guest systems between the tungsten complex and a Lewis base (L: DMF or pyridine) had dramatic effects on the mesomorphic behaviour of complexes **80** and **81**. **80**:L and **81**:L host–guest systems melted to the isotropic liquid at 115 and 84 °C respectively. Further heating (200–250 °C) led to the dissociation of the Lewis bases, and the columnar phases reappeared. These examples further confirmed this model, in which metal-ligand interactions (W=O—W=O—) are the main driving forces for mesophase formation.

6.5
Lyotropic Systems

While the study of thermotropic, metal-based materials has completely established itself as a proper branch in liquid crystals, lyotropic liquid crystals from amphiphilic metal complexes represent an area of growing interest, although there are still very few examples. As many metal complexes are ionic, they are potentially natural building units for the creation of amphiphilic materials, and in addition, all the properties specific to the metals can be brought to surfactants leading to new materials. It is thought that they might be good candidates for the generation of novel catalysts as well as playing an important role in biological systems, such as phospholipids and glycolipids [100].

In the past few years, several octahedral metal complexes have been reported to display lyotropic mesophases. The aquatetra(cyano)dodecyl-aminoferrate(II) complex [143] gives a hexagonal (H_1) mesophase in water; however, the complex slowly hydrolyses. The stability of such octahedral metal complexes is usually increased by the use of chelating ligands. Indeed, some cobalt(III) complexes of the general formula $[Co(L-L)(en)_2][PF_6]_2$ (L—L: dodecanoate, en: ethylenediamine), were stable to hydrolysis and gave a hexagonal (H_1) mesophase on contact with water at room temperature [144]. Lyotropic mesophases were also observed for some surfactant tris(bipyridine) [145], **82**, and bis(terpyridine) [146], **83**, complexes of Ru(II) and Rh(III).

In the first example, where one of the bipyridine ligand is monoalkylated (82: $R=Me$, $R'=C_nH_{2n+1}$, with $n=12$–31), the chloride salts showed a cubic (I_1) phase in water (and a hexagonal H_1 mesophase for one complex); the double-chained derivatives showed only H_1 hexagonal phases (82: $R=C_{12}H_{25}$, $R'=C_nH_{2n+1}$, with $n=12$, 19). For the bis(terpyridine) complexes, 83, both the Ru and Rh complexes were mesomorphic, but at higher temperature owing to their relative insolubility in water. Nevertheless, the ruthenium derivative showed a cubic (I_1) phase in water at temperature higher than 65 °C, while the longer chain homologue failed to be mesomorphic in water, but a cubic (I_1) phase was observed in ethylene glycol. The rhodium complex behaved in a similar manner. The relatively large spherical head of these surfactants, when compared to organic surfactants, does not appear to be a problem for the formation of lyotropic mesophases. Another driving force is also certainly the high number of charges found in these ionic complexes, which increase their solubility in polar solvents.

Many conceptually related organometallic amphiphiles which have been found to micellise at low concentration in water [147], or to form LB films [148]. It has also been shown that reversible vesicle formation could be achieved by oxidation of the head group of some complexes, and that it can be controlled by simple redox chemistry [149].

7
Mesophases Requiring Specific Intermolecular Interactions

7.1
Metal Carboxylates

It is now certain that metal carboxylates were the first metal-containing liquid crystals, and were first reported in 1855 with Heinz's work on magnesium tetradecanoate [4]. Then, many other mesomorphic metal carboxylates, with the general formula $M(O_2C-C_nH_{2n+1})_m$, were prepared [150]. They showed thermotropic lamellar and columnar mesophases, but also, when dissolved in water or alkanes, lyotropic mesophases.

Known for many years [151], dicopper tetraalkanoates have also been thoroughly investigated (84: M—Cu). Owing to their 'lantern' structures in the crystal state, it was expected that they would give rise to thermotropic liquid crystals, and in particular to columnar mesophases.

Indeed, all the copper alkanoates, 84, from $n=3$ to $n=23$, exhibit a mesophase at about 120 °C; the transitions to the isotropic liquid occur above 200 °C with extensive decomposition [152]. These materials have been studied by numerous experimental techniques, in the solid state and in the mesophase, including magnetic susceptibility [153], IR spectroscopy [154], dilatometry [155], EXAFS spectroscopy [156], incoherent quasielastic neutron scattering (IQENS) [157] and X-ray diffraction [152]. X-Ray diffraction identified the mesophase as hexagonal columnar ($n \geq 4$), and rectangular columnar ($n=3$). In the crystal state, the complexes form the core of the columns, with a 1-D, pseudo-polymeric chains or ladder structures (Fig. 8), in which each copper atom is surrounded by five oxygen atoms, in a square pyramidal environment; the Cu atoms have a square planar coordination with four oxygen atoms, and form a weak, axial intermolecular bond with one oxygen atom of the neighbouring molecule (axial ligation). These columns are then organised parallel to each other to form a lamellar structure. EXAFS clearly demonstrated that these columns (and the dimeric nature of the complexes) still exist in the mesophase, the columns being organised according to a 2D hexagonal lattice. During the transition between these two phases, a decrease in the stacking period is observed, attributed to a modification of the relative disposition of the dimers within the columns, in addition to a sharp increase in the molar volume (corresponding to an increase of the chain disorder). In the mesophase, the dimers stack on top of each other in the normal direction of a four fold helicoidal axis.

Subsequently, related dimetal tetracarboxylates but containing covalent metal–metal bonds were prepared and found to be mesomorphic. The first examples of this type are the dirhodium(II) complexes [158] with a Rh–Rh single bond (**84**: M = Rh), which formed Col_h mesophases between 100 and 200 °C, at which temperature they decomposed. Diruthenium tetracarboxylates (**84**: M = Ru), $[Ru_2(O_2CC_nH_{2n+1})_4]$ containing a Ru=Ru double bond, and the mixed-valence, paramagnetic $[Ru_2(O_2CC_nH_{2n+1})_5]$, also formed columnar mesophases [159]. The $Ru^{II}Ru^{II}$ complexes showed a Col_h mesophase above 100 °C, although the isotropic liquid could not be reached due to decomposition. The $Ru^{II}Ru^{III}$ complexes exhibited a Col_h phase with dodecanoate and a Col_r phase with nonanoate at 150 °C; the complexes $[Ru_2(O_2CC_nH_{2n+1})_4Cl]$ were not mesomorphic. X-Ray diffraction and EXAFS spectroscopy suggested the same model for the molecular organisation of the $Ru^{II}Ru^{II}$ complexes in the columnar phase as that proposed for their isostructural copper alkanoates (Fig. 8), although in the $Ru^{II}Ru^{III}$ complexes, the intermolecular Ru–O interactions are not present as the anionic group acts as a spacer between Ru_2 dimer units [160]. The use of branched and unsaturated carboxylate chains has

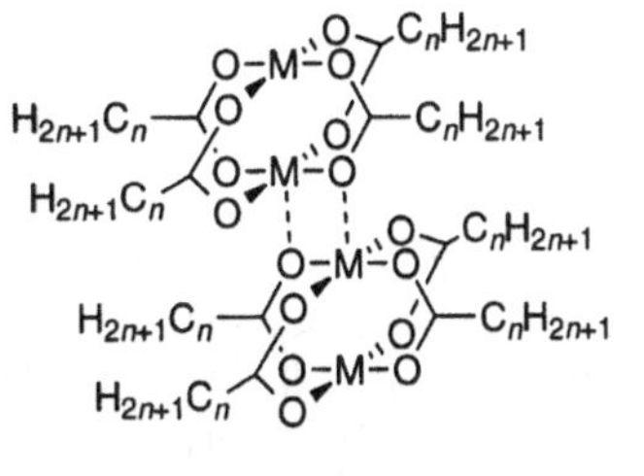

Fig. 8. Schematic diagram to show the arrangement in the columnar phase of dimetal tetracarboxylates

been successful in depressing the melting points of some copper and ruthenium complexes [161]; the mesophases observed were of two types: hexagonal columnar and lamello columnar.

Chisholm and co-workers prepared liquid-crystalline compounds containing Mo-Mo (84: M=Mo) and Cr—Cr (84: M=Cr) quadrupole bonds, which exhibited Col_h mesophases [162]. The molybdenum alkanoates were mesomorphic from $n=4$ to $n=10$, and were thermally stable up to 250 °C. The melting points decreased rapidly from $n=4$ (152 °C) to $n=5$ (108 °C), and then remained almost constant, whereas the clearing temperatures decreased almost linearly with increasing n, resulting in the loss of the mesophase from $n \geq 11$. The chromium complexes (84: M=Cr) behaved very much like their ruthenium analogues, while the tungsten complexes (84: M=W) were not mesomorphic. The similarity of the enthalpies of the crystal-to-columnar phase transition and of the melting temperatures for all the metal carboxylates of identical chain length implies that dimeric interactions, and thus, axial metal-oxygen interactions are preserved in all the systems despite the metal (and hence, that the same model that proposed for the copper complexes can be applied for Mo and Cr complexes). Furthermore, the differences in the thermal behaviour and in the clearing points are in agreement with the strength of the axial ligation, which follows the order Cr>>Mo > W, and Cu, Rh, Ru>>Mo.

7.2
The Complementary Shape Approach

A new concept that has been successfully used to design liquid crystals based on specific molecular interactions is the complementary molecular shapes approach. This consists of generating correlations between metal complexes, the resulting correlated structures being further organised to form columnar mesophase. Time-averaged disc-shaped structures can be the result of antiparallel or orthogonal correlation of nearest neighbours.

7.2.1
Semi-Disc-Shaped Compounds

Serrette and Swager reported the mesomorphic properties of a series of oxovanadium(IV) complexes of dimeric Schiff bases **85**, and **86** (Y=—CH$_2$—CH$_2$—, —CH$_2$—CH$_2$—CH$_2$—, —CH$_2$—C(Me)$_2$—CH$_2$—, R=OC$_{14}$ H$_{29}$) [163]. Despite the non-discoidal shape of these complexes, they all displayed columnar mesophases, some of the complexes showing several columnar-to-columnar phase transitions. The mesophases (hexagonal and rectangular) existed over very wide temperature-range, some of them being stable at ambient temperature.

IR spectroscopy revealed that the complexes form polymeric linear chain structures (··V=O···V=O···V=O··), and that the formation of the columnar phases would result from the stacking of correlated structures. A time-averaged, disc-shaped structure is generated when each molecule is rotated by 90° with respect to its neighbours, forming a time-averaged, disc-shaped structure (Fig. 9a).

A series of β-diketonate Schiff base complexes (**87**: M=Ni, Cu, Pd, VO; n=10, 12, 16) having a half disc-shape, was synthesised [164]. All the complexes, except the oxovanadium complexes, were mesomorphic (Col$_h$), at relatively low temperatures. This approach was also applied to half-disc-shaped homodinuclear and heterodinuclear 1,3,5-triketonate Schiff base complexes, **88**, and 1,3,5,7-tetraketonate Schiff base dicopper complexes, **89** [165]. The homodinuclear complexes **88**, with M=M'=Cu, X=OR, were mesomorphic showing Col$_h$ phases, whereas those with four chains (**88**: X=H) were not liquid-crystalline. Amongst the five heterodinuclear prepared (**88**: X=OR,n=14, Y=—CH$_2$—CH$_2$—, M=Cu, Pd, Ni, Mn, Co and M'=Ni), only those with M=Cu, M'=Ni and M=Pd, M'=Ni, showed a mesophase, namely a Col$_h$ phase. A Col$_h$ phase was also observed for a structurally related compound, **89**, (n=12, X=OR).

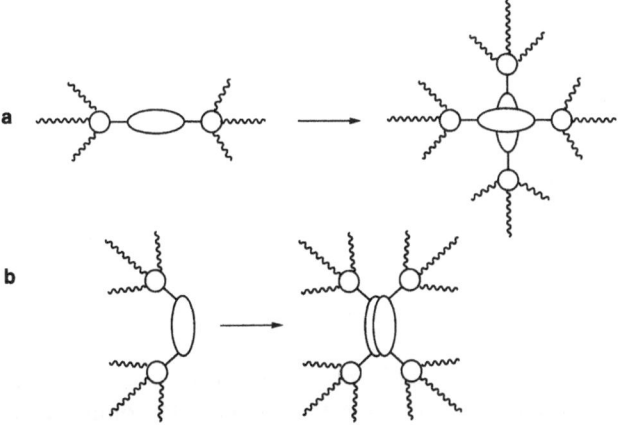

Fig. 9a,b. Possible arrangements of polycatenar mesogens in the columnar phase according to the complementary shape approach

R = OC_nH_{2n+1}, X = H, OR, Y = -CH_2-CH_2-, -CH_2-CH_2-CH_2-

The mesomorphism observed in these three series of semi-disc-shaped complexes (**87–89**) was explained on the basis of X-ray experiments. Indeed, a doubling of the periodicity along the columnar axis was measured, suggesting that the molecules are correlated in an antiparallel fashion (Fig. 9b), that is each molecule is rotated by 180 °C with respect to its neighbours, forming a time-averaged disc-shaped structure.

Liquid-crystalline 1,3,5,7-tetraketonate dicopper complexes **90** and **91** also showed Col$_h$ phases [163, 164]. In these cases, the neighbouring molecules must be rotated by 90° in respect to each other, in order to give rise to a time-averaged disc-shaped structure (Fig. 9a).

R = OC_nH_{2n+1}, X = H, OR, Y = CH_3, CF_3

Other groups are now successfully using this principle. For example, Szydlowska et al. reported non-discoidal copper(II) and nickel(II) binuclear complexes, forming Col$_r$ phases over very wide temperature-range [166], and recently, Lai et al. reported mesomorphic oxygen-bridged dicopper complexes [167], while Trzaska and Swager reported β-diketenato complexes of Ir^I [168].

7.2.2
Hemiphasmidic Compounds

As far as we are aware, three such structures have been realised, only two of them showing mesomorphic properties. The first series, the dioxomolybdenum complexes, **92**, described by Serrette and Swager [169], exhibited Col$_h$ mesophases, over an average temperature range of 40 °C, with melting temperatures between 96 °C (**92**: $n=12$, 14, and 16) and 106 °C (**92**: $n=10$). The octyloxy homologue was not liquid crystal and melted at 135 °C. The pyridinediyl-2,6-dimethanol ligands alone were not mesomorphic, clearly showing that the formation of the columnar mesophase arises from the MoO$_2$ unit and its tendency to form long chain polymeric structures. Indeed, variable temperature IR spectroscopy indicated that, in the mesophase, the complexes assemble in such a way that a polar polymeric ($\cdots$Mo$=$O$\cdots$ Mo$=$O$\cdots$) structure was formed. On the basis of both infrared and X-ray experiments, complexes **92** are very likely to be organised in such a way, that the polar chain ($\cdots$Mo$=$O$\cdots$Mo$=$O$\cdots$) structure is oriented along the columnar axis (Fig. 10).

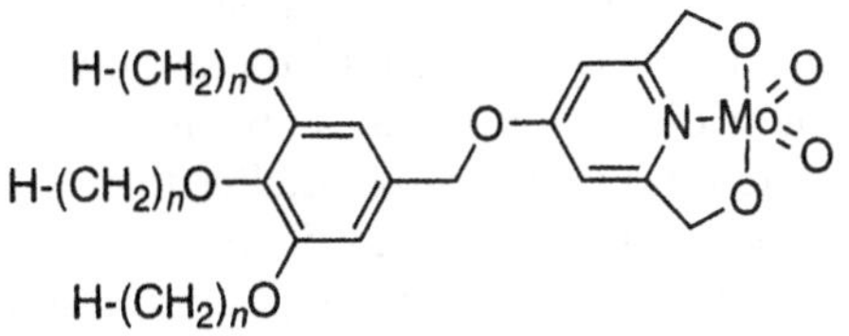

In the second series of materials, **93** [18], all the compounds with X=Cl, Br and I were mesomorphic, showing enantiotropic Col$_h$ mesophases from room temperature to 70–80 °C for the chloro-compounds, and 45 and 56 °C respectively for the bromo- and iodo-complexes. Here again, the ligands themselves were not mesomorphic. An intracolumnar distance of 4 Å was

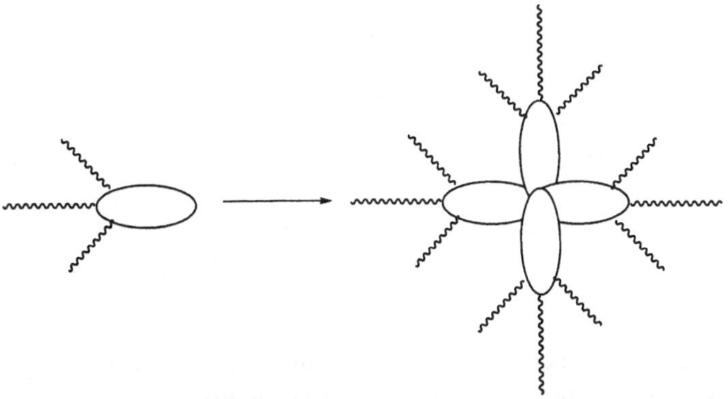

Fig. 10. Possible arrangement of hemiphasmidic complexes in a columnar mesophase

found (n=10, X=Cl), indicative of strong gold-ligand interactions, and the X-ray data were in agreement with the proposal of a discoid, dimeric molecular aggregate formed by two molecules of the complex in antiparallel disposition

$$C_nH_{2n+1}O$$

$$C_nH_{2n+1}O-\!\!\!\bigcirc\!\!\!-N{\equiv}C{-}Au{-}X$$

$$C_nH_{2n+1}O \qquad \textbf{93}$$

$$C_nH_{2n+1}O$$

$$C_nH_{2n+1}O-\!\!\!\bigcirc\!\!\!-CH{=}CH-\!\!\!\bigcirc\!\!\!-N{-}\underset{\underset{CO}{|}}{\overset{\overset{Cl}{|}}{Ir}}{-}CO$$

$$C_nH_{2n+1}O \qquad \textbf{94}$$

A series of structurally related iridium complexes, **94**, did not give rise to any mesomorphic materials [81]. The iridium fragment does not give rise to polymeric structures such as those found in the dioxomolybdenum complexes. The aromatic core of the iridium complexes is also more elongated and rigid, and the aliphatic/aromatic ratio is therefore smaller than in the molybdenum and gold complexes. The curvature radius is thus larger, thus excluding lamellar mesophases and the density of the aliphatic matrix is not sufficient enough to sustain the formation of columns.

Only two amongst these three structurally related materials were found to exhibit liquid crystalline behaviour. Two types of molecular organisation in the columnar mesophase are possible, both based on metal/ligand interactions, according to the principles expressed above on time-averaged disc-shaped structures.

8
Metallocrowns: A New Class of Metallomesogens!

Metallocrowns [170], as structural analogues of organic crown ethers in which the ethylene bridging groups are replaced by heteroatoms such as nitrogen and transition metals, represent a new direction of research in the field of metallomesogens. Serrano et al. reported the first liquid crystal metallocrown, derived from pyrazole ligands [171]. Two trinuclear gold complexes were prepared, differing only by the number of alkoxy chains (**95**: X=H or $OC_{10}H_{21}$). Compounds **95a** and **95b** displayed a Col$_h$ mesophase between 59 and 64 °C, and 36 and 59 °C, respectively. Thus the difference in mesomorphic temperature range between these two compounds seems to be essentially due to a different number of peripheral alkoxy chains.

H21C10O
H21C10O
OC10H21
OC10H21

X
X

N—N
Au   Au
N
N—Au—N
N

H21C10O
H21C10O

OC10H21
OC10H21

95a: X = H

95b: X = OC10H21

H21C10O
H21C10O
X
OC10H21
OC10H21

The mesophases were clearly identified by X-ray diffraction, with an average short-range correlation of 4.4 Å. This indicates that the macrocycles are mutually rotated in such a way that the gold atoms interact with the pyrazole rings of neighbouring macrocycle. An X-ray structure analysis of **95a** revealed that the columnar organisation of the molecules is already present in the solid state, with a repeat distance of 9.4 Å; yet the molecules crystallised in a orthorhombic cell. The solid-to-mesophase transition occurs with an increase of the chain volume, and the b cell parameter shrinks while the a parameter increases until $b/a = \sqrt{3}$, characteristic of the hexagonal lattice. In addition, clusters of two molecules must rotate to achieve a stacking distance of 4.4 Å. The reduction of the melting point of **95b** is due to the presence of two isomers as deduced for ^{1}H-NMR studies, on the one hand, but also to the off-plane distortion of the organic ligands in respect to the metallocrown which hinders an optimised interaction between the molecules in the solid state.

9
Conclusions

The range of metallomesogens has considerably expanded since the publication of the first review on the subject by Maitlis and Giroud-Godquin in 1991 [172]. These materials are noticeable for their huge variety and different shapes which have been achieved through the use of different geometries about the metal centre. Indeed, linear, square planar, five-coordinate and six-coordinate complexes have been found compatible with the formation of thermotropic lamellar and columnar mesomorphism, as well as lyotropic mesomorphism. Typical mesophases found in organic systems have been observed in metallomesogens, namely nematic, smectic, columnar and cubic phases. The field has benefited from the involvement of groups with a wide range of chemical backgrounds, ranging from physical chemists to synthetic chemists of the formally inorganic and organic 'persuasion', and including those who have long been involved in liquid crystals, and who see something new and exciting on offer. We have, in this subject, the whole wealth of inorganic chemistry which can be married with the range of organic chemistry and the existing knowledge of liquid crystals, to produce something which is potentially much greater than the sum of its parts. For example, it has been shown that metals can have important effects on the physical properties of liquid crystals [8, 9]. While a relatively small number of such studies have been carried out, and the rôle of the metal in relation to molecular structure and physical properties has not been fully elucidated, it is clear that electronic and magnetic properties can be greatly influenced to advantage. And no-one has really started to look at the reactivity of the metal centres yet. The future in this field full of surprises and excitement just waiting to be uncovered!

10
References

1. (a) Oriol L, Serrano JL (1995) Adv Mater 7: 348; (b) Oriol L, Pinol M, Serrano JL (1997) Prog Polym Sci 22: 873
2. Bruce DW, Heyns K, Vill V (1997) Liq Cryst 23: 813
3. Vorländer D (1923) Z Physik Chem 105: 211
4. Heintz W (1855) J Prakt Chem 66: 1
5. Malthête J, Billard J (1976) Mol Cryst, Liq Cryst 34: 117
6. (a) Giroud A-M, Mueller-Westerhoff UT (1977) Mol Cryst, Liq Cryst Lett 41: 11; (b) Giroud AM (1978) Ann Phys 3: 147; (c) Giroud AM, Nazzal A, Mueller-Westerhoff UT (1980) Mol Cryst, Liq Cryst Lett 56: 225
7. Mueller-Westerhoff UT, Nazzal A, Cox RJ, Giroud AM (1980) Mol Cryst, Liq Cryst Lett 56: 249
8. (a) Bruce DW (1996) Metal-containing liquid crystals. In: Bruce DW, O'Hare D (eds) Inorganic materials, 2nd edn, chap 8. Wiley; (b) Deschenaux R, Goodby JW (1995) Ferrocene-containing thermotropic liquid crystals. In: Togni A, Hayashi T (eds) Ferrocenes: homogeneous catalysis, organic synthesis, materials science, chap 9. VCH Verlagsgesellschaft, Weinheim; (c) Polishchuk AP, Timofeeva TV (1993) Russ Chem Rev 62: 291; (d) Espinet P, Esteruelas MA, Oro LA, Serrano JL, Sola E (1992) Coord Chem Rev 117: 215; (e) Hudson SA, Maitlis PM (1993) Chem Rev 93: 861
9. Serrano JL (ed) (1996) Metallomesogens. VCH Verlagsgesellschaft, Weinheim
10. Bruce DW (1994) Adv Mater 6: 699
11. Rourke JP, Fanizzi FP, Bruce DW, Dunmur DA, Maitlis PM (1992) J Chem Soc, Dalton Trans 3009
12. Esteruelas MA, Oro LA, Sola E, Ros MB, Serrano JL (1989) J Chem Soc, Chem Commun 55
13. Esteruelas MA, Sola E, Oro LA, Ros MB, Marcos M, Serrano JL (1990) J Organomet Chem 387: 103
14. Bruce DW, Dunmur DA, Esteruelas M, Hunt SE, Le Lagadec R, Maitlis PM, Marsden J, Sola E, Stacey JM (1991) J Mater Chem 1: 251
15. Adams H, Bailey NA, Bruce DW, Hudson SA, Marsden JR (1994) Liq Cryst 16: 643
16. Coco S, Espinet P, Falagan S, Martin-Alvarez JM (1995) New J Chem 19: 959
17. Benouazzane M, Coco S, Espinet P, Martin-Alvarez JM (1995) J Mater Chem 5: 441
18. Coco S, Espinet P, Martin-Alvarez JM, Levelut AM (1997) J Mater Chem 7: 19
19. Bruce DW, Hudson SA (1994) J Mater Chem 4: 479
20. Bayon R, Coco S, Espinet P, Fernandez-Mayordomo C, Martin-Alvarez J (1997) Inorg Chem 36: 2329
21. Kaharu T, Ishii R, Takahashi S (1994) J Chem Soc, Chem Commun 1349
22. Kaharu T, Ishii R, Adachi T, Yoshida T, Takahashi S (1995) J Mater Chem 5: 687
23. Bayon R, Coco S, Espinet P, Fernandez-Mayordomo C, Martin-Alvarez J (1997) Inorg Chem 36: 2329

24. Adams H, Albeniz A C, Bailey N A, Bruce D W, Cherodian A S, Dhillon R, Dunmur D A, Espinet P, Feijoo J L, Lalinde E, Maitlis P M, Richardson R M, Ungar G (1991) J Mater Chem 1: 843

25. (a) Ishii R, Kaharu T, Pirio N, Zhang SW, Takahashi S (1995) J Chem Soc, Chem Commun 1215; (b) Zhang SW, Ishii R, Takahashi S (1997) Organometallics 16: 20

26. Ghedini M, Pucci D (1990) J Organomet Chem 395: 112

27. Ghedini M, Pucci D, De Munno G, Viterbo D, Neve F, Armentano S (1991) Chem Mater 3: 65

28. (a) Baena MJ, Espinet P, Ros MB, Serrano JL (1991) Angew Chem Int Ed Engl 30: 711; (b) Buey J, Espinet P (1996) J Organomet Chem 507: 137

29. Ghedini M, Pucci D, Armentano S, Bartolino R, Versace C, Cipparrone G, Scaramuzza N (1992) It Pat No VE92,000,003

30. Lydon DP, Cave GWV, Rourke JP (1997) J Mater Chem 7: 403

31. Ghedini M, Morrone S, Neve F, Pucci D (1996) Gazz Chim It 126: 511

32. Ghedini M, Pucci D, Neve F (1996) Chem Commun 137

33. Baena MJ, Espinet P, Ros MB, Serrano JL, Ezcurra A (1993) Angew Chem Int Ed Engl 32: 1203

34. Thompson NJ, Iglesias R, Serrano JL, Baena MJ, Espinet P (1996) J Mater Chem 6: 1741

35. Thompson NJ, Serrano JL, Baena MJ, Espinet P (1996) Chem Eur J 2: 214

36. Usol'tseva N, Espinet P, Buey J, Serrano JL (1997) J Mater Chem 7: 215

37. Ghedini M, Morrone S, Francescangeli, Bartolino R (1992) Chem Mater 4: 1119

38. Ghedini M, Morrone S, Francescangeli, Bartolino R (1994) Chem Mater 6: 1971

39. Baena MJ, Buey J, Espinet P, Kitzerow HS, Heppke G (1993) Angew Chem Int Ed Engl 32: 1201

40. Buey J, Diez L, Espinet P, Kitzerow HS, Miguel JA (1996) Chem Mater 8: 2375

41. Espinet P, Lalinde E, Marcos M, Perez J, Serrano JL (1990) Organometallics 9: 555

42. Espinet P, Perez J, Marcos M, Ros MB, Serrano JL, Barbera J, Levelut AM (1990) Organometallics 9: 2028

43. Espinet P, Etxebarria J, Marcos M, Perez J, Remon A, Serrano JL (1989) Angew Chem Int Ed Engl 28: 1065

44. (a) Weissflog W, Demus D (1983) Cryst Res Technol 18: 21; (b) Weissflog W, Demus D (1984) 19: 55; (c) Demus D, Diele S, Hauser A, Latif I, Selbmann C, Weissflog W (1985) Cryst Res Technol 20: 1547; (d) Diele S, Roth K, Demus D (1986) Cryst Res Technol 21: 97

45. Weissflog W, Wiegeleben A, Demus D (1985) Mater Chem Phys 12: 461

46. Adams H, Bailey NA, Bruce DW, Dunmur DA, Lalinde E, Marcos M, Ridgway C, Smith AJ, Styring P, Maitlis PM (1987) Liq Cryst 2: 381

47. Kaharu T, Tanaka T, Sawada M, Takahashi S (1994) J Mater Chem 4: 859

48. (a) Kaharu T, Matsubara H, Takahashi S (1992) J Mater Chem 2: 43; (b) Kaharu T, Matsubara H, Takahashi S (1992) Mol Cryst, Liq Cryst 220: 191

49. Bruce DW, Lea MS, Marsden JR (1996) Mol Cryst, Liq Cryst 275: 183

50. Salt NJS (1990) PhD Thesis, University of Sheffield

51. Marcos M, Romero P, Serrano JL (1989) J Chem Soc, Chem Commun 1641

52. Hoshino N, Hayakawa R, Shibuya T, Matsunaga Y (1990) Inorg Chem 29: 5129

53. Serrano JL, Romero P, Marcos M, Alonso PJ (1990) J Chem Soc, Chem Commun 859

54. Caruso U, Roviello A, Sirigu A (1990) Liq Cryst 7: 421

55. Caruso U, Roviello A, Sirigu A (1990) Liq Cryst 7: 431

56. Hoshino N, Kodama A, Shibuya T, Matsunaga Y, Miyajima S (1991) Inorg Chem 30: 3091

57. Caruso U, Roviello A, Sirigu A (1991) Liq Cryst 10: 85

58. Galyametdinov YG, Ivanova GI, Ovchinnikov IV (1989) Izv Akad SSSR, Ser Khim 1931

59. Campillos E, Marcos M, Serrano JL (1993) J Mater Chem 3: 1049

60. Marcos M, Romero P, Serrano JL, Barbera J, Levelut AM (1990) Liq Cryst 7: 251

61. Barbera J, Levelut AM, Marcos M, Romero P, Serrano JL (1991) Liq Cryst 10: 119

62. (a) Pyzuk W, Galyametdinov Y (1993) Liq Cryst 15: 265; (b) Griesar K, Galyametdinov Y, Athanassopoulou M, Ovchinnikov I, Haase W (1994) Adv Mater 6: 381
63. Bulkin BJ, Rose RK, Santoro A (1977) Mol Cryst, Liq Cryst 43: 53
64. Ohta K, Yokoyama M, Kusabayashi S, Mikava H (1980) J Chem Soc, Chem Commun 392
65. Mühlberger B, Haase W (1989) Liq Cryst 5: 251
66. (a) Ohta K, Takenaka O, Hasebe H, Morizumi Y, Fujimoto T, Yamamoto I (1991) Mol Cryst, Liq Cryst 195: 123; (b) Ohta K, Akimoto H, Fujimoto T, Yamamoto I (1994) J Mater Chem 4: 61; (c) Ohta K, Takenaka O, Hasebe H, Morizumi Y, Fujimoto T, Yamamoto I (1991) Mol Cryst, Liq Cryst 195: 135; (d) Ohta K, Morizumi Y, Akimoto H, Takenaka O, Fujimoto T, Yamamoto I (1992) Mol Cryst, Liq Cryst 214: 143
67. Szydlowska J, Pyzuk W, Krowczynski A, Bikchantaev I (1996) J Mater Chem 6: 733
68. (a) Pyzuk W, Gorecka E, Krowczynski A (1992) Liq Cryst 11: 797; (b) Pyzuk W, Gorecka E, Krowczynski A, Przedmojki J (1993) Liq Cryst 14: 773; (c) Pyzuk W, Gorecka E, Krowczynski A (1994) Mol Cryst, Liq Cryst 249: 17
69. Berdagué P, Perez F, Judeinstein P, Bayle JP (1995) New J Chem 19: 293
70. Bayle JP, Bui E, Perez F, Courtieu J (1989) Bull Soc Chim Fr 4: 532
71. Perez F, Judeinstein P, Bayle JP (1995) New J Chem 19: 1015
72. Lydon DP, Rourke JP (1997) Chem Commun 1741
73. (a) Gray GW, Goodby JW (1984) Smectic liquid crystals: textures and structures. Leonard Hill Publishers, Glasgow; (b) Zorkii PM, Timofeeva TV, Polishchuk AP (1989) Russ Chem Rev 58: 1971
74. (a) Destrade C, Gasparoux H, Foucher P, Nguyen HT, Malthête J (1983) J Chim Phys 80: 137; (b) Levelut AM (1983) J Chim Phys 80: 149; (c) Destrade C, Foucher P, Gasparoux H, Nguyen HT, Levelut AM, Malthête J (1984) Mol Cryst, Liq Cryst 106: 121; (d) Chandrasekhar S (1993) Liq Cryst 14: 3
75. Demus D (1989) Liq Cryst 5: 75
76. (a) Prade H, Miethchen R, Vill V (1995) J Prakt Chem 337: 427; (b) Tschierske C (1996) Prog Polym Sci 21: 775
77. (a) Date RW, Imrie CT, Luckhurst GR, Seddon JM (1992) Liq Cryst 12: 203; (b) Attard GS, Date RW, Imrie CT, Luckhurst GR, Roskilly SJ, Seddon JM, Taylor L (1994) Liq Cryst 16: 529
78. Dehne H, Roger A, Demus D, Diele S, Kresse H, Pelzl G, Wedler W, Weissflog W (1989) Liq Cryst 6: 47
79. (a) Malthête J, Nguyen HT, Destrade C (1993) Liq Cryst 13: 171; (b) Nguyen HT, Destrade C, Malthête J (1997) Adv Mater 9: 375
80. (a) Hendrikx Y, Levelut AM (1988) Mol Cryst, Liq Cryst 165: 233; (b) Skoulios A, Guillon D (1988) Mol Cryst, Liq Cryst 165: 317; (c) Tschierscke C (1998) J Mater Chem 8: 1485
81. Donnio B, Bruce DW (1997) J Chem Soc, Dalton Trans 2745
82. Guillon D, Heinrich B, Ribeiro AC, Cruz C, Nguyen HT (1998) Mol Cryst, Liq Cryst; in press
83. Berdagué P, Perez F, Courtieu J, Bayle JP (1994) Bull Soc Chim Fr 131: 335
84. Krowczynski A, Szydlowska J, Pociecha, Gorecka E (1996) Polish J Chem 70: 32
85. Chandrasekhar S (1992) Liquid Crystals, 2nd edn. Cambridge University Press, Cambridge, p 414, chap 6, Sect. 6.6
86. Praefcke K, Singer D, Gündoğan B (1992) Mol Cryst, Liq Cryst 223: 181
87. Praefcke K, Bilgin B, Pickardt J, Borowski M (1994) Chem Ber 127: 1543
88. Barbera J, Espinet P, Lalinde E, Marcos M, Serrano JL (1987) Liq Cryst 2: 833
89. Singer D, Liebmann A, Praefcke K, Wendorff JH (1993) Liq Cryst 14: 785
90. Praefcke K, Singer D (1994) Mol Mater 3: 265
91. (a) Praefcke K, Singer D, Gündoğan B, Gutbier K, Langner M (1993) Ber Bunsenges Phys Chem 97: 1358; (b) Praefcke K, Singer D, Gündoğan B, Gutbier K, Langner M (1993) 98: 118; (c) Usol'tseva N, Praefcke K, Singer D, Gündoğan B (1994) Liq Cryst 16: 601; (d) Praefcke K, Diele S, Pickardt J, Gündoğan B, Nütz U, Singer D (1995) Liq Cryst

18: 857; (e) Praefcke K, Bilgin B, Usol'tseva N, Heinrich B, Guillon D (1995) J Mater Chem 5: 2257

92. Heinrich B, Praefcke K, Guillon D (1997) J Mater Chem 7: 1363
93. Usol'tseva N, Hauck G, Koswig HD, Praefcke K, Heinrich B, Guillon D (1996) Liq Cryst 20: 731
94. Usol'tseva N, Praefcke K, Singer D, Gündoğan B (1994) Liq Cryst 16: 617
95. Usol'tseva N, Praefcke K, Singer D, Gündoğan B (1994) Mol Mater 4: 253
96. Praefcke K, Holbrey JD (1996) J Inclusion Phenom Mol Recog Chem 24: 19
97. Mitchell DJ, Tiddy GJT, Waring L, Bostock T, McDonald MP (1983) J Chem Soc, Faraday Trans 1 79: 975
98. (a) Mariani P, Luzzati V, Delacroix H (1988) J Mol Biol 204: 165; (b) Seddon JM, Hogan JL, Warrender NA, Pebay-Peyrovla E (1990) Prog Colloid Polym Sci 81: 189
99. Seddon JM, Templer RH (1993) Phil Trans R Soc Lond A 344: 377
100. Seddon JM (1996) Ber Bunsenges Phys Chem 100: 380
101. Levelut AM, Clerc M (1997) Liq Cryst 24: 105
102. Bruce DW, Dunmur DA, Hudson SA, Maitlis PM, Styring P (1992) Adv Mater Opt Electron 1: 37
103. Bruce DW, Dunmur DA, Hudson SA, Lalinde E, Maitlis PM, McDonald MP, Orr R, Styring P, Cherodian AS, Richardson RM, Feijoo JL, Ungar G (1991) Mol Cryst, Liq Cryst 206: 79
104. Adams H, Bailey NA, Bruce DW, Davis SC, Dunmur DA, Hempstead PD, Hudson SA, Thorpe S (1992) J Mater Chem 2: 395
105. Bruce DW; unpublished results
106. Bruce DW, Donnio B, Hudson SA, Levelut AM, Megtert S, Petermann D, Veber M (1995) J Phys II (France) 5: 289
107. Donnio B, Bruce DW, Delacroix H, Gulik-Krzywicki T (1997) Liq Cryst 23: 147
108. Levelut AM, Donnio B, Bruce DW (1997) Liq Cryst 22: 753
109. Bruce DW, Donnio B, Guillon, D, Heinrich B, Ibn-Elhaj M (1995) Liq Cryst 19: 537
110. Donnio B, Bruce DW, Heinrich B, Guillon D, Delacroix H, Gulik-Krzywicki T (1997) Chem Mater 9: 2251
111. Kékicheff, P (1991) Mol Cryst, Liq Cryst 198: 131
112. Baena MJ, Espinet P, Lequerica MC, Levelut AM (1992) J Am Chem Soc 114: 4182
113. Cotton FA, Wilkinson G (1988) Advanced Inorganic Chemistry, 5th edn. Wiley, New York
114. Ziminski L, Malthête J (1990) J Chem Soc, Chem Commun 1495
115. Jacq P, Malthête J (1996) Liq Cryst 21: 291
116. Huang D, Yang J, Wan W, Ding F, Zhang L (1996) Mol Cryst, Liq Cryst 281: 43
117. Campillos E, Deschenaux R, Levelut AM (1996) J Chem Soc, Dalton Trans 2533
118. Yang J, Huang D, Ding F, Zhao W Zhang L (1996) Mol Cryst, Liq Cryst 281: 51
119. Bhatt J, Fung BM, Nicholas KM, Poon CD (1988) J Chem Soc, Chem Commun 1439
120. Loubser C, Imrie C, Rooyen PH van (1993) Adv Mater 5: 45
121. Imrie C, Loubser C (1994) J Chem Soc, Chem Commun 2159
122. Werner A, Friedrichsen W (1994) J Chem Soc, Chem Commun 365
123. Deschenaux R, Kosztics I, Nicolet B (1995) J Mater Chem 5: 2291
124. Thompson NJ, Goodby JW, Toyne KJ (1993) Liq Cryst 13: 381
125. Andersch J, Diele S, Tschierske C (1996) J Mater Chem 6: 1465
126. Deschenaux R, Schweissguth M, Levelut AM (1996) Chem Commun 1275
127. Neve F (1996) Adv Mater 8: 277
128. (a) Bruce DW, Liu XH (1994) J Chem Soc, Chem Commun 729; (b) Bruce DW, Liu XH (1995) Liq Cryst 18: 165; (c) Liu X-H, Manners I, Bruce DW (1998) J Mater Chem 8: 1555
129. Liu XH, Abser MN, Bruce DW (1998) J Organomet Chem 551: 271
130. Morrone S, Bruce DW, Guillon D (1996) Inorg Chem 35: 7041
131. Rowe KE, Bruce DW (1996) J Chem Soc, Dalton Trans 3913
132. Bruce DW, Rowe KE (1995) Liq Cryst 18: 161
133. Rowe KE, Bruce DW (1996) Liq Cryst 20: 183

134. Deschenaux R, Donnio B, Rheinwald G, Stauffer F, Süss-Fink G, Velker J (1997) J Chem Soc, Dalton Trans 4351
135. Bruneau C, Dixneuf PH (1997) Chem Commun 507
136. Lahoz FJ, Martin A, Esteruelas MA, Sola E, Serrano JL, Oro LA (1991) Organometallics 10: 1794
137. Giroud-Godquin AM, Rassat A (1982) C R Sceances Acad Sci Ser 294: 2
138 (a) Zheng H, Swager TM (1994) J Am Chem Soc 116: 761; (b) Swager TM, Zheng H (1995) Mol Cryst, Liq Cryst 260: 301
139. Schmidt S, Lattermann G, Kleppinger R, Wendorff JH (1994) Liq Cryst 16: 693
140. Lattermann G, Schmidt S, Kleppinger R, Wendorf JH (1992) Adv Mater 4: 30
141. Stebani U, Lattermann G, Wittenbger M, Wendorff JH (1996) Angew Chem Int Ed Engl 35: 1859
142. Xu B, Swager TM (1993) J Am Chem Soc 115: 1159
143. Bruce DW, Dunmur DA, Maitlis PM, Watkins J. M (1992) Liq Cryst 11: 127
144. Bruce DW, Denby IR, Tiddy GJT, Watkins JM (1993) J Mater Chem 3: 911
145. Bruce DW, Holbrey JD, Tajbakhsh AR, Tiddy GJT (1993) J Mater Chem 3: 905
146. Holbrey JD, Tiddy GJT, Bruce DW (1995) J Chem Soc, Dalton Trans 1769
147. Iida M, Yamamoto M, Fujita N (1996) Bull Chem Soc Jpn 69: 3217
148. (a) Yamagishi A, Sasa N, Taniguchi M, Endo A, Sakamoto M, Shimizu K (1997) Langmuir 13: 1689; (b) Deschenaux R, Megert S, Zumbrunn C, Ketterer J, Steiger R (1997) Langmuir 13: 2363
149. (a) Munoz S, Abel E, Wang K, Gokel GW (1995) Tetrahedron 51: 423; (b) Munoz S, Gokel GW (1996) Inorg Chim Acta 259: 59; (c) Wang K, Munoz S, Zhang L, Castro R, Kaifer AE, Gokel GW (1996) J Am Chem Soc 118: 6707
150. (a) Skoulios AE, Luzzati V (1961) Acta Cryst 14: 278; (b) Spegt PA, Skoulios AE (1963) Acta Cryst 16: 301; (c) Spegt PA, Skoulios AE (1963) Acta Cryst 17: 198; (d) Busico V, Ferraro A, Vacatello M (1984) J Phys Chem 88: 4055
151. Grant RF (1964) Can J Chem 42: 951
152. (a) Abied H, Guillon D, Skoulios A, Weber P, Giroud-Godquin AM, Marchon JC (1987) Liq Cryst 2: 269; (b) Ibn-Elhaj M, Guillon D, Skoulios A, Giroud-Godquin AM, Maldivi P Liq Cryst (1992) 11: 731; (c) Guillon D, Skoulios A (1996) J Phys IV (France) C4: 41; (d) Ibn-Elhaj M, Guillon D, Skoulios A, Giroud-Godquin A-M, Marchon J-C (1992) J Phys II (France) 2: 2197; (e) Seghrouchni R, Skoulios A (1995) J Phys II (France) 5: 1385
153. Giroud-Godquin AM, Latour JM, Marchon JC (1985) Inorg Chem 24: 4452
154. Strommen DP, Giroud-Godquin AM, Maldivi P, Marchon JC (1987) Liq Cryst 2: 689
155. (a) Abied H, Guillon D, Skoulios A, Giroud-Godquin AM, Maldivi P, Marchon JC (1988) Colloid Polym Sci 266: 579; (b) Ibn-Elhaj M, Guillon D, Skoulios A (1992) Phys Rev 46: 7643
156. Abied H, Guillon D, Skoulios A, Dexpert H, Giroud-Godquin AM, Marchon JC (1988) J Phys France 49: 345
157. (a) Giroud-Godquin AM, Maldivi P, Marchon JC, Bée M, Carpentier L (1989) Mol Phys 68: 1353; (b) Carpentier L, Bée M, Giroud-Godquin AM, Maldivi P, Marchon JC (1989) Mol Phys 68: 1367; (c) Bée M, Giroud-Godquin AM, Maldivi P, Williams J (1994) Mol Phys 81: 57
158. (a) Giroud-Godquin AM, Marchon JC, Guillon D, Skoulios A (1986) J Phys Chem 90: 5502; (b) Poizat O, Strommen DP, Maldivi P, Giroud-Godquin AM, Marchon JC (1990) Inorg Chem 29: 4853; (c) Ibn-Elhaj M, Guillon D, Skoulios A, Maldivi P, Giroud-Godquin AM, Marchon JC (1992) J Phys II France 2: 2237
159. (a) Maldivi P, Giroud-Godquin AM, Marchon JC, Guillon D, Skoulios A (1989) Chem Phys Lett 157: 552; (b) Cukiernik FD, Maldivi P, Giroud-Godquin AM, Marchon JC, Ibn-Elhaj M, Guillon D, Skoulios A (1991) Liq Cryst 9: 903; (c) Bonnet L, Cukiernik FD, Maldivi P, Giroud-Godquin AM, Marchon JC, Ibn-Elhaj M, Guillon D, Skoulios A (1994) Chem Mater 6: 31

160. Cukiernik FD, Ibn-Elhaj M, Chaia ZD, Marchon JC, Giroud-Godquin AM, Guillon D, Skoulios A, Maldivi P (1998) Chem Mater 10: 83

161. (a) Attard GS, Cullum PR (1990) Liq Cryst 8: 299; (b) Attard GS, Templer RH (1993) J Mater Chem 3: 207; (c) Maldivi P, Bonnet L, Giroud-Godquin AM, Ibn-Elhaj M, Guillon D, Skoulios A (1993) Adv Mater 5: 909

162. (a) Cayton RH, Chisholm MH, Darrington FD (1990) Angew Chem Int Ed Engl 29: 1481; (b) Baxter DV, Cayton RH, Chisholm MH, Huffman JC, Putilina EF, Tagg SL, Wesemann JL, Zwanziger JW, Darrington FD (1994) J Am Chem Soc 116: 4551

163. Serrette AG, Swager TM (1993) J Am Chem Soc 115: 8879

164. Zheng H, Lai CK, Swager TM (1994) Chem Mater 6: 101

165. (a) Lai CK, Serrette AG, Swager TM (1992) J Am Chem Soc 114: 7948; (b) Serrette AG, Lai CK, Swager TM (1994) Chem Mater 6: 2252

166. Krowczynski A, Pociecha D, Szydlowska J, Przedmojski J, Gorecka E (1996) Chem Commun 2731

167. Lai CK, Lu MY, Lin FJ (1997) Liq Cryst 23: 313

168. Trzaska ST, Swager TM (1998) Chem Mater 10: 438

169. Serrette AG, Swager TM (1994) Angew Chem Int Ed Engl 33: 2342

170. Pecoraro VL, Stemmler AJ, Gibney BR, Bodwin JJ, Wang H, Kampf JW, Barwinski A (1997) Metallocrowns: a new class of molecular recognition agents. In: Karlin K (ed) Prog Inorg Chem. Pergamon, vol 45, chap 2

171. Barbera J, Elduque A, Gimenez R, Oro LA, Serrano JL (1996) Angew Chem Int Ed Engl 35: 2832

172. Giroud-Godquin AM, Maitlis PM (1991) Angew Chem Int Ed Engl 30: 375

Author Index Volumes 1–95

Drago RS (1973) Quantitative Evaluation and Prediction of Donor-Acceptor Interactions. *15*: 73–139

Dreher C, see Buchler JW (1995) *84*: 1–70

Drillon M, Darriet J (1992) Progress in Polymetallic Exchange-Coupled Systems, some Examples in Inorganic Chemistry. *79*: 55–100

Duffy JA (1977) Optical Electronegativity and Nephelauxetic Effect in Oxide Systems. *32*: 147–166

Dunham WR, see Bearden AJ (1970) *8*: 1–52

Dunn MF (1975) Mechanisms of Zinc Ion Catalysis in Small Molecules and Enzymes. *23*: 61–122

Eatough DJ, see Izatt RM (1973) *16*: 161–189

Eller PG, see Ryan RR (1981) *46*: 47–100

Emmerling A, see Fricke J (1991) *77*: 37–88

Emsley E (1984) The Composition, Structure and Hydrogen Bonding of the β-Diketones. *57*: 147–191

Englman R (1981) Vibrations in Interaction with Impurities. *43*: 113–158

Epstein IR, Kustin K (1984) Design of Inorganic Chemical Oscillators. *56*: 1–33

Ermer O (1976) Calculations of Molecular Properties Using Force Fields. Applications in Organic Chemistry. *27*: 161–211

Ernst RD (1984) Structure and Bonding in Metal-Pentadienyl and Related Compounds. *57*: 1–53

Erskine RW, Field BO (1976) Reversible Oxygenation. *28*: 1–50

Evain M, Brec R (1992) A New Approach to Structural Description of Complex Polyhedra Containing Polychalcogenide Anions. *79*: 277–306

Fackler PH, see Clarke MJ (1982) *50*: 57–58

Fajans K (1967) Degrees of Polarity and Mutual Polarization of Ions in the Molecules of Alkali Fluorides, SrO and BaO. *3*: 88–105

Fan M-F, see Lin Z (1997) *87*: 35–80

Fee JA (1975) Copper Proteins – Systems Containing the "Blue" Copper Center. *23*: 1–60

Feeney RE, Komatsu SK (1966) The Transferrins. *1*: 149–206

Fehlner TP (1997) Metalloboranes. *87*: 111–136

Felsche J (1973) The Crystal Chemistry of the Rare-Earth Silicates. *13*: 99–197

Ferreira R (1976) Paradoxical Violations of Koopmans' Theorem, with Special Reference to the 3d Transition Elements and the Lanthanides. *31*: 1–21

Fichtinger-Schepman AMJ, see Reedijk J (1987) *67*: 53–89

Fidelis IK, Mioduski T (1981) Double-Double Effect in the Inner Transition Elements. *47*: 27–51

Field BO, see Erskine RW (1976) *28*: 1–50

Figlar J, see Maroney MJ (1998) *92*: 1–66

Fischer J, see Mathey F (1984) *55*: 153–201

Fischer S, see Tytho KH (1999) *93*: 125–317

Follmann H, see Lammers M (1983) *54*: 27–91

Fonticella-Camps JC (1998) Biological Nickel. *91*: 1–30

Fournier JM, Manes L (1985) Actinide Solids. 5f Dependence of Physical Properties. *59/60*: 1–56

Fournier JM (1985) Magnetic Properties of Actinide Solids. *59/60*: 127–196

Fraga S, Valdemoro C (1968) Quantum Chemical Studies on the Submolecular Structure of the Nucleic Acids. *4*: 1–62

Frasinski LJ, see Codling K (1996) *85*: 1–26

Fraústo da Silva JJR, Williams RJP (1976) The Uptake of Elements by Biological Systems. *29*: 67–121

Frenking G, see Jørgensen CK (1990) *73*: 1–16

Labarre JF (1978) Conformational Analysis in Inorganic Chemistry: Semi-Empirical Quantum Calculation vs. Experiment 35: 1–35

Lammers M, Follmann H (1983) The Ribonucleotide Reductases: A Unique Group of Metallo-enzymes Essential for Cell Proliferation. 54: 27–91

Le Brun NE, Thomson AJ, Moore GR (1997) Metal Centres of Bacterioferritins or Non-Heam-Iron-Containing Cytochromes b_{557}. 88: 103–138

Leciejewicz J, Alcock NW, Kemp TJ (1995) Carboxylato Complexes of the Uranyl Ion: Effects of Ligand Size and Coordinat. Geometry Upon Molecular and Crystal Structure. 82: 43–84

Lecomte C, see Guilard R (1987) 64: 205–268

Lee YJ, see Scheidt WR (1987) 64: 1–70

Lehmann H, see Schultz H (1991) 74: 41–146

Lehn J-M (1973) Design of Organic Complexing Agents. Strategies Towards Properties. 16: 1–69

Li H, see Sun H (1997) 88: 71–102

Lioccia S, Paolesse R (1995) Metal Complexes of Corroles and Other Corrinoids. 84: 71–134

Lin Z, Fan M-F (1997) Metal-Metal Interactions in Transition Metal Clusters with π-Donor Ligands. 87: 35–80

Linarès C, Louat A, Blanchard M (1977) Rare-Earth Oxygen Bonding in the $LnMO_4$ Xenotime Structure. 33: 179–207

Lindskog S (1970) Cobalt(II) in Metalloenzymes. A Reporter of Structure-Function Relations. 8: 153–196

Ling JH, see Hall DI (1973) 15: 3–51

Liu A, Neilands JB (1984) Mutational Analysis of Rhodotorulic Acid Synthesis in Rhodotorula philimanae. 58: 97–106

Livage J, see Henry M (1991) 77: 153–206

Livorness J, Smith T (1982) The Role of Manganese in Photosynthesis. 48: 1–44

Llinás M (1973) Metal-Polypeptide Interactions: The Conformational State of Iron Proteins. 17: 135–220

Louat A, see Linarès C (1977) 33: 179–207

Luchinat C, see Banci L (1990) 72: 113–136

Luchinat C, see Bertini I (1982) 48: 45–91

Luchinat C, see Bertini I (1995) 83: 1–54

Luchinat C, see Capozzi F (1998) 90: 127–160

Lucken EAC (1969) Valence-Shell Expansion Studied by Radio-Frequency Spectroscopy. 6: 1–29

Luckhurst GR, see also Bates MA (1999) 94: 65–137

Ludi A, Güdel HU (1973) Structural Chemistry of Polynuclear Transition Metal Cynaides. 14: 1–21

Lutz HD (1988) Bonding and Structure of Water Molecules in Solid Hydrates. Correlation of Spectroscopic and Structural Data. 69: 125

Lutz HD (1995) Hydroxide Ions in Condensed Materials – Correlation of Spectroscopy and Structural Data. 82: 85–104

Maaskant WJA (1995) On Helices Resulting from a Cooperative Jahn-Teller Effect in Hexagoal Perovskites. 83: 55–88

Maggiora GM, Ingraham LL (1967) Chlorophyll Triplet States. 2: 126–159

Magyar B (1973) Salzebullioskopie III. 14: 111–140

Makovicky E, Hyde BG (1981) Non-Commensurate (Misfit) Layer Structures. 46: 101–170

Manes L, see Fournier JM (1985) 59/60: 1–56

Manes L, Benedict U (1985) Structural and Thermodynamic Properties of Actinide Solids and Their Relation to Bonding. 59/60: 75–125

Mann S (1983) Mineralization in Biological Systems. 54: 125–174

March NH (1993) The Ground-State Energy of Atomic and Molecular Ions and Its Variation with the Number of Electrons. 80: 71–86